Astrophysics and Space Science Proceedings

Volume 49

More information about this series at http://www.springer.com/series/7395

Tiago L. Campante • Nuno C. Santos •
Mário J.P.F.G. Monteiro
Editors

Asteroseismology and Exoplanets: Listening to the Stars and Searching for New Worlds

IVth Azores International Advanced School in Space Sciences

Springer

Editors
Tiago L. Campante
School of Physics and Astronomy
University of Birmingham
Birmingham, United Kingdom

Nuno C. Santos
Instituto de Astrofísica e Ciências do Espaço
Universidade do Porto
Porto, Portugal

Mário J.P.F.G. Monteiro
Instituto de Astrofísica e Ciências do Espaço
Universidade do Porto
Porto, Portugal

ISSN 1570-6591 ISSN 1570-6605 (electronic)
Astrophysics and Space Science Proceedings
ISBN 978-3-319-59314-2 ISBN 978-3-319-59315-9 (eBook)
DOI 10.1007/978-3-319-59315-9

Library of Congress Control Number: 2017946182

Printed on acid-free paper

This Springer imprint is published by Springer Nature
The registered company is Springer International Publishing AG
The registered company address is: Gewerbestrasse 11, 6330 Cham, Switzerland

As armas e os barões assinalados,
Que da ocidental praia Lusitana,
Por mares nunca de antes navegados,
Passaram ainda além da Taprobana,
Em perigos e guerras esforçados,
Mais do que prometia a força humana,
E entre gente remota edificaram
Novo Reino, que tanto sublimaram;

— Luís de Camões, in *Os Lusíadas*

Foreword to Part I

Vibrations of an object depend on the overall characteristics, such as size, of the object and its more detailed internal properties. The sound of a violin is very different from that of a double bass, but at a more subtle level there are also differences between a violin built by Stradivarius and one bought in a cheap store. This principle is the basis for the use of waves and vibrations to probe the properties of objects, from machinery over seismic studies of the Earth to stars.

Stars may support a broad range of oscillations. Whether or not these are useful as diagnostics of the stars depends on their excitation to observable levels. Here I concentrate on the so-called solar-like oscillators where, as in the Sun, the modes are intrinsically stable and would damp out if not excited by some external source. That source is the vigorous near-surface convective motions, with speed approaching the local speed of sound, found in relatively cool stars with outer convection zones.

Solar oscillations were first detected in the early 1960s, but it was only in 1975 that their nature as normal modes of the Sun was definitely established. The modes initially identified had short horizontal wavelengths, but the presence of modes in the 5-min range of very large scale including radial (spherically symmetric) oscillations was established a few years later. The diagnostic potential of these observations for the study of the solar interior was immediately obvious, and indeed even the early observations led to interesting inferences on, for example, the depth of the solar convective envelope. Also, early results strongly indicated that models of the solar core were essentially correct, ruling out attempts to explain the low observed flux of electron neutrinos from the Sun in terms of modifications to solar modelling and supporting the idea of neutrino oscillations. As a result of these early successes, several projects were developed to carry out detailed observations of solar oscillations, culminating in the ground-based GONG network of observing stations and instruments on the *SOHO* satellite, both starting operations in the middle of the 1990s.

Analysis of the helioseismic observations has resulted in remarkably detailed inferences of structure and rotation in most of the solar interior. The latitudinal differential rotation observed at the solar surface has been found to extend throughout the convection zone, with a sharp transition to nearly solid-body rotation in the radiative interior. The resulting rotational shear is believed to play an important role in the generation of the solar magnetic activity. The sound speed in the solar interior was determined with very high accuracy, agreeing within a fraction of a per cent with existing models. Revised determinations of the solar surface composition have, however, caused changes in solar models, greatly increasing the discrepancy. The origin of this discrepancy is still to be found.

The solar excitation mechanism is expected to be active in all stars with substantial outer convection zones. However, the very small amplitudes of these modes make the observations challenging, and it was only in the late 1990s that definite detections of solar-like oscillations in other stars were made. Subsequent observing campaigns, in many cases using some of the largest telescopes in the World to observe bright stars, resulted in analysis of the oscillations in a few main-sequence and red-giant stars. However, a major breakthrough in the field of solar-like asteroseismology came with space-based photometry from the *CoRoT* and *Kepler* missions, launched in 2006 and 2009, respectively, with the combined goals to search for exoplanets and carry out asteroseismology. These missions have provided high-quality asteroseismic data for hundreds of main-sequence and subgiant stars, and tens of thousands of red giants.

Early results have focused on the determination of overall stellar parameters, such as mass, radius and age. This has been particularly important in the cases where asteroseismic characterisation was possible for hosts of exoplanets. A second important application of such asteroseismic surveys is the use in galactic archaeology, i.e., the investigation of the structure and evolution of the Milky Way Galaxy based on characterising the present distribution, ages and composition of stars. Here asteroseismology of red giants plays a crucial role in providing determinations of distances and ages of stars over a large part of the Galaxy.

From the point of view of stellar astrophysics the asteroseismic data provide detailed tests of the internal structure and in many cases rotation of the stars. An important example is the study of the rotation of subgiant and red-giant stars which has demonstrated the need for so far unidentified mechanisms of angular-momentum transport to explain the comparatively slow rotation of the stellar cores. Detailed analysis of the data in terms of stellar internal structure is just starting. Indeed, the already available data from *CoRoT* and *Kepler* will remain an invaluable resource for stellar astrophysics for a long time to come. More data will come from the NASA *TESS* mission scheduled for launch in 2017 and the ESA *PLATO* mission with expected launch in 2025.

The fields of helio- and asteroseismology remain extremely rich in the potential for further scientific investigations and breakthroughs, towards the goal of obtaining a proper physically based understanding of stellar structure and evolution. I hope that the school, and this volume, will inspire a new generation of young scientists to work with the marvellous data now available and expected, and tackle these issues. You will surely have fun!

Aarhus, Denmark
January 2017

Jørgen Christensen-Dalsgaard

Foreword to Part II

Since the Copernican revolution, philosophers and astronomers have interpreted the nearly circular and coplanar orbits of the Sun's family of planets as clues to their origins. The Sun contains over 99% of the solar system's mass, but 98% of its angular momentum resides in the planets. In the eighteenth century, Kant and Laplace hypothesised that if the Sun had condensed from a contracting cloud of gas, any residual angular momentum must necessarily result in the formation of a flattened disc of nebular material. If planets condensed within such a nebular disc, near-circular orbits and coplanarity would be assured. Others were less convinced. In the following century, James Clerk Maxwell argued that Keplerian shear within the disc would inhibit condensation of planetary bodies. Early in the twentieth century, James Jeans proposed instead that planet formation was a rare process occurring in tidal streams resulting from close stellar encounters. Planets were either almost impossibly rare products of close stellar encounters or commonplace natural by-products of the star formation process itself.

The advent of infrared astronomy in the 1980s produced the first indirect evidence that protoplanetary systems might be common features of young stars in star-forming regions like Taurus-Auriga. The spectral energy distributions of T Tauri stars were found to have multiple components: the stellar spectrum itself, an ultraviolet component arising from accretion-driven heating of the outer stellar atmosphere, and an extended infrared tail emitted by circumstellar dust at temperatures from 1000 K down to a few tens of K. Links between emission-line behaviour and the infrared luminosity suggested active disc accretion might be present. These conclusions were vindicated spectacularly by early Hubble Telescope images showing protoplanetary discs clearly silhouetted against the background of the Orion Nebula.

The prospects of detecting mature planets remained bleak in the face of angular separations smaller than the seeing limitations on ground-based observations, and contrast ratios ranging from billions for reflected light to thousands for gas giants viewed at thermal-infrared wavelengths. In the end, the breakthrough came not from advances in direct imaging but from clever use of indirect dynamical effects. As long ago as 1952, Otto Struve had proposed that if Jupiter-mass planets existed in close orbits about their parent stars, it should be possible to detect the Doppler effect of the host star's reflex motion about the system's centre of mass. Moreover, some 10% of planets in few-day orbits would transit the face of the host star, yielding a periodic, temporary reduction in stellar flux of 1 or 2% for a Jupiter-like gas giant. The 1992 detection by Alex Wolczcan and Dale Frail of a family of planets around pulsar PSR J1300+1240 was achieved through modulation of the pulse period by reflex orbital motion. Three years later, Michel Mayor and Didier Queloz detected the 4.2-day, 56 m s^{-1} reflex motion of a normal Sun-like star, 51 Peg, around an unseen object with about half the mass of Jupiter.

Encouraged by this result, other teams carrying out high-precision radial-velocity searches announced further discoveries of Jupiter-mass planets in short-period orbits around F, G and K main-sequence stars. Their gas-giant nature was confirmed with the detection of transits in HD 209458b by David Charbonneau and Tim Brown. Their instrument was the prototype for a new type of planet search: wide-field transit surveys using small off-the-shelf camera lenses on robotic mounts, backed by science-grade CCDs. Over the ensuing decade, transit surveys such as TrES, WASP, HAT and XO eventually amassed hundreds of hot-Jupiter detections, while high-precision, long-duration radial-velocity surveys pushed towards ever longer orbital periods and lower planet masses. Space-based surveys such as *MOST*, *CoRoT* and *Kepler* revealed dozens, then hundreds, then thousands of transiting planets with radii too small to detect from the ground. The exquisite photometric performance and uninterrupted observations of these missions also made asteroseismic characterisation of exoplanet host stars possible for the first time. Radial-velocity surveys reached the threshold of 1 Earth mass at about the same time as *Kepler* detected the first transits of Earth-sized planets, in 2012.

We are now moving from an era of discovery into one of characterisation. Many stars possess compact systems of multiple coplanar planets in orbits closer than Mercury's orbit about the Sun. Others possess hot Jupiters and little else; others still have gas-giant planets in highly eccentric orbits. The orbital planes of a significant fraction of hot Jupiters are strongly inclined with respect to their stars' rotation axes. Some are even retrograde. Although it is now clear that Kant and Laplace were on the right track after all, we still have a long way to go in understanding the detailed physical processes that take place in protoplanetary discs, and the compositions and orbital properties of the planetary bodies that eventually emerge.

In only 20 years, exoplanetary science has grown into one of the most challenging and rewarding areas of research in observational and theoretical astrophysics. It is a rich mix of superbly engineered instrumentation, large-scale data handling, orbital dynamics, gas and dust chemistry, and stellar physics. The latter is vital: the properties of the host star are intimately connected to the composition of a planetary system, and its surface activity presents a challenge to our efforts to characterise small planets. The techniques presented at the school and in this volume are the foundations on which our explorations of worlds around other stars are based. I hope that it will prove useful and inspirational to all who read it. With the *TESS*, *CHEOPS* and *PLATO* missions on the near horizon, the adventure is only just beginning.

St Andrews, Scotland
January 2017

Andrew Collier Cameron

Preface

This volume is a collection of original review articles resulting from the lectures presented at the *IVth Azores International Advanced School in Space Sciences* on

ASTEROSEISMOLOGY AND EXOPLANETS:
LISTENING TO THE STARS AND SEARCHING FOR NEW WORLDS
17–27 July 2016, Horta, Faial, Azores Islands, Portugal
Website: http://www.iastro.pt/research/conferences/faial2016/

This Advanced School was jointly organised by the *Instituto de Astrofísica e Ciências do Espaço – Universidade do Porto*, the *Universidade dos Açores* and the *University of Birmingham*. Its main goal was to address the topics at the forefront of scientific research being conducted in the fields of stellar physics and exoplanetary science, being mainly aimed at PhD and MSc students in any field of Astrophysics. The School was an excellent opportunity for the young researchers to network with fellow students and lecturers, thereby promoting awareness of areas outside the main specialisation of the student, and potential cross-fertilisation of techniques and concepts.

The School covered two scientific topics that share many synergies and resources: Asteroseismology and Exoplanets. Therefore, the program was defined with a clear strategy of building opportunities for cooperation and sharing of methods that will benefit both communities. This cooperation has experienced great success in the context of past space missions such as *CoRoT* and *Kepler*. Upcoming photometry and astrometry from space, as well as complementary data from ground-based networks, will continue to foster this cooperation. Observations of bright stars and clusters in the ecliptic plane are being made by the repurposed K2 mission, and NASA's *TESS* and ESA's *CHEOPS* missions will soon start obtaining similar data over the entire sky, thus allowing the detection and precise characterisation of planets around nearby stars. ESA's *PLATO* mission will then build upon these successes by providing photometric light curves on a wealth of stars. Ground-based spectroscopy from state-of-the-art instruments will complement the satellite data for the brightest stars in the sky. This includes projects such as the Stellar Observations Network Group (SONG) and a whole new generation of high-precision spectrographs being developed for the ESO, like the Echelle SPectrograph for Rocky Exoplanets and Stable Spectroscopic Observations (ESPRESSO).

Lectures at the School included both a teaching and hands-on component, respectively consisting of a series of theoretical courses and tutorials. These were presented by a group of young, dynamic lecturers, who have already established themselves as leaders in their respective fields of research. This volume is then the collection of these lectures, covering in detail several critical methods and descriptions that are central to the School's two main thematic lines. As such, this volume constitutes a valuable and timely review that should prove useful to a new generation of PhD students and young postdocs in the fields of Asteroseismology and Exoplanets. We would like to thank all lecturers for accepting the challenge to take part in this School and for submitting the manuscripts for inclusion in this volume.

We are very grateful for the hard work and dedication invested by all participants in the School, in particular by the students, who have contributed to a very pleasant and friendly atmosphere (the evenings spent at Peter's shall never be forgotten!). A special thanks goes to the Chair of the Local Organising Committee, João Miguel Ferreira (Universidade dos Açores), for his dedication and thorough planning, and to Elsa Silva for her invaluable support over the entire duration of the School.

The organisation of the School and the publication of the present volume were supported by the *Instituto de Astrofísica e Ciências do Espaço – Universidade do Porto* (IA–U.Porto) through funds from European Commission's SPACEINN Project on 'Exploitation of Space Data for Innovative Helio- and Asteroseismology' (FP7-SPACE-2012-312844), as well as by the *Fundação para a Ciência e a Tecnologia* (FCT) through national funds (UID/FIS/04434/2013, PTDC/FIS-AST/1526/2014) and by FEDER through COMPETE2020 (POCI-01-0145-FEDER-007672, POCI-01-0145-FEDER-016886). The School also received support from the *Governo dos Açores* (81-9/168-1355), the *Stellar Astrophysics Centre – Aarhus University* through Grant DNRF106 from the *Danish National Research Foundation*, the *University of Birmingham* through funds from the UK *Science and Technology Facilities Council* (STFC), and from a private donation made by Eng.º Adelino Campante.

Birmingham, UK — Tiago L. Campante
Porto, Portugal — Mário J.P.F.G. Monteiro
Porto, Portugal — Nuno C. Santos
March 2017

Contents

Part I
Asteroseismology

Stellar Evolution and Modelling Stars

Víctor Silva Aguirre

Abstract In this chapter I give an overall description of the structure and evolution of stars of different masses, and review the main ingredients included in state-of-the-art calculations aiming at reproducing observational features. I give particular emphasis to processes where large uncertainties still exist as they have strong impact on stellar properties derived from large compilations of tracks and isochrones, and are therefore of fundamental importance in many fields of astrophysics.

1 Stars, Why Bother?

A comprehensive view of how stars evolve and interact with their surrounding medium is crucial for many fields in astrophysics. Stars are the main sources of chemical evolution in the Universe (e.g., Matteucci 2016), and their physical, chemical, and kinematic characteristics preserve information about their birth environment and subsequent evolution to present age. For these reasons, the integrated properties of stellar populations can be used as tools to understand the evolution of distant galaxies (see, e.g., Conselice 2014, and references therein). When individual stars can be resolved by observations, as is the case in our Galaxy, accurately characterising them can help unveiling the main processes responsible for the formation history and evolution of the Milky Way (for a review, see Bland-Hawthorn and Gerhard 2016).

Stars are also the progenitors of objects and events of high astrophysical importance, such as supernovae, gamma-ray bursts, planetary nebulae, and stellar black holes (e.g., Herwig 2005; Camenzind 2007; Smartt 2009; Maoz et al. 2014; Kumar and Zhang 2015, and references therein). Last, and definitely not least, stars are the hosts of exoplanets, and our ability to characterise planetary systems (from size and mass to their atmospheric composition) depends critically on having an accurate representation of the parent star's properties (e.g., Winn and Fabrycky 2015). You only

V.S. Aguirre (✉)
Department of Physics and Astronomy, Stellar Astrophysics Centre (SAC), Aarhus University, Ny Munkegade 120, DK-8000 Aarhus C, Denmark
e-mail: victor@phys.au.dk

T.L. Campante et al. (eds.), *Asteroseismology and Exoplanets: Listening to the Stars and Searching for New Worlds*, Astrophysics and Space Science Proceedings 49, DOI 10.1007/978-3-319-59315-9_1

know your planet as well as you know your star. From these examples it is clear that stars are the building blocks of astrophysics, and understanding them holds the key to unveiling the relevant physical processes in many areas of astronomical research. We must discover what are the main drivers of stellar evolution and the impact in observable properties that can help us decoding the wonderful nature of stars.

The luminosity flux radiated from the surface of a star is a natural consequence of the intricate processes taking place in the stellar interior. How this flux evolves and changes with time, the field concerning with stellar structure and evolution, is routinely described to a considerable extent as a 'solved' problem in astrophysics. However, and although the main ingredients of the theory of stellar evolution are well established, large uncertainties still remain in cases where empirical evidence is lacking or the physical descriptions are not well constrained and thus parametrised to a simpler form. How stars produce, transport, and emit their energy; how they burn and mix chemical elements; how is the stellar matter constituted and what is its state etc., are some of the many relevant topics where still many uncertainties exist.

There are several excellent textbooks on stellar structure and evolution written by notable astronomers, and I encourage the reader to explore titles such as Cox and Giuli (1968) (recently updated by Weiss et al. 2005), Clayton (1983), Hansen et al. (2004), and Kippenhahn et al. (2012), just to name a few. Based on these references, I will introduce the basic concepts of stellar structure and evolution emphasising those relevant to asteroseismic studies, and refer the reader when necessary to textbooks where the relevant (and interesting!) details can be found.

2 Modelling Stars

Stars owe their brightness to a delicate interplay between gravitational contraction and thermonuclear reactions as energy generation sources, both mechanisms not always acting simultaneously. This interaction produces changes in the star on different timescales, which become relevant at different stages of stellar evolution.

When nuclear reactions are not efficient in producing energy, the physical conditions in stellar interiors evolve with the rate of change of the gravitational potential and internal energy. We can consider that the star contracts gradually while maintaining sphericity and that this is the sole responsible for the stellar luminosity. The corresponding timescale a star can shine through this mechanism is called the *Kelvin–Helmholtz* timescale,

$$\tau_{\rm kh} \simeq \frac{GM^2}{2RL}\,, \tag{1}$$

where G is the Newtonian constant of gravitation, M is the mass of the star, R its radius, and L the associated luminosity produced by the contraction. For a $1\,M_\odot$ star, this is approximately 15 Myr.

Stellar matter is naturally victim of interacting forces such as gravitation and pressure, but if no acceleration of material takes place a stage of mechanical equilibrium is maintained. This equilibrium in a star, referred to as hydrostatic equilibrium, is one of the pillars of stellar structure studies. The shortest relevant timescale describes the time a star needs to recover its equilibrium when the balance between gravitational forces and pressure is disturbed by some dynamical process, for instance a pressure wave. If a star is close to hydrostatic equilibrium, it corresponds to the free-fall timescale of the star, or the *dynamical* timescale,

$$\tau_{\rm dyn} \simeq \sqrt{\frac{R^3}{GM}} \simeq \sqrt{\frac{1}{G\langle\rho\rangle}}, \tag{2}$$

where $\langle\rho\rangle$ corresponds to the mean density of the star. In the case of the Sun, the free-fall timescale is of the order of 20 min.

Finally, the longest relevant timescale involved is the *nuclear* timescale,

$$\tau_{\rm nuc} = \frac{\epsilon q M c^2}{L}, \tag{3}$$

where c is the speed of light, q the fraction of the total stellar mass involved in the nuclear burning, and ϵ the amount of mass that is converted into energy as a result of the nuclear reaction processes. Essentially, this timescale describes how long a star can shine with nuclear fusion as its sole source of energy. If the Sun was made of pure hydrogen and the central 10% of its mass would contribute to nuclear reactions, it could shine through this mechanism for approximately 10 Gyr.

There are large differences between the relevant timescales for stellar evolution, where $\tau_{\rm nuc} \gg \tau_{\rm kh} \gg \tau_{\rm dyn}$. If we intent on following a large portion of the evolution of a star, the processes occurring in the shortest timescales must be consistently parametrised over larger periods. Moreover, some processes are usually neglected due to the lack of a consistent theory describing their effects in the overall stellar evolution, such as rotation and magnetic fields. Therefore, many assumptions and simplifications about the nature of stellar matter and complex physics must be used when modelling stars.

2.1 *The Main Equations*

The vast majority of stars are currently in long-lasting phases of their evolution, in which the timescales involved for appreciable change to occur are too large to be observed. Thus, the evolutionary changes are described by the four basic differential equations of stellar structure which are dubbed the macrophysics in modern theory of stellar studies:

$$\frac{\partial r}{\partial m} = \frac{1}{4\pi r^2 \rho}, \tag{4}$$

$$\frac{\partial P}{\partial m} = -\frac{Gm}{4\pi r^4}, \tag{5}$$

$$\frac{\partial L}{\partial m} = \epsilon - \epsilon_\nu + \epsilon_g, \tag{6}$$

$$\frac{\partial T}{\partial m} = -\frac{GmT}{4\pi r^4 P}\nabla, \tag{7}$$

where

$$\nabla = \frac{\partial \ln T}{\partial \ln P}. \tag{8}$$

In these equations, r is the distance from the centre of the star, and m the mass contained within this distance. P, T and ρ are the thermodynamic variables pressure, temperature and density respectively, while L is the luminosity at the corresponding position of r (or m). The ϵ term corresponds to the energy rate per unit mass generated by nuclear reactions, ϵ_ν to the energy rate lost (in form of neutrinos), and ϵ_g to the work that is performed on the gas during any expansion or contraction of the star. These equations are, in order of appearance, the *mass conservation equation*, the *hydrostatic equilibrium equation*, the *energy conservation equation*, and the *energy transport equation*.

The solutions to the equations are not stationary but rather evolve with time as a consequence of contraction and nuclear reactions taking place, and the resulting changes in the chemical composition and mean molecular weight brought about by them. The first two equations define the mass profile in the stellar interior, while the latter two equations determine the thermal profile inside of the star. In fact, Eq. (8) is simply the definition of ∇, whose value must be derived from a theory of energy transport.

Under normal circumstances there is a steady flow of energy from the deep stellar interior, where the nuclear reactions take place, to the outermost layers of the star, where energy is radiated to the interstellar medium. Depending on the thermodynamical properties of matter in the stellar interior, this energy transport can occur via *radiative transfer*, *convective motions*, or *conductive transfer*. The latter transport method becomes very efficient under degenerate matter conditions.

2.1.1 Energy Transport by Radiation

In the time-independent, three-dimensional case, the equation of radiative transfer can be written as:

$$\mu_i \frac{\partial \mathbf{I}}{\partial x_i} = -(\kappa_{\rm ab} + \kappa_{\rm sc})\, \rho \mathbf{I} + \kappa_{\rm ab}\, \rho B + \kappa_{\rm sc}\, \rho J\,, \tag{9}$$

where $\mathbf{I}(\mathbf{x}, \boldsymbol{\mu}, t)$ is the specific intensity at $\mathbf{x}$ in direction $\boldsymbol{\mu}$, $\kappa_{\rm ab}$ the mean absorption opacity, $\kappa_{\rm sc}$ the scattering opacity, $B = (ac/4\pi)T^4$ is the integrated Planck intensity and J the mean intensity. In order to estimate the flux carried by radiation, it is customary to use the *Eddington approximation* (e.g., Unno and Spiegel 1966). Assuming the intensity to be isotropic, a relation between the zeroth- and first-order moments of Eq. (9) can be obtained, that yields the radiation flux as

$$\mathbf{F} = -\frac{4\pi}{3\rho\,(\kappa_{\rm ab} + \kappa_{\rm sc})} \nabla J\,. \tag{10}$$

The near-isotropy of the radiation intensity is usually associated with a short photon mean free path, where radiation is efficiently trapped (i.e., $\rho\kappa_{\rm ab} \to \infty$). In this case, also referred to as the optically thick case, a diffusive mechanism takes place for radiative energy transport leading to the following expression for the energy flux:

$$\mathbf{F} = -\frac{4\pi}{3\rho\kappa} \nabla B = -\frac{4acT^3}{3\rho\kappa} \nabla T\,, \tag{11}$$

where $\kappa = |\kappa_{\rm ab} + \kappa_{\rm sc}|$. This is known as the *diffusion approximation.*

The radiation flux then clearly depends on the opacities κ, of which we have so far neglected their natural frequency dependence. This comes from the idea that we can replace the problem of frequency dependence (the *non-grey atmosphere* problem) through some sort of mean opacities. In fact, one can show that a particular average of the opacities can be found by imposing the Eddington approximation in the equation of radiative transfer (Eq. (9)), thus making the problem frequency-independent. These are called *Rosseland* mean opacities, and are defined as

$$\kappa_{\rm ross}^{-1} = \frac{\int_0^\infty \frac{1}{\kappa_\nu} \frac{\partial B_\nu}{\partial T}\, {\rm d}\nu}{\int_0^\infty \frac{\partial B_\nu}{\partial T}\, {\rm d}\nu} = \frac{\pi}{acT^3} \int_0^\infty \frac{1}{\kappa_\nu} \frac{\partial B_\nu}{\partial T}\, {\rm d}\nu\,, \tag{12}$$

where a is the radiation constant, and B_ν is the monochromatic Planck function.

When essentially all energy is transported outwards by photons (condition of radiative equilibrium), it can be shown that the temperature gradient in Eq. (8) takes the form:

$$\nabla_{\rm rad} = \frac{3}{16\pi acG} \frac{\kappa LP}{mT^4}\,, \tag{13}$$

where κ is the Rosseland mean opacity of the stellar matter. A very clear description of the radiation theory in stellar interiors and its connection with stellar atmospheres can be found in Mihalas (1970) (recently updated by Hubeny and Mihalas 2014).

2.1.2 Energy Transport by Convection

When the temperature gradient indicated in Eq. (13) is too steep, radiation is not able to carry all the energy outwards and convective instabilities set in. A theory of convective transport includes a stability criterion for convection to take place, and a consistent description of how energy is transported outwards by convective motions.

Based on the displacement analysis of random bubbles inside of the star, a criterion for the onset of convective stabilities can be derived. In a nutshell, let us consider a slight temperature fluctuation in a gas element with respect to its surroundings. Assuming that the element remains in pressure equilibrium with the medium, a temperature increase translates into a density decrease if we consider that the stellar matter obeys an ideal gas law. Thus, this lighter bubble will be lifted upwards by the force of buoyancy. The gas element will travel until it becomes unstable due to turbulence and dissolves into the surrounding gas. For adiabatic motions of bubbles, it can be shown that a layer will remain stable if

$$\nabla_{\rm rad} < \nabla_{\rm ad} + \frac{\varphi}{\delta}\nabla_{\mu} , \tag{14}$$

where $\nabla_{\rm ad}$ is the temperature gradient introduced in Eq. (8) when the displacement of the bubble takes place adiabatically, and

$$\varphi = \left(\frac{\partial \ln \rho}{\partial \ln \mu}\right)_{\rm P,\mathit{T}} , \qquad \delta = -\left(\frac{\partial \ln \rho}{\partial \ln T}\right)_{\rm P,\mu} , \qquad \nabla_{\mu} = \left(\frac{d \ln \mu}{d \ln P}\right) . \tag{15}$$

Equation (14) is the *Ledoux* criterion for convection (Ledoux 1947), which takes into account variations in the molecular weight μ to define the boundaries of convective regions. However, in regions of homogeneous composition one has simply the *Schwarzschild* criterion (Schwarzschild and Härm 1958), according to which a region remains stable against convection as long as

$$\nabla_{\rm rad} < \nabla_{\rm ad} . \tag{16}$$

This is by far the most commonly used criterion when modelling stellar evolution. One reason for this is that once convection sets in a given region of the star, the chemical composition gradient is annihilated by convective mixing and Eq. (14) simplifies to (Eq. (16)). However, the use of a different criterion can have large effects in the size of the convective regions, with a subsequent impact on, e.g., the resulting luminosity, effective temperature, and main-sequence lifetime of the star.

In a radiative (dynamically stable) layer, a displaced element is pushed back by buoyancy forces. This interaction imprints a certain momentum in the gas element, which will overshoot from its original position when descending, becoming lighter than its surroundings and thus ascending again. Such oscillations of gas elements could occur in the form of thin needles, and when they take place adiabatically they are characterised by the *Brunt–Väisälä* frequency:

$$N^2 = \frac{g\delta}{H_p}\left(\nabla_{\rm ad} - \nabla + \frac{\varphi}{\delta}\nabla_\mu\right), \tag{17}$$

where H_p is the pressure scale height given by

$$H_p{}^{-1} = -\frac{\mathrm{d}\ln P}{\mathrm{d}r}. \tag{18}$$

For the displacement to be oscillatory the condition for the frequency $N^2 > 0$ must be fulfilled, which is the case for convectively stable (radiative) regions (cf. Eq. (14)). From this simple analysis of the stability of a fluid element against local perturbations it transpires that internal gravity waves cannot occur in convective regions. It is interesting to note that, for stars in the subgiant and red-giant phase of evolution, N^2 reaches very high values in the core of the star due to the strong central condensation and resulting high value of the local gravity (e.g., Christensen-Dalsgaard 2004). Stars in these evolutionary phases are expected to show a rich spectrum of gravity-dominated pulsation frequencies, as beautifully confirmed by asteroseismic observations (e.g., Kjeldsen et al. 2003; Bedding et al. 2010; Deheuvels and Michel 2011).

Once a region is found by any stability criterion to be convective, the temperature gradient of that zone needs to be defined (Eq. (8)). The usual way to do this is by using the *mixing-length theory* for convection (MLT) in any of its flavours, most commonly the original formulation proposed by Böhm-Vitense (1958). The critical free parameter involved in the formulation is the so-called mixing-length parameter $\alpha_{\rm MLT} = l/H_p$, where l is the distance a bubble will traverse before dissolving into the surrounding medium. Other formulations, such as the full spectrum of turbulent eddies by Canuto and Mazzitelli (1991, 1992), also require the definition of a convective efficiency to properly determine the temperature gradient in convective regions. The most commonly adopted procedure to calibrate this free parameter is using the solar properties, as described in Sect. 2.3.2.

2.2 Solving the Equations

Up to now, I have discussed the problem of stellar evolution calculations based on the structure of stars and physical processes taking place inside of them. The correct description of these processes depends critically on the properties of stellar matter, which are termed the microphysics of stellar evolution.

Inspection of the structure equations (Eqs. (4)–(7)) easily reveals that we are trying to solve a problem for five explicitly shown unknowns (r, ρ, P, L, and T) through a set of four equations. The missing relation is given by the Equation Of State (EOS), which provides one of the thermodynamic quantities in terms of the others (for instance, $\rho = \rho(P, T, \mu)$, where μ is just an indicator of the general chemical composition). It is customary to refer to the EOS as one of the *constitutive equations* of stellar structure; the other quantities that enter the equations and need to be defined form the set of constitutive equations. These can be written as:

$$\rho = \rho\,(P, T, \mu)\,, \tag{19}$$

$$c_{\rm P} = c_{\rm P}\,(P, T, \mu)\,, \tag{20}$$

$$\kappa_\nu = \kappa_\nu\,(P, T, \mu)\,, \tag{21}$$

$$r_{jk} = r_{jk}\,(P, T, \mu)\,, \tag{22}$$

$$\epsilon_\nu = \epsilon_\nu\,(P, T, \mu)\,, \tag{23}$$

where $c_{\rm P}$ is the specific heat at constant pressure, κ_ν the monochromatic opacity of stellar matter (a particular average of it was introduced in Sect. 2.1.1), r_{jk} the thermonuclear reaction rate transforming nuclei j into nuclei k, with the corresponding energy generation rate ϵ_{jk} given by the product of r_{jk} and the energy released when the transformation takes place. Time evolution of a certain chemical species X_i when only nuclear reactions create or destroy it is given by

$$\frac{\partial X_i}{\partial t} = \frac{m_i}{\rho}\left(\sum_j r_{ji} - \sum_k r_{ik}\right)\,, \tag{24}$$

with the constraint that $\sum_i X_i = 1$. If exchange of mass occurs between different stellar layers, diffusive processes also affect the evolution of chemical species.

2.2.1 Nuclear Reactions

Thermonuclearly fuelled reactions in stellar interiors produce energy that is carried out through radiation, convection or conduction (see Sect. 2.1). I briefly mention here the main channels for burning hydrogen into helium, the longest evolutionary phase in the lifetime of a star, and the reaction network to transform helium into heavier elements.

There are two reaction chains that transform four protons into one ^{4}He nucleus, namely the *p-p chain* and the *CNO cycle* (see Table 1). Presence of C, N, or O isotopes is necessary for the CNO cycle to begin, and since they are both destroyed and produced during the cycle they act as catalysts for the reactions. The p-p chain and the CNO cycle usually take place simultaneously in a star but with different efficiencies depending on the total stellar mass.

Table 1 Reaction networks involved in the *p-p chain* and the *CNO cycle*

p-p chain		
pp I	pp II	pp III
$^{1}H + ^{1}H \rightarrow ^{2}D + e^{+} + \nu_e$	$^{3}He + ^{4}He \rightarrow ^{7}Be + \gamma$	$^{3}He + ^{4}He \rightarrow ^{7}Be + \gamma$
$^{2}D + ^{1}H \rightarrow ^{3}He + \gamma$	$^{7}Be + e^{-} \rightarrow ^{7}Li + \nu_e$	$^{7}Be + ^{1}H \rightarrow ^{8}B + \gamma$
$^{3}He + ^{3}He \rightarrow ^{4}He + ^{1}H + ^{1}H$	$^{7}Li + ^{1}H \rightarrow ^{4}He + ^{4}He$	$^{8}B \rightarrow ^{8}Be + e^{+} + \nu_e$
		$^{8}Be \rightarrow ^{4}He + ^{4}He$
CNO cycle		
CN cycle	NO cycle	
$^{12}C + ^{1}H \rightarrow ^{13}N + \gamma$	$^{15}N + ^{1}H \rightarrow ^{16}O + \gamma$	
$^{13}N \rightarrow ^{13}C + e^{+} + \nu_e$	$^{16}O + ^{1}H \rightarrow ^{17}F + \gamma$	
$^{13}C + ^{1}H \rightarrow ^{14}N + \gamma$	$^{17}F \rightarrow ^{17}O + e^{+} + \nu_e$	
$^{14}N + ^{1}H \rightarrow ^{15}O + \gamma$	$^{17}O + ^{1}H \rightarrow ^{14}N + ^{4}He$	
$^{15}O \rightarrow ^{15}N + e^{+} + \nu_e$		
$^{15}N + ^{1}H \rightarrow ^{12}C + ^{4}He$		

The nuclear energy generation rate (ϵ) of these channels has different temperature sensitivities, meaning that the conditions in the stellar interior will define the efficiency with which each one of them operates. The p-p chain has an average relation of the order of $\epsilon_{\rm pp} \propto T^4$ at $T \approx 15 \times 10^6$ K, while the CNO cycle has a higher value of $\epsilon_{\rm CNO} \propto T^{18}$ at $T \approx 20 \times 10^6$ K. As an example, in the centre of the Sun $T \approx 15 \times 10^6$ K and more than 90% of the energy budget corresponds to the p-p chain. An important consequence of the temperature sensitivities of the nuclear reaction chains is that, if the H-burning process is dominated by the CNO cycle, it will be confined towards the very central regions of the star. This results in a larger energy flux arising from the innermost regions which favours the presence of a convective core.

Following the exhaustion of hydrogen in the centre, stars begin burning helium as soon as the central temperature increases enough to produce the *triple alpha* (3α) reaction:

$$^{4}He + ^{4}He \rightarrow ^{8}Be$$
$$^{8}Be + ^{4}He \rightarrow ^{12}C + \gamma\,.$$

The temperature sensitivity is quite strong for the 3α reaction: $\epsilon_{3\alpha} \propto T^{40}$ at $T \approx 10^8$ K. Thus, for the same physical reason as in the case of the CNO cycle, stars burning helium via the 3α mechanism have extended convective cores. For the sake of completeness, I mention the other important nuclear reactions involved in the helium burning process:

$$^{12}C + ^{4}He \rightarrow ^{16}O + \gamma$$
$$^{16}O + ^{4}He \rightarrow ^{20}Ne + \gamma\,.$$

It is clear by looking at these reactions that helium burning transforms ^{4}He particles mainly into ^{12}C, ^{16}O and ^{20}Ne. These elements have burning processes in more advanced stages of stellar evolution. I refer the reader to the monograph by Clayton (1983) for a detailed description of the above mentioned reactions, and further explanations on the reaction networks of elements heavier than helium.

2.3 Choosing the Ingredients

Based on the descriptions of the previous section, we now turn our attention to the practicalities of modelling a particular star or stellar population. The target(s) in question will have some set of observed properties that we aim at reproducing with our stellar evolution model, for example colour (or effective temperature), magnitude (or gravity), surface chemical composition, and oscillation frequencies, just to name a few. The task is to select the appropriate ingredients in our evolutionary code that will produce the most realistic representation possible of the star, and this requires some assumptions on the relevant physical processes to be included in the modelling procedure.

We start by defining the microphysics applied in the calculation as described in Sect. 2.2. In the case of the Equation Of State, several compilations relevant for stellar calculations are available such as those computed by the OPAL group (Rogers et al. 1996; Rogers and Nayfonov 2002) and the FreeEOS (Cassisi et al. 2003). These may need to be complemented at the low-temperature regime by other compilations of EOS, such as the MHD EOS (Hummer and Mihalas 1988) or that of Saumon et al. (1995).

Regarding the opacity of stellar matter, this requires different treatments for the radiative and conductive cases. Compilations of Rosseland mean radiative opacities (see Eq. (12)) from the OPAL group (Iglesias and Rogers 1996) and Opacity Project (OP; Badnell et al. 2005) are available, normally complemented at low temperatures by molecular opacities from Alexander and Ferguson (1994) and Ferguson et al. (2005). For the conductive opacities, relevant in the cases of degenerate matter, calculations by Itoh et al. (1983) or the more updated Cassisi et al. (2007) are commonly adopted.

For the nuclear reaction rates, the most frequently used large compilations of cross sections are those by the NACRE collaboration (Angulo et al. 1999) and the Solar Fusion (Adelberger et al. 1998, 2011). We note in passing that the cross sections of two important astrophysical factors (S_{34} and $S_{1,14}$) have been updated by Marta et al. (2008) and Costantini et al. (2008), and is highly recommended that the chosen set of thermonuclear reactions includes these latest values. Energy losses via neutrino emission (Eq. (23)) are determined following prescriptions by, e.g., Haft et al. (1994) and Itoh et al. (1996).

Once the desired microphysics of the evolutionary calculation is defined, a few more considerations must be made before we can model our targets. First, one must select a criterion for defining the boundaries of convective regions in the stellar interior, which as described in Sect. 2.1.2 is normally done using the Ledoux or Schwarzschild criterion. Second, we must define the value of an efficiency parameter describing the gas motions in zones found to be convective to determine the real temperature gradient (Eq. (8)). Then, we must define the initial chemical composition used in our stellar modelling exercise based on some knowledge of the surface abundances of the target. In the following I describe how these parameters are normally obtained in evolutionary calculations, and mention some of the additional mixing processes included in some evolutionary calculations depending on type of star considered.

Before closing this section, I note in passing that not all the aforementioned sets of opacities, EOS, nuclear reactions etc. are available in a given evolutionary code, and each modeller should carefully ensure that the combination of microphysics is consistent for each calculation (i.e., the used opacities are calculated for the adopted solar mixture).

2.3.1 Chemical Composition

In stellar evolution calculations abundances are represented with the letters X, Y, Z, defining the mass fractions of hydrogen, helium, and all elements heavier than helium ('metals'), respectively. These three quantities must be provided to the evolutionary code performing the calculation, and I review in this section the assumptions adopted in their determination.

First, we need the observed surface abundance of the star we aim at modelling as obtained by, e.g., photometry or spectroscopy. In most cases the observed chemical composition is given in terms of logarithmic abundance ratios with respect to the solar value. For the species i and j, this is expressed as

$$\log(i/j) - \log(i/j)_{\odot} \equiv [i/j]\,, \tag{25}$$

where the dependence of the observed abundances on the solar reference value is clearly established. Normally, observations provide a measurement of the bulk stellar metallicity in terms of the iron abundance [Fe/H], and to transform it into an estimate of the mass fraction of heavy elements required in evolutionary calculations it is assumed that

$$\log(Z/X) - \log(Z/X)_{\odot} \simeq [\mathrm{Fe/H}]\,. \tag{26}$$

As we mentioned above, Z in evolutionary calculations includes every element heavier than helium and thus this transformation has two important consequences. First it assumes that iron is by large the most abundant of the heavy elements in a star, which is normally the case unless excess of α elements is present (such as in metal-poor stars; see, e.g., Sneden 1985). Second, since a solar reference value $\log(Z/X)_{\odot}$ must be defined, it implies that the fraction of each element comprising Z is distributed accordingly to the chosen set of solar abundance ratios (or solar 'mixture').

The topic of the atmospheric chemical composition of the Sun has been one of hot debate in the past years. The most recent determinations of the solar individual abundances based on realistic 3D hydrodynamical simulations of stellar atmospheres have considerably decreased the total metallicity of the Sun compared to the older predictions based on simpler 1D atmospheric models. However, these new sets of solar abundances are not yet completely accepted by the astrophysics community mainly due to the discrepancy of the predicted internal sound speed profile with the results of helioseismology (see, e.g., Serenelli et al. 2009), and therefore both the old (Grevesse and Noels 1993; Grevesse and Sauval 1998) and new (Asplund et al. 2009; Caffau et al. 2011) determinations are widely applied according to personal preference. The reader is referred to Bahcall et al. (2005) and Basu and Antia (2008) for further explanations on the topic and its implications for helioseismology.

In terms of total metallicity determinations, the ratio $(Z/X)_{\odot}$ varies from 0.0245 to 0.0183 among the different compilations, which according to Eq. (26) implies that the value of [Fe/H] in a stellar model computed with the same Z and X can vary by $\sim$0.13 dex just by the choice of solar mixture. This must be kept in mind when comparing evolutionary calculations with spectroscopic or photometric metallicities, which can quote uncertainties well below this level.

Equation (26) is not sufficient to estimate the three necessary components X, Y, and Z, of the stellar composition. Moreover, it is not possible to directly measure helium abundances in stars of temperatures lower than $\sim 10^4$ K. Thus, a so-called 'galactic chemical evolution law' is applied (Peimbert and Torres-Peimbert 1976), which basically relates the amount of fresh helium supplied by stars to the interstellar medium relative to their supply of heavy elements:

$$\frac{\Delta Y}{\Delta Z} = \frac{Y - Y_{\rm ref}}{Z - Z_{\rm ref}}. \tag{27}$$

The existence and value of the $\Delta Y/\Delta Z$ relation is still a matter of debate, and it is usually considered to lie in the range $1.0 \leq \Delta Y/\Delta Z \leq 3.0$ (e.g., Pagel and Portinari 1998; Jimenez 2003; Casagrande et al. 2007). In Eq. (27) one must provide a reference point for Y and Z; big bang nucleosynthesis values can be considered ($Z_{\rm ref} = 0$ and $Y_{\rm ref} \sim 0.2488$; Steigman 2010), or the initial Z and Y abundances obtained in the Sun from a solar calibration (see Sect. 2.3.2 below). In the latter case, the chosen set of today's surface abundances in the Sun plays an important role, although it has been suggested that the initial helium abundance of the Sun could be independent of them (Serenelli and Basu 2010). Finally, the third equation comes from the obvious fact that $X + Y + Z = 1$.

2.3.2 Convective Efficiency

As mentioned in Sect. 2.1.2 the value of the convective efficiency (i.e., $\alpha_{\rm MLT}$ under the mixing-length formulation) ultimately defines the value of ∇ in a convective region, but unfortunately it cannot be obtained from first principles. Instead it is normally determined via a standard solar calibration, an iterative procedure in which three input parameters are tuned to reproduce the properties of the Sun. This is an optimisation process where the codes vary the initial composition (two parameters, e.g., Y and Z) and the convective efficiency to reproduce the solar radius and luminosity, at present solar age, for a chosen surface composition of the Sun. In the case of the mixing-length theory, the resulting value usually ranges between $1.5 < \alpha_{\rm MLT} < 2.5$ depending on the input physics and evolutionary code employed. There is a vast amount of literature devoted to the results of solar calibrations where the interested reader can find more details (see, e.g., Christensen-Dalsgaard et al. 1996; Bahcall et al. 2001, and references therein).

In recent years, a new possibility has appeared for determining convective efficiency values from 3D simulations of stellar envelopes. These simulations solve the time-dependent hydrodynamic equations of mass, momentum, and energy conservation and are by design free from adjustable parameters such as the $\alpha_{\rm MLT}$ (e.g., Nordlund et al. 2009; Kupka and Muthsam 2017, and references therein). Using suitable averages, it is possible to match the atmospheric stratification of the 3D model with the equivalent from one-dimensional calculations and extract a calibrated convective efficiency (e.g., Trampedach et al. 2014; Magic et al. 2015). These results have been recently implemented in standard stellar evolution codes at solar metallicity (Salaris and Cassisi 2015; Rørsted et al. 2016) and are being extended to other chemical compositions.

2.3.3 Additional Mixing

The last layer of complications in the macrophysics of stellar evolution comes from physical processes that are likely to take place in stars and are still poorly understood. Among these we can mention the cases of atomic diffusion of helium and heavier elements, radiative levitation, rotational mixing, the influence of magnetic fields, stellar winds etc. (for a review, see Pinsonneault 1997). All these processes require the inclusion of some additional free parameter controlling their efficiency as a function of, e.g., mass, temperature, luminosity etc., and are relevant in different regimes of stellar evolution to accurately reproduce the observational results. As an example, the inclusion of atomic diffusion in standard solar models greatly improved the agreement with helioseismic data via inversion techniques (Christensen-Dalsgaard et al. 1993) and is now regarded as a necessary ingredient to accurately reproduce the properties of the Sun. However, diffusion as it operates in the solar case requires the inclusion of additional processes to counteract some of its effects in late-type stars via radiative accelerations (e.g., Turcotte et al. 1998). Similarly, overshoot from the convective core is necessary to reproduce the shape of the colour-magnitude diagram of open clusters (e.g., Maeder and Meynet 1991), but despite extensive theoretical studies there is still no firm verdict on the amount of additional mixing required and its dependence on parameters such as mass or metallicity (e.g., Zahn 1991).

Asteroseismology is beginning to make substantial contributions to our understanding of such additional mixing processes, and among these we can mention the first measurements of integrated mass loss in the red-giant branch (Miglio et al. 2012), constraints on the amount of rotational mixing in subgiant and red-giant stars (Eggenberger et al. 2012), detection of convective cores and constraints on the overshoot efficiency during the core hydrogen- and helium-burning phases (Silva Aguirre et al. 2013; Constantino et al. 2015; Deheuvels et al. 2016), as well as the inversion of the rotation profile in *Kepler* targets (Deheuvels et al. 2014). It is expected that these results will further our understanding of these fundamental physical processes and allow us to produce even more realistic models of stellar interiors.

3 Overview of Stellar Evolution

The ultimate fate of a star depends mostly on its initial mass and chemical composition, properties that are related to the place and time where the star was born and possible interactions with the medium surrounding it. It is customary to analyse the main phases of stellar evolution following the path described by the surface luminosity, L, and effective temperature, $T_{\rm eff}$, throughout the star's lifetime. This is the so-called *Hertzsprung–Russell Diagram* (HRD), shown for several masses in Fig. 1 at a fixed chemical composition. I will mostly focus on the evolution of stars with masses below $\sim$2.5–3.0 $M_\odot$, from the beginning of the hydrogen

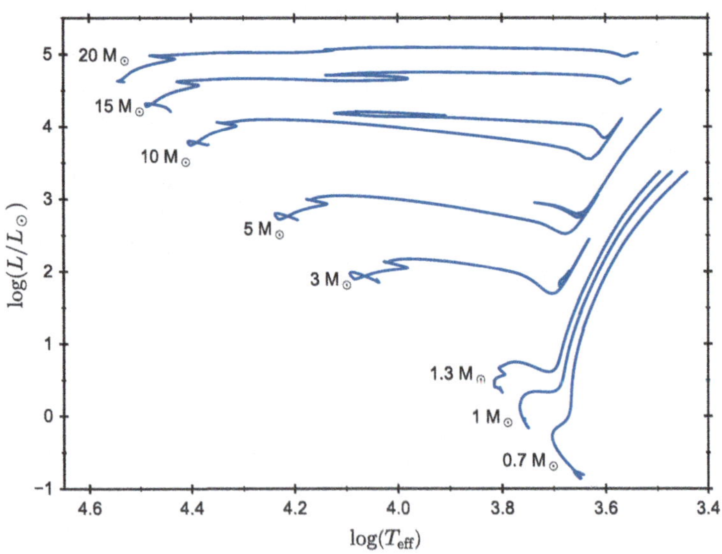

Fig. 1 Hertzsprung–Russell Diagram plotted for different masses at a fixed chemical composition of $Y = 0.28$ and $Z = 0.02$. The tracks cover the evolution from the beginning of the main sequence to the red-giant phase or, for $M \gtrsim 3.0\,M_\odot$, also the helium burning phase

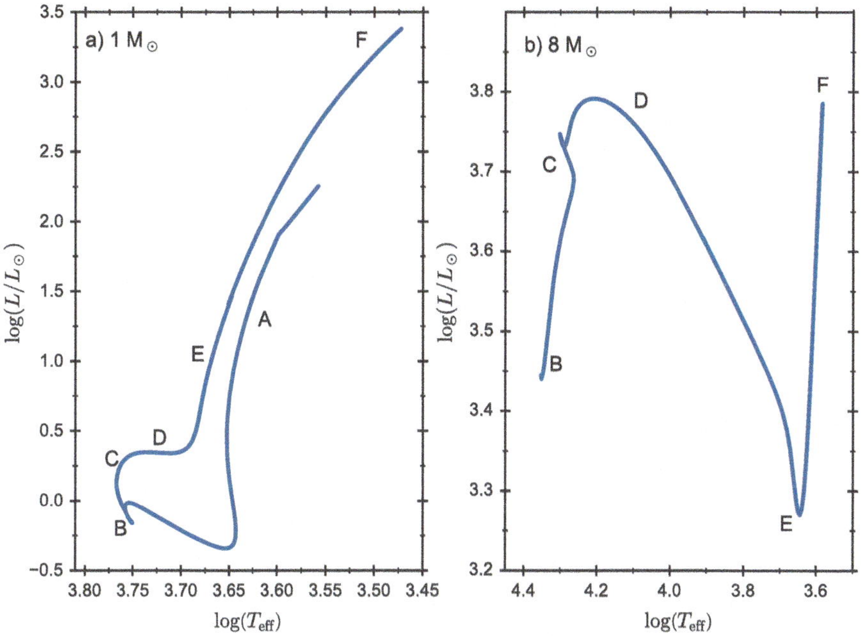

Fig. 2 Hertzsprung–Russell Diagram for stars at the same metallicity as in Fig. 1. *Left:* Evolution of a 1 $M_{\odot}$ star from the pre-main sequence to the tip of the red-giant branch. *Right:* Similar evolution of an 8.0 $M_{\odot}$ star, where the pre-main sequence has been removed for better visualization

burning phase until helium is exhausted in the centre. Nevertheless, I will also broadly describe the evolution of more massive stars. For better guidance through the different evolutionary stages, Fig. 2 presents two HRDs of different evolutionary phases for stars of the same metallicity as that of the tracks in Fig. 1. The left panel shows a 1 $M_{\odot}$ star evolving from the pre-main sequence until helium ignition, while the right panel depicts an 8 $M_{\odot}$ star in similar evolutionary phases.

3.1 Pre-main Sequence Phase

Protostars form through condensation of interstellar matter at low temperatures in hydrostatic equilibrium. Giant molecular clouds with masses large enough to undergo gravitational collapse fragment into smaller subunits due to the presence of inhomogeneities. Each of these subregions forms a hydrostatic core and accretes free-falling gas from its surroundings. Once this accretion process is complete, the protostar collapses again until hydrostatic equilibrium is restored, giving birth to a new star.

A star in this phase is fully convective, has a low temperature, a large radius and a high luminosity. It evolves almost vertically in the HRD (at roughly constant temperature) along the so-called *Hayashi track* (see point 'A' in the left panel of Fig. 2), its exact location depending on the initial mass and chemical composition. Once a radiative core appears, the star leaves the Hayashi track and moves to higher effective temperatures while its convective envelope slowly retreats. The star contracts, increasing its central temperature and density, until fusion of hydrogen to helium becomes efficient and releases enough energy to counteract the gravitational force. At this point, the star has reached the Zero Age Main Sequence (ZAMS; point 'B' in both panels of Fig. 2).

Due to the temperature sensitivity of the reaction networks, stars of masses $M \leq 1.3\,\mathrm{M}_\odot$ burn hydrogen mainly through the p-p chain (see Table 1), the exact transition mass value depending on the initial chemical composition (it is $M \approx 1.3\,\mathrm{M}_\odot$ at solar metallicity). Before reaching the ZAMS, some nuclear burning occurs in these stars, such as transforming deuterium into ^{3}He. While this reaction takes place and the abundance of ^{3}He is not high enough as to complete the pp I branch of the p-p chain, the star is forced to reach higher temperatures and densities in order to satisfy its energy needs. This higher temperature also induces the first three reactions of the CNO cycle, burning ^{12}C into ^{14}N. A small convective core appears as a consequence of an energy generation more concentrated toward the centre; it only survives until the ^{12}C abundance decreases and the amount of ^{3}He increases enough for the p-p reactions to become more important and the energy generation to be redistributed over a larger area.

3.2 *Main Sequence Evolution*

The main sequence corresponds to the phase in which a star transforms hydrogen into helium at its centre, and it is the longest of the evolutionary phases in a star's lifetime (points 'B' to 'C' in Fig. 2). Its duration is mainly controlled by the mass of the star, while in comparison the chemical composition and mixing processes play a secondary role. From homology relations (see, e.g., Kippenhahn et al. 2012), there exists a mass-luminosity relation that gives the dependence between these parameters in different evolutionary stages. For stars on the main sequence, at a given chemical composition, it is of the order of

$$\left(\frac{L}{\mathrm{L}_\odot}\right) \propto \left(\frac{M}{\mathrm{M}_\odot}\right)^{3.5} . \tag{28}$$

Recalling the nuclear timescale given by Eq. (3), and replacing L with the mass-luminosity relation, it is clear that the time a star can shine with nuclear burning as its energy source decreases for increasing stellar mass. To give some numbers, at solar metallicity, a $1.0\,\mathrm{M}_\odot$ star burns hydrogen for approximately 9 Gyr, while a $20\,\mathrm{M}_\odot$ star does it for 8 Myr only.

Stars more massive than $\sim 1.7\,\mathrm{M}_\odot$ have a convective core and a radiative envelope, while their less massive counterparts have convective envelopes on top of their radiative interiors and can host either radiative or convective cores. The presence of a convective envelope has important consequences for the pulsation properties of stars, as it stochastically produces excitation of modes.

The main effect of a larger mass is a significant increase of the interior temperature, resulting in different efficiencies of the H-burning reaction networks. Depending on the resulting mechanism employed to burn hydrogen, stars are usually classified in lower main-sequence stars (masses below $\sim 1.3\,M_\odot$ where the p-p chain is the main mechanism) and upper main-sequence stars (more massive stars where the CNO cycle plays the leading role). When the CNO cycle is the dominant source of energy production, the centre of the star becomes convective due to a very high energy flux in the innermost regions. As the mass increases, convective core size also grows as a consequence of the higher temperature in the interior leading to a larger flux.

It is assumed that the convective core is homogeneously mixed, and its size determines the amount of available fuel for hydrogen burning. For a star of a given mass and chemical composition, this defines the total time it will spend burning hydrogen, the size of its helium core once the hydrogen in the centre is exhausted, and the exact position of the star in the HRD. If the star had a convective core during its main-sequence evolution, its disappearance once the star reaches the *turn-off* point leaves a characteristic hook-like feature in the HRD (compare points 'C' in both panels of Fig. 2). The existence of this feature in the Colour Magnitude Diagram (CMD) of clusters is used to calibrate the amount of mixing beyond the convective core and directly impacts age determinations via isochrone fitting (e.g., Vandenberg et al. 2007). Despite its importance, the exact extension of the convective core is still an open problem due to the uncertainty in the 'true' convective boundary definition, and the contribution of the different physical processes that mix material beyond this formal boundary (e.g., Silva Aguirre et al. 2011; Gabriel et al. 2014).

3.3 Subgiants, Giants, and Clump Giants

At the turn-off point (related to hydrogen exhaustion in the centre), hydrogen burning ceases to be a central process and becomes a shell-burning process in a layer outside of the He-rich core. At the same time, the stellar envelope expands cooling down the star and moving it to the right in the HRD. This constitutes the *subgiant* phase (point 'D' in Fig. 2), where stars evolve roughly at the pace set by the Kelvin–Helmholtz timescale (Eq. (1)).

For masses above $\sim 2.5\text{–}3.0\,M_\odot$, the core contracts since it cannot counteract the pressure exerted by the layers above it (the exact relation between core and envelope mass is given by the *Schönberg–Chandrasekhar* limit; Schönberg and Chandrasekhar 1942). A convective envelope develops due to the cooling down of the outer layers, which marks the beginning of the *red giant* (RG) phase (point 'E' in Fig. 2b). From here onwards, the star evolves at a roughly constant temperature, burning hydrogen in a shell while increasing its luminosity and radius and further contracting the core. Eventually, the central temperature reaches values high enough as to ignite helium in the centre, marking the end of the RG phase (point 'F' in Fig. 2b). This occurs under non-degenerate conditions, as the central density is low enough to prevent the onset of electron degeneracy. For increasing stellar mass, the time needed to reach He-burning central temperatures is very short and the RG might even disappear.

The case of masses lower than $\sim 2.5\,M_{\odot}$ is slightly different from their more massive counterparts. The gas in the He-rich core is electron-degenerate, providing enough pressure to support the envelope above it and at the same time grow from the production of helium in the H-burning shell. During the subgiant phase, the cooling down of the outer layers results in an inward penetration of the already existing convective envelope, which drags partially processed nuclear material to the surface. Point 'E' in Fig. 2a shows the position in the HRD where maximum inward penetration of the envelope occurs, a phenomenon called the *first dredge-up*. This is observationally witnessed by the change in CNO abundances due to mixing of former nuclearly processed material dredged-up to the surface.

After this process, the convective envelope begins to retreat and the star continues its vertical ascent along the *red-giant branch* (RGB; the portion of the HRD populated by stars evolving through the RG phase). One must keep in mind that, during the evolution up the RGB, stars lose some amount of mass through stellar winds due to their increasingly larger envelopes. Nevertheless, the mass-loss rates have not been tightly constrained by either observations or theory, and ad hoc parameterizations are used to reproduce this phenomenon, such as the one given by Reimers (1975, 1977).

Although helium ignition occurs quietly in massive stars, the case for stars below $\sim 2.5\,M_{\odot}$ is somewhat different. For better visualisation, Fig. 3 depicts the evolution from the tip of the red-giant branch until centre helium exhaustion of a $1.0\,M_{\odot}$ star at solar metallicity. In the late stages of the RGB, stars lose large amounts of energy in the form of neutrinos, this form of dissipation being most efficient where the stellar matter is more dense (its centre). An inversion of the thermal profile occurs in the He core, the hottest place being a layer located off-centre within the He-rich core. When temperatures high enough to start helium burning are reached, the ignition takes place off-centre in a sort of thermonuclear runaway, called the *core helium flash* (point 'F' in Figs. 2a and 3).

The reason for this phenomenon has to do with a property of the electron-degenerate gas, which is the decoupling of the temperature dependence from the $P-\rho$ relation (the EOS; see Sect. 2.2). For such a case, the energy input to the medium due to the nuclear reactions when helium ignites increases the local temperature, but no compensating increase of the pressure takes place. As a consequence, the region where burning takes place does not expand to cool down the material, leading to an increase in the thermonuclear reaction rates, which leads to a further increase in the local temperature, and so forth. The thermodynamical runaway is terminated once the increasing temperature removes degeneracy and the EOS becomes temperature-dependent. During this phase, the large amounts of energy

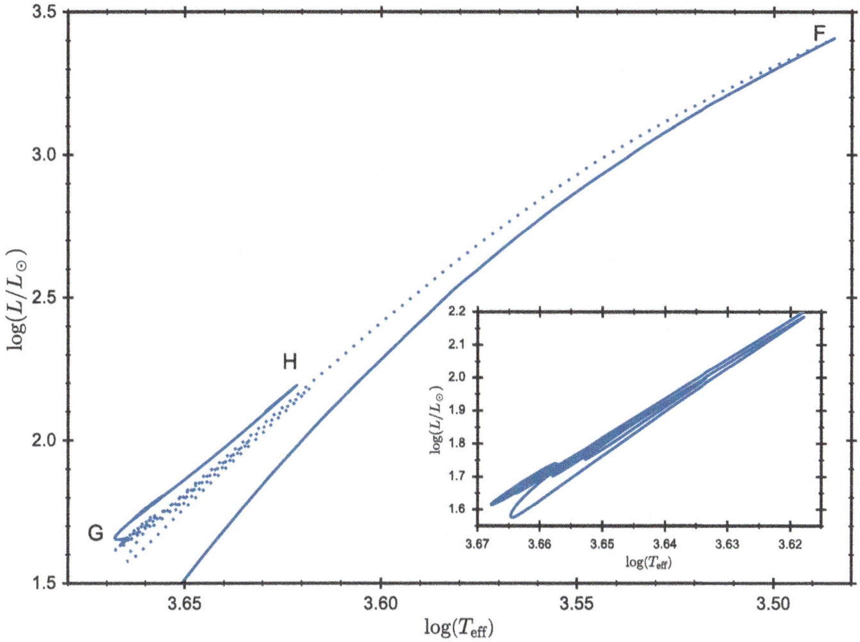

Fig. 3 Hertzsprung–Russell Diagram of a $1.0\,M_\odot$ star from the tip of the red-giant branch until centre helium exhaustion at low metallicity. The evolution during the helium flash is marked with a *dotted line* in the main panel for clarity, and is also depicted in the inset (*solid line*)

produced by the ignition are used to lift up the degeneracy in the core, decreasing considerably the luminosity; in fact, secondary flashes take place increasingly closer to the centre until degeneracy has been lifted throughout the He core, producing loops in the HRD (see inset in Fig. 3).

The star is now able to burn helium quiescently in a convective core and hydrogen in a shell, marking the end of the RG phase. This evolutionary stage is called the *horizontal branch* phase for stars in Globular Clusters (low-mass, metal-poor stars; see point 'G' in Fig. 3), while it is usually known as the *red clump* for composite populations since their location in the HRD can be closer to the RGB. In this latter case, a secondary (and less luminous) red clump can also be present, comprising the more massive stars that started the He-burning in non-degenerate conditions (Girardi 1999).

The horizontal branch or clump is the second longest evolutionary phase in the life of a star, in which helium is burned in a homogeneously mixed convective core through the 3α mechanism (see Sect. 2.2.1), surrounded by a H-burning shell. Evolution until exhaustion of helium in the centre is shown in Fig. 3 (points 'G' to 'H'), which for this case lasts approximately 80 Myr.

3.4 Advanced Stages of Evolution

For the sake of completeness, I briefly mention the evolutionary phases beyond the helium-burning stage. Once helium is exhausted in the core, stars with initial masses $\lesssim 8.0\,M_\odot$ shine by helium- and hydrogen-shell burning. They are said to ascend the *asymptotic giant branch* (AGB), a phase where extinction and re-ignition of helium leads to the occurrence of thermal pulses. During this evolutionary stage, stars undergo large amounts of mixing and complex nuclear reactions, such as slow-neutron capture and carbon burning. Large amounts of mass are lost due to dust-driven winds and large-amplitude pulsations. Once hydrogen is largely exhausted in the burning shell, the remaining envelope is rapidly lost and shines due to ionisation by the bare core of the star as a *planetary nebula*. This exposes the central star (a *white dwarf*), which consequently evolves down the white dwarf cooling curve over a time scale of billions of years.

The value of $\sim 8.0\,M_\odot$ given above is a very loose approximation of the maximum initial mass a star should have to end its life as a white dwarf. Its ultimate fate will depend on its capability of losing enough mass by stellar winds throughout evolution to have a final mass smaller than the so-called *Chandrasekhar* limiting mass. This limit gives the highest possible final core mass a star can have in order to be a stable white dwarf, and it is usually considered to be approximately $\sim 1.45\,M_\odot$.

Naturally, there are stars that have core masses far higher than the *Chandrasekhar* limiting mass once they have exhausted helium in the centre. These stars will ignite carbon under non-degenerate conditions. As they go through several cycles of nuclear burning, they produce shells of heavier elements in a so-called *onion skin* model. At last, the core, consisting mostly of ^{56}Fe (or a neighbouring nucleus), becomes dynamically unstable and core collapse sets in, resulting in a supernova explosion.

4 Closing Remarks

The picture of stellar structure and evolution sketched in this chapter is an overwhelmingly simplified one and I encourage the reader to consult the extensive available literature for all the interesting details about the birth, life, and fate of stars. I have reviewed the main components and assumptions entering stellar evolution calculations in different phases and regimes, highlighting the areas where large uncertainties still remain and how additional tools such as asteroseismology can help constraining these processes.

Ultimately, our knowledge of the formation and evolution of planetary systems critically depends on our understanding of stars. Several compilations of stellar tracks and isochrones including state-of-the-art macrophysics and microphysics have been computed throughout the years by different groups, and are perfectly suited for a wide range of applications in astrophysical research. Some of the most commonly used include the BASTI isochrones (Pietrinferni et al. 2004; Cordier et al. 2007; Salaris et al. 2010), the Darthmouth stellar evolution database (Dotter et al. 2008), and the MESA (Choi et al. 2016) and PARSEC stellar tracks and isochrones (Bressan et al. 2012). These are routinely used in the characterisation of exoplanet-host stars (e.g., Schlaufman 2010; Huber et al. 2013; Silva Aguirre et al. 2015), and can be combined with sophisticated Bayesian schemes to extract stellar properties given a set of observations (e.g., Da Silva et al. 2006; Serenelli et al. 2013; Silva Aguirre et al. 2017).

Acknowledgements The author would like to thank the organisers for their invitation to be an instructor at this Summer School, giving me the opportunity of sharing my care for stars with an outstanding group of avid young astronomers. The author also thanks Jørgen Christensen-Dalsgaard, Achim Weiss, Santi Cassisi, and Aldo Serenelli for many stimulating discussions that helped shaping the content of this lecture.

References

Adelberger, E.G., Austin, S.M., Bahcall, J.N., et al.: Rev. Mod. Phys. **70**, 1265 (1998)
Adelberger, E.G., García, A., Robertson, R.G.H., et al.: Rev. Mod. Phys. **83**, 195 (2011)
Alexander, D.R., Ferguson, J.W.: Astrophys. J. **437**, 879 (1994)
Angulo, C., Arnould, M., Rayet, M., et al.: Nucl. Phys. A **656**, 3 (1999)
Asplund, M., Grevesse, N., Sauval, A.J., Scott, P.: Annu. Rev. Astron. Astrophys. **47**, 481 (2009)
Badnell, N.R., Bautista, M.A., Butler, K., et al.: Mon. Not. R. Astron. Soc. **360**, 458 (2005)
Bahcall, J.N., Pinsonneault, M.H., Basu, S.: Astrophys. J. **555**, 990 (2001)
Bahcall, J.N., Basu, S., Pinsonneault, M., Serenelli, A.M.: Astrophys. J. **618**, 1049 (2005)
Basu, S., Antia, H.M.: Phys. Rep. **457**, 217 (2008)
Bedding, T.R., Huber, D., Stello, D., et al.: Astrophys. J. Lett. **713**, L176 (2010)
Bland-Hawthorn, J., Gerhard, O.: Annu. Rev. Astron. Astrophys. **54**, 529 (2016)
Böhm-Vitense, E.: Z. Astrophys. **46**, 108 (1958)
Bressan, A., Marigo, P., Girardi, L., et al.: Mon. Not. R. Astron. Soc. **427**, 127 (2012)
Caffau, E., Ludwig, H.-G., Steffen, M., Freytag, B., Bonifacio, P.: Sol. Phys. **268**, 255 (2011)
Camenzind, M.: Compact Objects in Astrophysics: White Dwarfs. Springer, Berlin (2007)
Canuto, V.M., Mazzitelli, I.: Astrophys. J. **370**, 295 (1991)
Canuto, V.M., Mazzitelli, I.: Astrophys. J. **389**, 724 (1992)
Casagrande, L., Flynn, C., Portinari, L., Girardi, L., Jimenez, R.: Mon. Not. R. Astron. Soc. **382**, 1516 (2007)
Cassisi, S., Salaris, M., Irwin, A.W.: Astrophys. J. **588**, 862 (2003)
Cassisi, S., Potekhin, A.Y., Pietrinferni, A., Catelan, M., Salaris, M.: Astrophys. J. **661**, 1094 (2007)
Choi, J., Dotter, A., Conroy, C., et al.: Astrophys. J. **823**, 1 (2016)
Christensen-Dalsgaard, J.: Sol. Phys. **220**, 137 (2004)
Christensen-Dalsgaard, J., Proffitt, C.R., Thompson, M.J.: Astrophys. J. **403**, L75 (1993)
Christensen-Dalsgaard, J., Dappen, W., Ajukov, S.V., et al.: Science **272**, 1286 (1996)
Clayton, D.D.: Principles of Stellar Evolution and Nucleosynthesis. University of Chicago Press, Chicago (1983). http://adsabs.harvard.edu/abs/1983psen.book.....C
Conselice, C.J.: Annu. Rev. Astron. Astrophys. **52**, 291 (2014)
Constantino, T., Campbell, S.W., Christensen-Dalsgaard, J., Lattanzio, J.C., Stello, D.: Mon. Not. R. Astron. Soc. **452**, 123 (2015)
Cordier, D., Pietrinferni, A., Cassisi, S., Salaris, M.: Astron. J. **133**, 468 (2007)
Costantini, H., Bemmerer, D., Confortola, F., et al.: Nucl. Phys. A **814**, 144 (2008)

Cox, J.-P., Giuli, R.T.: Principles of Stellar Structure. Gordon and Breach, New York (1968)
Da Silva, L., Girardi, L., Pasquini, L., et al.: Astron. Astrophys. **458**, 609 (2006)
Deheuvels, S., Michel, E.: Astron. Astrophys. **535**, 91 (2011)
Deheuvels, S., Doğan, G., Goupil, M.J., et al.: Astron. Astrophys. **564**, A27 (2014)
Deheuvels, S., Brandão, I., Silva Aguirre, V., et al.: Astron. Astrophys. **589**, A93 (2016)
Dotter, A., Chaboyer, B., Jevremović, D., et al.: Astrophys. J. Suppl. Ser. **178**, 89 (2008)
Eggenberger, P., Montalbán, J., Miglio, A.: Astron. Astrophys. **544**, L4 (2012)
Ferguson, J.W., Alexander, D.R., Allard, F., et al.: Astrophys. J. **623**, 585 (2005)
Gabriel, M., Noels, A., Montalbán, J., Miglio, A.: Astron. Astrophys. **569**, A63 (2014)
Girardi, L.: Mon. Not. R. Astron. Soc. **308**, 818 (1999)
Grevesse, N., Noels, A.: In: Hauck, B., Paltani, S., Raboud, D. (eds.) 35ème cours de perfectionnement de l'Association Vandoise des Chercheurs en Physique, p. 205. AVCP, Lausanne (1993)
Grevesse, N., Sauval, A.J.: Space Sci. Rev. **85**, 161 (1998)
Haft, M., Raffelt, G., Weiss, A.: Astrophys. J. **425**, 222 (1994)
Hansen, C.J., Kawaler, S.D., Trimble, V.: Stellar Interiors: Physical Principles, Structure, and Evolution, 2nd edn. Springer, New York (2004). http://adsabs.harvard.edu/abs/2004sipp.book.....H
Herwig, F.: Annu. Rev. Astron. Astrophys. **43**, 435 (2005)
Hubeny, I., Mihalas, D.: Theory of Stellar Atmospheres. Princeton University Press, Princeton (2014)
Huber, D., Chaplin, W.J., Christensen-Dalsgaard, J., et al.: Astrophys. J. **767**, 127 (2013)
Hummer, D.G., Mihalas, D.: Astrophys. J. **331**, 794 (1988)
Iglesias, C.A., Rogers, F.J.: Astrophys. J. **464**, 943 (1996)
Itoh, N., Mitake, S., Iyetomi, H., Ichimaru, S.: Astrophys. J. **273**, 774 (1983)
Itoh, N., Hayashi, H., Nishikawa, A., Kohyama, Y.: Astrophys. J. Suppl. Ser. **102**, 411 (1996)
Jimenez, R. Science **299**, 1552 (2003)
Kippenhahn, R., Weigert, A., Weiss, A.: Stellar Structure and Evolution. Springer, Berlin (2012). http://adsabs.harvard.edu/abs/2012sse..book.....C
Kjeldsen, H., Bedding, T.R., Baldry, I.K., et al.: Astron. J. **126**, 1483 (2003)
Kumar, P., Zhang, B.: Phys. Rep. **561**, 1 (2015)
Kupka, F., Muthsam, H.J.: Modelling of stellar convection. Living Rev. Comput. Astrophys. **3**(1), Article ID. 1, 159 pp. (2017). http://adsabs.harvard.edu/abs/2017LRCA....3....1K
Ledoux, W.P.: Astrophys. J. **105**, 305 (1947)
Maeder, A., Meynet, G.: Astron. Astrophys. Suppl. **89**, 451 (1991)
Magic, Z., Weiss, A., Asplund, M.: The Stagger-grid: a grid of 3D stellar atmosphere models. III. The relation to mixing length convection theory. Astron. Astrophys. **573**, ID.A89, 17 pp. (2015). http://adsabs.harvard.edu/abs/2015A%26A...573A..89M
Maoz, D., Mannucci, F., Nelemans, G.: Annu. Rev. Astron. Astrophys. **52**, 107 (2014)
Marta, M., Formicola, A., Gyürky, G., et al.: Phys. Rev. C **78**, 22802 (2008)
Matteucci, F.: J. Phys. Conf. Ser. **703**, 012004 (2016)
Miglio, A., Brogaard, K., Stello, D., et al.: Mon. Not. R. Astron. Soc. **419**, 2077 (2012)
Mihalas, D.: Stellar Atmospheres, Series of Books in Astronomy and Astrophysics. W.H. Freeman & Co. Ltd., San Francisco (1970)
Nordlund, Å., Stein, R.F., Asplund, M.: Solar surface convection. Living Rev. Sol. Phys. **6**(1), 2, 117 pp. (2009). http://adsabs.harvard.edu/abs/2009LRSP....6....2N
Pagel, B.E.J., Portinari, L.: Mon. Not. R. Astron. Soc. **298**, 747 (1998)
Peimbert, M., Torres-Peimbert, S.: Astrophys. J. **203**, 581 (1976)
Pietrinferni, A., Cassisi, S., Salaris, M., Castelli, F.: Astrophys. J. **612**, 168 (2004)
Pinsonneault, M.: Annu. Rev. Astron. Astrophys. **35**, 557 (1997)
Reimers, D.: Mem. Soc. R. Liège 6 Sér. **8**, 369 (1975)
Reimers, D.: Astron. Astrophys. **57**, 395 (1977)
Rogers, F.J., Nayfonov, A.: Astrophys. J. **576**, 1064 (2002)
Rogers, F.J., Swenson, F.J., Iglesias, C.A.: Astrophys. J. **456**, 902 (1996)

Rørsted, M.J., Silva, A.V., Weiss, A., Christensen-Dalsgaard, J., Trampedach, R.: Improving 1D Stellar Models with 3D Atmospheres (2016). arXiv:1610.07323. http://adsabs.harvard.edu/abs/2016arXiv161007323R

Salaris, M., Cassisi, S.: Stellar models with mixing length and T(τ) relations calibrated on 3D convection simulations. Astron. Astrophys. **577**, ID.A60, 6 pp. (2015). http://adsabs.harvard.edu/abs/2015A%26A...577A..60S

Salaris, M., Cassisi, S., Pietrinferni, A., Kowalski, P.M., Isern, J.: Astrophys. J. **716**, 1241 (2010)

Saumon, D., Chabrier, G., van Horn, H.M.: Astrophys. J. Suppl. **99**, 713 (1995)

Schlaufman, K.C.: Astrophys. J. **719**, 602 (2010)

Schönberg, M., Chandrasekhar, S.: Astrophys. J. **96**, 161 (1942)

Schwarzschild, M., Härm, R.: Astrophys. J. **128**, 348 (1958)

Serenelli, A.M., Basu, S.: Astrophys. J. **719**, 865 (2010)

Serenelli, A.M., Basu, S., Ferguson, J.W., Asplund, M.: Astrophys. J. Lett. **705**, L123 (2009)

Serenelli, A.M., Bergemann, M., Ruchti, G., Casagrande, L.: Mon. Not. R. Astron. Soc. **429**, 3645 (2013)

Silva Aguirre, V., Ballot, J., Serenelli, A.M., Weiss, A.: Astron. Astrophys. **529**, 63 (2011)

Silva Aguirre, V., Basu, S., Brandão, I.M., et al.: Astrophys. J. **769**, 141 (2013)

Silva Aguirre, V., Davies, G.R., Basu, S., et al.: Mon. Not. R. Astron. Soc. **452**, 2127 (2015)

Silva Aguirre, V., Lund, M.N., Antia, H.M., et al.: Astrophys. J. **835**, 1 (2017)

Smartt, S.J.: Annu. Rev. Astron. Astrophys. **47**, 63 (2009)

Sneden, C.: ESO Workshop on Production and Distribution of C, vol. 21, p. 1 (1985)

Steigman, G.: J. Cosmol. Astropart. Phys. **2010**, 29 (2010)

Trampedach, R., Stein, R.F., Christensen-Dalsgaard, J., Nordlund, Å., Asplund, M.: Improvements to stellar structure models, based on a grid of 3D convection simulations - II. Calibrating the mixing-length formulation. Mon. Not. R. Astron. Soc. **445**(4), 4366–4384 (2014). http://adsabs.harvard.edu/abs/2014MNRAS.445.4366T

Turcotte, S., Richer, J., Michaud, G.: Astrophys. J. **504**, 559 (1998)

Unno, W., Spiegel, E.A.: Publ. Astron. Soc. Jpn. **18**, 85 (1966)

Vandenberg, D.A., Gustafsson, B., Edvardsson, B., Eriksson, K., Ferguson, J.: Astrophys. J. **666**, L105 (2007)

Weiss, A., Hillebrandt, W., Thomas, H.-C., Ritter, H. Adv. Astron. Astrophys. (2005)

Winn, J.N., Fabrycky, D.C. Annu. Rev. Astron. Astrophys. **53**, 409 (2015)

Zahn, J.-P. Astron. Astrophys. **252**, 179 (1991)

Theory of Stellar Oscillations

Margarida S. Cunha

Abstract In recent years, astronomers have witnessed major progresses in the field of stellar physics. This was made possible thanks to the combination of a solid theoretical understanding of the phenomena of stellar pulsations and the availability of a tremendous amount of exquisite space-based asteroseismic data. In this context, this chapter reviews the basic theory of stellar pulsations, considering small, adiabatic perturbations to a static, spherically symmetric equilibrium. It starts with a brief discussion of the solar oscillation spectrum, followed by the setting of the theoretical problem, including the presentation of the equations of hydrodynamics, their perturbation, and a discussion of the functional form of the solutions. Emphasis is put on the physical properties of the different types of modes, in particular acoustic (p-) and gravity (g-) modes and their propagation cavities. The surface (f-) mode solutions are also discussed. While not attempting to be comprehensive, it is hoped that the summary presented in this chapter addresses the most important theoretical aspects that are required for a solid start in stellar pulsations research.

1 Introduction

The study of stellar pulsations is revolutionizing our knowledge of the internal structure and dynamics of stars and, as a consequence, also our understanding of stellar evolution. This is made possible through the combination of a solid theoretical understanding of the phenomena of stellar pulsations and the availability of a tremendous amount of high-quality data, in particular that acquired from space with satellites such as *SOHO* (Domingo et al. 1995), observing the Sun for over 20 years, and *CoRoT* (Baglin et al. 2006) and *Kepler* (Gilliland et al. 2010; Koch et al. 2010).

In this chapter I review basic aspects of the theory of stellar pulsations. Given the limited space available, options had to be made on what to discuss. A more detailed

M.S. Cunha (✉)
Instituto de Astrofísica e Ciências do Espaço, Universidade do Porto, CAUP, Rua das Estrelas, 4150-762 Porto, Portugal
e-mail: mcunha@astro.up.pt

T.L. Campante et al. (eds.), *Asteroseismology and Exoplanets: Listening to the Stars and Searching for New Worlds*, Astrophysics and Space Science Proceedings 49, DOI 10.1007/978-3-319-59315-9_2

view of the aspects considered here, as well as discussions of the issues that have been left out can be found, e.g., in published books (Unno et al. 1989; Aerts et al. 2010), lecture notes (Gough 1993), as well as in other long reviews (Cunha et al. 2007; Basu 2016).

While this chapter is dedicated to the theory of stellar pulsations, it is interesting and motivating to start by inspecting one of the main observational results in this context, namely, the oscillation power density spectrum of the Sun. This is shown in the upper panel of Fig. 1. The first aspect that catches the eye is that the oscillation spectrum is composed of a number of discrete frequencies, whose power is modulated over frequency, showing a close to Gaussian shape. This is typical of oscillation spectra of solar-like pulsators in which modes are intrinsically stable (meaning that small perturbations are damped) and continuously excited

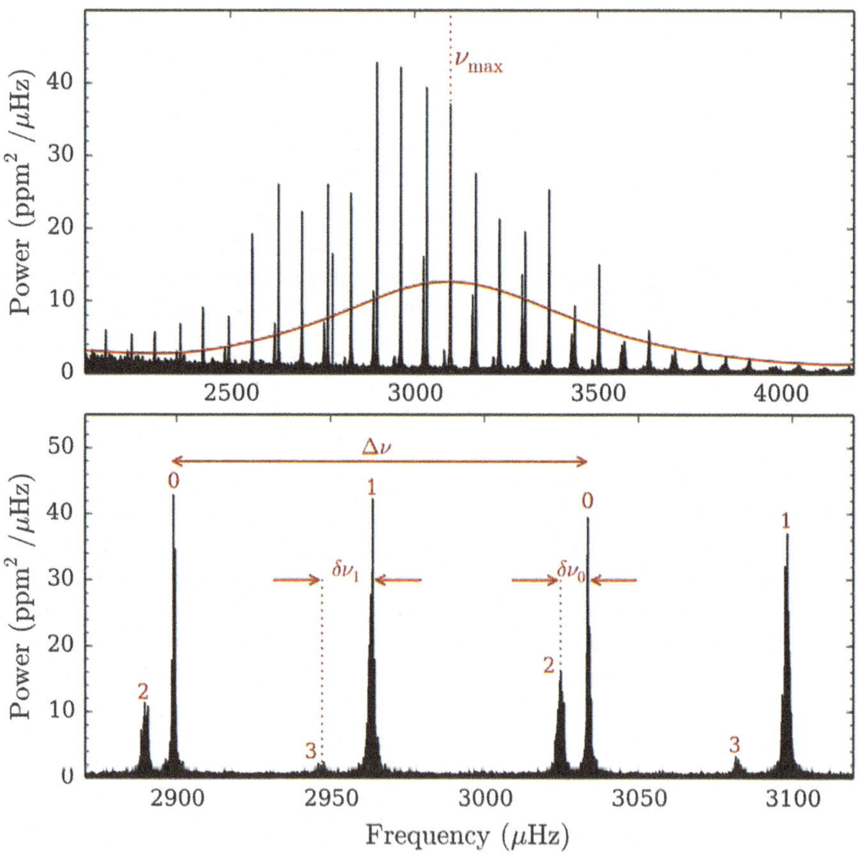

Fig. 1 Power density spectrum of the Sun obtained from data acquired with VIRGO/SPM onboard the *SOHO* satellite (Fröhlich et al. 1995; Jiménez et al. 2002). *Top:* The *red line* shows the power spectrum density smoothed by $3\Delta\nu$ and multiplied by 50, used to estimate ν_{max}. *Bottom:* A zoom of the upper panel illustrating a few modes, identified by mode degree. The large frequency separation, $\Delta\nu$, and the small frequency separations between pairs of modes with degrees $l = 0, 2$, $\delta\nu_0$, and pairs of modes with degrees $l = 1, 3$, $\delta\nu_1$, are also shown (figure courtesy of A. Santos)

stochastically by convection. Other stellar pulsators, in which oscillations are intrinsically unstable, with small perturbations growing due to some sort of coherent excitation mechanism, will still show oscillation spectra composed of discrete frequencies, but often a less regular pattern as a consequence of not all possible frequencies being excited or observed. An important observable for solar-like pulsators is the frequency of maximum power, $\nu_{\rm max}$, shown in Fig. 1. There are different approaches to derive it (Verner et al. 2011, and references therein) that usually involve considering a heavily-smoothed version of the oscillation power spectrum.

The lower panel of Fig. 1 shows a close-up of the regular peak structure seen in the upper panel. Here each mode is identified by a positive integer, the mode degree, which will be discussed in detail in Sect. 2.3. Two main separations are identified in the figure, namely, the large separation, $\Delta\nu$, between consecutive modes of the same degree and the small separation, $\delta\nu$, between modes of similar frequency and degree differing by two.

The large separation has been shown to scale as $\Delta\nu \propto \sqrt{\overline{\rho}}$ (Tassoul 1980), where $\overline{\rho}$ is the mean density of the star. Moreover, the frequency of maximum power has been suggested to scale with the surface gravity and effective temperature as $\nu_{\rm max} \propto gT_{\rm eff}^{-1/2}$ (Brown et al. 1991; Kjeldsen and Bedding 1995). Together, these scaling relations provide two equations that can be used for a first estimate of the stellar mass and radius, once the effective temperature is known, namely,

$$\frac{R}{\mathrm{R}_\odot} \approx \left(\frac{\nu_{\rm max}}{\nu_{\rm max,\odot}}\right)\left(\frac{\overline{\Delta\nu}}{\overline{\Delta\nu_\odot}}\right)^{-2}\left(\frac{T_{\rm eff}}{T_{\rm eff,\odot}}\right)^{1/2},$$

$$\frac{M}{\mathrm{M}_\odot} \approx \left(\frac{\nu_{\rm max}}{\nu_{\rm max,\odot}}\right)^{3}\left(\frac{\overline{\Delta\nu}}{\overline{\Delta\nu_\odot}}\right)^{-4}\left(\frac{T_{\rm eff}}{T_{\rm eff,\odot}}\right)^{3/2}, \tag{1}$$

where the overbar stands for a suitable average taken over the different pairs of modes and solar values are marked by the index '$\odot$'.

Finally, the small separation depends strongly on the sound speed in the stellar core and is, thus, very sensitive to stellar age. These quantities shall be discussed further in Sect. 5. Before that, I will introduce the pulsation equations in Sect. 2, and discuss the corresponding solutions in Sects. 3–4.

2 Equations for Linear, Adiabatic Stellar Pulsations

In this section I set the problem of linear, adiabatic stellar pulsations. I start from the equations of hydrodynamics for an inviscid fluid and then consider small, adiabatic perturbations about a spherically symmetric, static equilibrium. Finally, I discuss the functional form of the solutions on the sphere and the boundary conditions.

Let us assume that a gas can be treated as a continuum with thermodynamic properties well defined at each position in space, $\boldsymbol{r}$. Let f be a scalar property of the gas. There are two ways of looking at the time evolution of f: (1) at fixed position

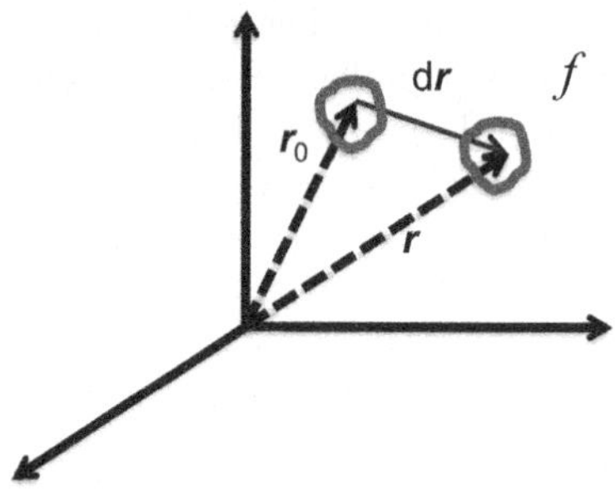

Fig. 2 Eulerian versus Lagrangian descriptions. In the Eulerian description the evolution of the property, f, of the gas is considered at fixed position, $\boldsymbol{r}_0$, by comparing $f(\boldsymbol{r}_0, t_0)$ with $f(\boldsymbol{r}_0, t_1)$, while in the Lagrangian description the evolution is considered following the motion by comparing $f(\boldsymbol{r}_0, t_0)$ with $f(\boldsymbol{r}, t_1)$

$\boldsymbol{r}_0$ and (2) following the motion (see Fig. 2). The first corresponds to an Eulerian description and the second to a Lagrangian description. Both perspectives are useful and commonly used in the study of stellar pulsations. The two descriptions are related by

$$\frac{\mathrm{d}f}{\mathrm{d}t} = \frac{\partial f}{\partial t} + \nabla f \cdot \frac{\mathrm{d}\boldsymbol{r}}{\mathrm{d}t} \equiv \frac{\partial f}{\partial t} + \boldsymbol{v} \cdot \nabla f \,, \tag{2}$$

where $\boldsymbol{v}$ is the velocity, $\mathrm{d}/\mathrm{d}t$ is the time derivative following the motion (Lagrangian description) and $\partial/\partial t$ is the time derivative at fixed position (Eulerian description). Likewise, for a vector quantity, $\boldsymbol{F}$, the two derivatives are related by

$$\frac{\mathrm{d}\boldsymbol{F}}{\mathrm{d}t} = \frac{\partial \boldsymbol{F}}{\partial t} + (\boldsymbol{v} \cdot \nabla)\,\boldsymbol{F} \,. \tag{3}$$

2.1 The Conservation Laws

The evolution of the properties of a fluid is described by a set of equations that translate conservation laws. In what follows these equations are summarized under particular conditions that will be discussed below. Conservation of mass, linear momentum, and energy are expressed, respectively, by

$$\begin{aligned}
\frac{\mathrm{d}\rho}{\mathrm{d}t} &= -\rho \nabla \cdot \boldsymbol{v} \,, \\
\rho \frac{\mathrm{d}\boldsymbol{v}}{\mathrm{d}t} &= -\nabla p + \rho \boldsymbol{g} + \boldsymbol{F}_{\mathrm{oth}} \,, \\
\frac{\mathrm{d}q}{\mathrm{d}t} &= \frac{\mathrm{d}E}{\mathrm{d}t} + p \frac{\mathrm{d}\,(1/\rho)}{\mathrm{d}t} \,,
\end{aligned} \tag{4}$$

where ρ and p are, respectively, the fluid density and pressure, $\boldsymbol{g}$ is the acceleration of gravity, $\boldsymbol{F}_{\rm oth}$ are other body forces, expressed per unit volume, that may act on the fluid, besides gravity (e.g., the Lorentz force, if a magnetic field is present), and E and q are, respectively, the internal energy and heat supplied to the system, both per unit mass.

The first of these equations, known as the continuity equation, expresses that the rate of change of the mass within a given volume must equal, with opposite sign, the mass crossing the surface that encloses that volume, per unit time. The second, the equation of motion, expresses that the change in linear momentum of an element of fluid must equal the force acting on it by its surroundings. It is written under the assumption that the fluid is inviscid, which is a good approximation under stellar conditions. The third equation translates the first law of thermodynamics and it states that the change in the internal energy of a system must equal the heat supplied to the system minus the work done by the system on its surroundings. This equation can be written in different forms. A useful one, adopted below, is

$$\frac{{\rm d}q}{{\rm d}t} = \frac{1}{\rho\,(\Gamma_3 - 1)}\left(\frac{{\rm d}p}{{\rm d}t} - \frac{\Gamma_1 p}{\rho}\frac{{\rm d}\rho}{{\rm d}t}\right), \tag{5}$$

where Γ_1 and Γ_3 are adiabatic exponents defined by the adiabatic derivatives,

$$\Gamma_1 = \left(\frac{\partial \ln p}{\partial \ln \rho}\right)_{\rm ad}, \qquad \Gamma_3 - 1 = \left(\frac{\partial \ln T}{\partial \ln \rho}\right)_{\rm ad}, \tag{6}$$

and T is the temperature of the fluid. From Eq. (5) one can further define the adiabatic sound speed, c. Making the left-hand side equal to zero one finds:

$$c^2 \equiv \frac{{\rm d}p}{{\rm d}\rho} = \frac{\Gamma_1 p}{\rho}. \tag{7}$$

Finally, it should be noted that the thermodynamic variables T, ρ, and p are not all independent, but rather are related by the equation of state that can be expressed as $\mathscr{F}\,(T, p, \rho) = 0$, where $\mathscr{F}$ is a function that depends on the conditions of the fluid. Since in this chapter only adiabatic oscillations will be considered, the explicit form of the equation of state will not be needed. However, this equation will still be required if the reader is interested in deriving the temperature fluctuations associated to the perturbations in the density and pressure.

2.2 *Perturbative Analysis*

Consider an equilibrium state that is: (1) static, meaning that there are no velocities and all derivatives at fixed position are null ($\partial/\partial t = 0$) and (2) spherically symmetric, implying, e.g., that there is no rotation or magnetic fields. Then, in the equilibrium, one has:

$$\nabla p_0 = \rho_0 \mathbf{g}_0 \equiv -\rho_0 g_0 \hat{\mathrm{a}}_\mathrm{r} \,, \tag{8}$$

where the index '0' is used to identify the equilibrium quantities and $\hat{\mathrm{a}}_\mathrm{r}$ is the unit vector in the radial direction pointing outwardly from the centre of the star, making the scalar g_0 a positive quantity.

Now let us assume that the equilibrium is perturbed under the following conditions: (1) the perturbations are adiabatic and (2) they are small, in the sense that non-linear terms in the perturbations can be neglected.

The adiabatic condition implies the assumption that no heat is exchanged with the element of fluid during the perturbation, a condition that is very closely satisfied almost everywhere in the star. This can be seen by comparing the characteristic timescales of pulsations, typically found in the range of minutes to a few days, with the timescale for radiation that, except very close to the stellar surface, has characteristic values many orders of magnitude larger than the pulsation period (e.g., exceeding a million years in the Sun, when the Sun is considered as a whole).

Under the above conditions, let f be a scalar property of the gas, and f' and δf be, respectively, the Eulerian and Lagrangian perturbations to it. Then $f = f_0 + f'$ and $\delta f = f' + \boldsymbol{\xi} \cdot \nabla f_0$, where $\boldsymbol{\xi}$ is the displacement vector ($\equiv \boldsymbol{r} - \boldsymbol{r}_0$). Moreover, since the perturbations are linear, the velocity of a given element of fluid is

$$\boldsymbol{v} \equiv \frac{\mathrm{d}\boldsymbol{\xi}}{\mathrm{d}t} \approx \frac{\partial \boldsymbol{\xi}}{\partial t} \,. \tag{9}$$

Perturbing the system of equations (4) [with energy conservation expressed as in Eq. (5)], using Eqs. (8) and (9), and integrating in time the equations of continuity and energy, one finds that linear adiabatic perturbations about a static spherically symmetric equilibrium are described by the following set of equations:

$$\begin{aligned}
\rho' &= -\nabla \cdot (\rho_0 \boldsymbol{\xi}) \,, \\
\rho_0 \frac{\partial^2 \boldsymbol{\xi}}{\partial t^2} &= -\nabla p' - \rho_0 \nabla \phi' - \rho' \nabla \phi_0 \,, \\
\nabla^2 \phi' &= 4\pi G \rho' \,, \\
p' + \boldsymbol{\xi} \cdot \nabla p_0 &= \frac{\Gamma_{1,0} p_0}{\rho_0} \left(\rho' + \boldsymbol{\xi} \cdot \nabla \rho_0 \right) \,.
\end{aligned} \tag{10}$$

To reach the system of equations above, I have further defined the acceleration of gravity in terms of the gravitational potential ϕ, such that $\mathbf{g} = -\nabla \phi$ and, accordingly, considered in addition the Poisson equation that relates the gravitational potential to the fluid density.

Taking the equilibrium quantities as known, one can identify four variables in the system above (three scalars and one vector), namely ρ', p', ϕ', and $\boldsymbol{\xi}$. These four equations thus form a closed system that can be solved, with adequate boundary conditions.

2.3 Solutions on a Sphere

Consider the spherical coordinate system (r, θ, φ) such that the variables $\rho', p', \phi', \boldsymbol{\xi}$ are expressed as functions of r, θ, φ and t.

It can be shown by substitution (or derived by the technique of separation of variables) that the system of equations (10) admits solutions of the type

$$f'(r, \theta, \varphi, t) = \Re\left\{f'(r)\, Y_l^m(\theta, \varphi)\, \mathrm{e}^{-i\omega t}\right\}, \tag{11}$$

$$\boldsymbol{\xi}(r, \theta, \varphi, t) = \Re\left\{\left[\xi_r(r)\, Y_l^m(\theta, \varphi)\, \hat{\mathrm{a}}_r + \xi_h(r)\left(\frac{\partial Y_l^m}{\partial \theta}\hat{\mathrm{a}}_\theta + \frac{1}{\sin\theta}\frac{\partial Y_l^m}{\partial \varphi}\hat{\mathrm{a}}_\varphi\right)\right]\mathrm{e}^{-i\omega t}\right\},$$

where f' stands for any of the scalar perturbations, ξ_r and ξ_h are, respectively, the depth-dependent amplitudes of the radial and horizontal components of the displacement and $\hat{\mathrm{a}}_i$ are the components of the unit vectors of the spherical coordinate system.

The time dependence of the solution is associated to the angular oscillation frequency ω. The sign in the exponential is arbitrary. Here it is chosen to be negative to guarantee that in cases when ω is complex the growth rate (i.e., the imaginary part of the frequency) is positive when the perturbation grows. The possible values of ω are determined by imposing the boundary conditions that shall be discussed later. In practice, since the equations were derived under the assumption that the perturbations are adiabatic, so far as the boundary conditions are fully reflective (i.e., no energy is lost through the boundary), ω is real. That will be the only case discussed in this chapter.

The angular dependence of the solutions is given by the spherical harmonic functions Y_l^m, characterized by the angular degree l (a non-negative integer), and the azimuthal order m, an integer that takes values between $-l$ and l. The angular degree defines the number of surface nodes and the absolute value of the azimuthal order defines the subset of those that cross the equator. This means that $|m|$ defines the orientation of that solution on the sphere, something that will be relevant for the discussion below. An example of low-degree spherical harmonic functions with identified values of l and $|m|$ is shown in Fig. 3. Since $Y_l^m \propto P_l^m(\cos\theta)\mathrm{e}^{im\varphi}$, where P_l^m are the associated Legendre functions, the sign of m defines whether the associated solution is travelling eastwardly or westwardly in the chosen reference frame. Given the negative sign adopted for the time-dependent part of the solution the perturbations are found to vary as $\mathrm{e}^{i(m\varphi - \omega t)}$. This means that in this case a positive m corresponds to a solution travelling eastwardly. I note, however, that not all literature adopts the same definition, since sometimes the opposite sign is chosen for the exponent in the time-dependent exponential. Finally, from the properties of the spherical harmonics, one has that

$$\nabla_h^2 Y_l^m = -\frac{l(l+1)}{r^2} Y_l^m \equiv -\kappa_h^2 Y_l^m, \tag{12}$$

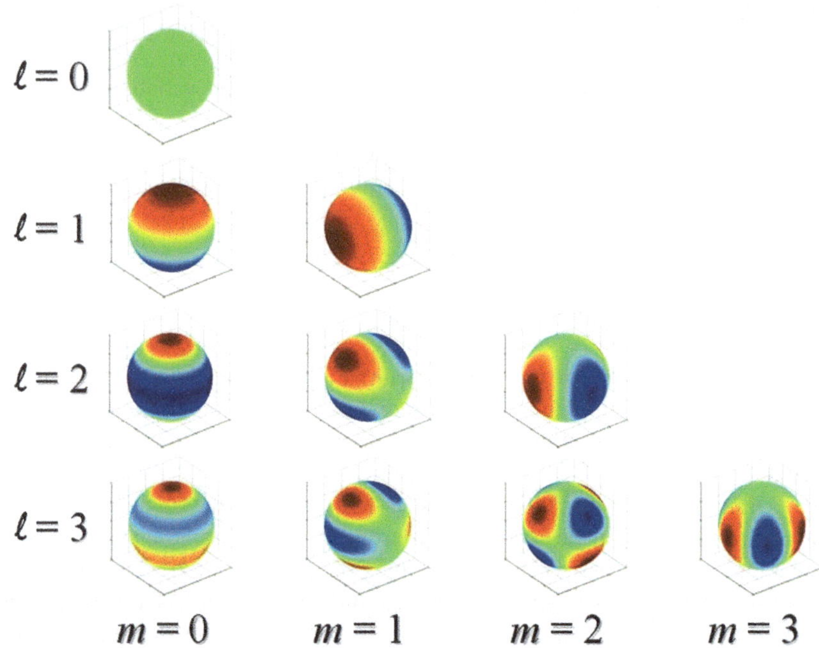

Fig. 3 Examples of spherical harmonic functions, Y_l^m, for mode degrees in the range $l = 0$–3. For each mode degree, all possible non-negative values of m are shown. *Red* and *blue* show perturbations of opposite sign

where κ_h has been identified as the horizontal wavenumber of the perturbation when the latter is interpreted, locally, as a plane wave.

The last part of the solutions in Eq. (11) are the radial-dependent amplitudes. For simplicity, radial-dependent parts of scalar perturbations have been named with the same symbol as the full solutions. Using the full solutions separated as in Eq. (11) in the system of equations (10), one can derive a set of equations governing these amplitude functions. After eliminating the horizontal component of the displacement by combining the continuity equation and the horizontal divergence of the perturbed momentum equation, and eliminating the Eulerian perturbation to the density through the adiabatic relation, one finds that the radial-dependent amplitudes $p'(r)$, $\phi'(r)$, and $\xi_r(r)$, obey the following system of equations:

$$\frac{1}{r^2}\frac{\mathrm{d}}{\mathrm{d}r}\left(r^2\xi_r\right) - \frac{g_0}{c_0^2}\xi_r - \left(\frac{S_l^2}{\omega^2} - 1\right)\frac{1}{c_0^2\rho_0}p' = \frac{l(l+1)}{r^2\omega^2}\phi' ,$$

$$\frac{\mathrm{d}p'}{\mathrm{d}r} + \frac{g_0}{c_0^2}p' - \rho_0\left(\omega^2 - N_0^2\right)\xi_r = -\rho_0\frac{\mathrm{d}\phi'}{\mathrm{d}r} ,$$

$$\frac{1}{r^2}\frac{\mathrm{d}}{\mathrm{d}r}\left(r^2\frac{\mathrm{d}\phi'}{\mathrm{d}r}\right) - \frac{l(l+1)}{r^2}\phi' = 4\pi G\left(\frac{p'}{c_0^2} + \frac{\rho_0 N_0^2}{g_0}\xi_r\right) , \quad (13)$$

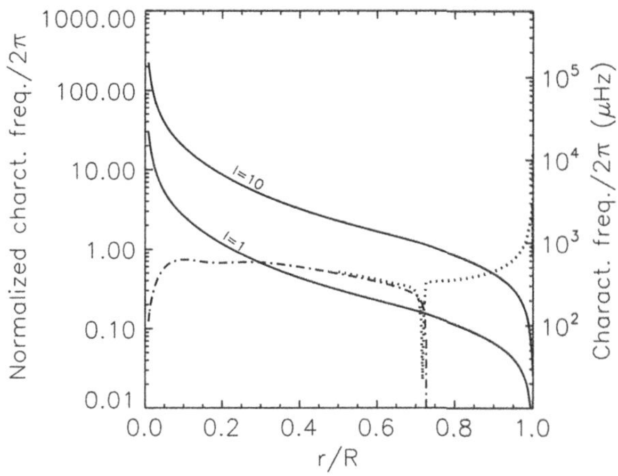

Fig. 4 Lamb frequency, S_l (*continuous lines*), for $l = 1$ and $l = 10$, buoyancy frequency, N_0 (*dashed-dotted line*), and critical frequency, ω_c (*dotted line*, displayed in the outer 50% of the stellar radius only), for a model of the Sun, all divided by 2π. The *left vertical axis* shows dimensionless values of these characteristic frequencies obtained by multiplying them by $t_{\rm dyn}$. The *right vertical axis* indicates the true physical values

where two characteristic frequencies have been defined: the Lamb frequency, S_l; and the buoyancy (or Brunt–Väisälä) frequency, N_0. The squares of these quantities are given, respectively, by

$$S_l^2 = \frac{l(l+1)\,c_0^2}{r^2}\,, \qquad N_0^2 = g_0\left[\frac{1}{\Gamma_{1,0}}\frac{{\rm d}\ln p_0}{{\rm d}r} - \frac{{\rm d}\ln \rho_0}{{\rm d}r}\right]. \tag{14}$$

Examples of the Lamb frequency and buoyancy frequency are shown in Figs. 4 and 5 for a model of the Sun and a model of a star in the red-giant branch, respectively. The buoyancy frequency is seen only where $N^2 > 0$, which marks stellar layers that are stable to convection.

A significant difference is found in the characteristic values of the Lamb frequency in the two stars, as seen by comparing the scales on the right-hand side vertical axes in Figs. 4 and 5. This is because this frequency scales approximately with the inverse of the dynamical timescale of the star, $t_{\rm dyn} = (R^3/GM)^{1/2}$, hence depending significantly on stellar radii. The vertical scale on the left-hand side of each figure shows the characteristic frequencies in units of $1/t_{\rm dyn}$, illustrating the similarity of the Lamb frequency in the two stars once the scaling is accounted for. The second aspect that calls attention in these figures is that the buoyancy frequency in the more evolved star shows a much more significant contrast, increasing significantly towards the stellar centre. This is a consequence of the increasing density gradient in the innermost layers as the star evolves and the core contracts. These structural differences between main-sequence and red-giant stars have significant impact on the properties of their oscillations, as will be discussed in Sect. 4.

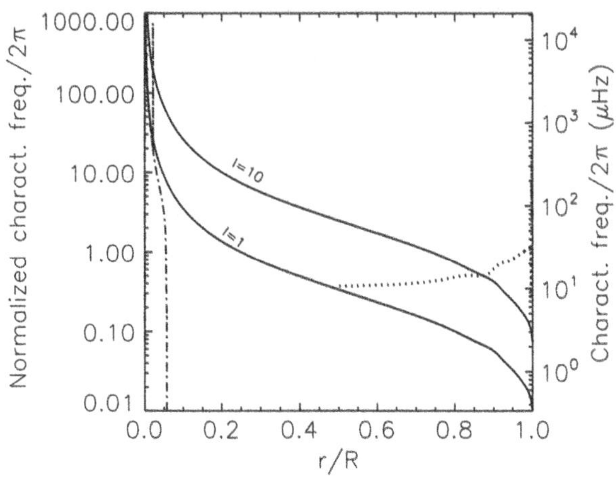

Fig. 5 Same as Fig. 4 but for a model of a star in the red-giant branch

There are a number of points that should be stressed in relation to the system of equations (13):

- First, the system only contains total derivatives, reminding us that the variables are the depth-dependent parts of the solutions only. In fact, the system of equations (13) forms a system of linear, total differential equations of fourth order for four unknown functions, namely, the depth-dependent amplitude functions, ξ_r, p', ϕ', and $\mathrm{d}\phi'/\mathrm{d}r$. We have, thus, at this point, reduced the original 3-dimensional problem into a 1-dimensional problem.
- The second point is that in the case of spherically symmetric perturbations ($l = 0$), the perturbed Poisson equation can be integrated, reducing the system to second order for the variables p' and ξ_r. An important consequence, that will not be explored here, is that it is then possible to combine the two first-order differential equations to obtain a single second-order differential equation for the displacement, that can then be cast in the form of a standard wave equation. Moreover, Takata (2005, 2016) has shown that in the case of dipolar modes ($l = 1$), momentum conservation can be used to derive an integral that again reduces the system to second order. In the same work, the author has also argued that integrals such as those found for the case of radial and dipolar modes do not exist for any other mode degree.
- For other mode degrees, reduction of the system of equations (13) to a second order system can be achieved by performing the *Cowling approximation*, which consists in neglecting the perturbation to the gravitational potential. This approximation is adequate for perturbations that vary on relatively short scales (much smaller than the radius of the star). Since the integral solution of the Poisson equation relates ϕ' to the integral of ρ', the cancellation effect that results from the integration when ρ' varies on short scales, leads to ϕ' being small in that case. While this reduction of the order of the system requires an approximation, it has

proven to be extremely useful for the asymptotic analysis of the equations and, in that way, a better understanding of the physical picture involved. I will get back to this in Sect. 3.

- The fourth, and last point, that should be made is that while the coefficients in the system of equations (13) depend on the mode degree, l, they are independent of the mode azimuthal order, m. This is a consequence of our assumption that the equilibrium state is spherically symmetric. In fact, under that assumption, there is no preferential direction in the star and, so far as the boundary conditions (to be discussed in Sect. 2.4) are also independent of m, the solutions must be independent of the reference about which the spherical harmonic functions are defined. Since m defines the orientation of the spherical harmonic on the sphere, the equations and, hence, the solutions, must not depend on m. This means that under this assumption the solutions will be degenerate in the azimuthal order. That degeneracy, however, is broken (partially broken), in the presence of agents that break (partially break) the spherical symmetry, such as, e.g., rotation or magnetic fields.

2.4 *Boundary Conditions*

Of the four boundary conditions required to solve the problem set by the system of equations (13), two will be defined at the stellar centre ($r = 0$) and two at the stellar surface ($r = R$).

2.4.1 At the Stellar Centre

The boundary conditions at the stellar centre are derived by imposing that the solutions are regular (do not diverge) there. By expanding the equations near $r = 0$ one finds that the regular solutions require $p' \sim \mathcal{O}(r^l)$, $\phi' \sim \mathcal{O}(r^l)$, and $\xi_r \sim \mathcal{O}(r^\alpha)$, where $\alpha = 1$ for $l = 0$ and $\alpha = l - 1$ for $l > 0$. This means that, as $r \to 0$,

$$\frac{\mathrm{d}\phi'}{\mathrm{d}r} - \frac{l}{r}\phi' \to 0 \ , \qquad \frac{\mathrm{d}p'}{\mathrm{d}r} - \frac{l}{r}p' \to 0 \ , \qquad \frac{\mathrm{d}\xi_r}{\mathrm{d}r} - \frac{\alpha}{r}\xi_r \to 0 \, . \tag{15}$$

The three conditions above are not all independent. In fact, the condition on the displacement can be derived from the other two by first noting that the gradient of the thermodynamic variables in the equilibrium structure must be zero at the centre of the star and, then, applying that knowledge to the perturbed equations. Moreover, in that process one also finds that for non-radial modes $\xi_r = l\xi_h$, at $r = 0$. The regularity of the solutions thus provides us with two independent boundary conditions to apply at $r = 0$. One interesting point to notice is that the displacement at the centre of the star is non-zero only for dipolar modes ($l = 1$). In all other cases the centre of the star does not move. That such is the case can also be seen from symmetry arguments, as only for $l = 1$ modes one would recover the same non-zero displacement vector at $r = 0$ independently from where the centre is approached.

2.4.2 At the Stellar Surface

As the density vanishes outside the star, at the surface the perturbation to the gravitational potential must match continuously onto the physically-meaningful solution of the Poisson equation for a vacuum field (vanishing at infinity). That implies that $\phi' \sim \mathcal{O}(r^{-l-1})$ and thus, at $r = R$,

$$\frac{\mathrm{d}\phi'}{\mathrm{d}r} = -\frac{l+1}{r}\phi' . \tag{16}$$

The second boundary condition to be applied at the surface depends on how one treats the atmosphere of the star. If one assumes a free boundary, then one must consider that the pressure at the boundary is constant and, hence, that the Lagrangian pressure perturbation there is zero. In that case, a second boundary condition at $r = R$ is found in the form:

$$p' + \frac{\mathrm{d}p_0}{\mathrm{d}r}\xi_r = 0 . \tag{17}$$

This condition is reasonable for low-frequency waves, as will be seen later. However, as the frequency of the waves increases, the details of the atmosphere become more important for the solution. It is therefore common, when solving the pulsation equations, to adopt a more adequate boundary condition, such as that derived from the matching of the radial displacement solution onto the physically meaningful analytical solution derived for an isothermal atmosphere. The analytical solution can be derived assuming a plane-parallel isothermal equilibrium composed of an ideal gas and with constant adiabatic exponent and mean molecular weight. Under these assumptions, the sound speed is constant and so are the density and pressure scale heights, i.e., the characteristic lengths associated with the variations of density and pressure, given respectively by $H = -\mathrm{d}r/\mathrm{d}\log\rho_0$ and $H_p = -\mathrm{d}r/\mathrm{d}\log p_0$. Moreover, in this case, the equilibrium pressure and density decrease exponentially with height in the atmosphere $\rho, p \propto \exp[(R-r)/H)]$. Then, considering the system of equations (13) under the Cowling approximation, the displacement is found to have the form $\xi_r \propto \exp(\kappa r)$, where, for all cases that will be of interest to us,[1]

$$\kappa = \frac{1}{2H}\left[1 \pm \left(1 - \frac{4\omega^2 H^2}{c_0^2}\right)^{1/2}\right] \equiv \frac{1}{2H}\left[1 \pm \left(1 - \frac{\omega^2}{\omega_c^2}\right)^{1/2}\right] . \tag{18}$$

[1]Here the term that would dominate in the case of atmospheric gravity waves is being neglected, as those will not be discussed in these lectures.

Combining the perturbed equation of mass conservation and the adiabatic condition (first and fourth equations in the system of Eqs. (10)) with the solution for ξ_r, and noting that in the isothermal atmosphere $H = H_p$, one finds an alternative boundary condition to be applied at $r = R$ [replacing Eq. (17)], namely,

$$p' \approx -\left(\Gamma_{1,0} p_0 \kappa + \frac{\mathrm{d}p_0}{\mathrm{d}r}\right)\xi_r = \frac{p_0}{H}\left(1 - \Gamma_{1,0}\kappa H\right)\xi_r\,, \tag{19}$$

where $\nabla \cdot \boldsymbol{\xi}$ has been approximated by $\mathrm{d}\xi_r/\mathrm{d}r$, implicitly assuming that the perturbation varies much more rapidly in the radial than in the horizontal direction.

In Eq. (18) the critical frequency, ω_c, has been introduced, which in the isothermal atmosphere is constant and equal to $c_0/(2H)$. It is important to note that when $\omega < \omega_c$, κ is real and the physically meaningful solution corresponds to choosing the negative sign, ensuring that the energy density, $\propto \rho\xi_r^2$, decreases outwardly. When $\omega > \omega_c$, κ is complex and assuming no waves are being sent into the atmosphere from outside, the imaginary part must be chosen to guarantee that the wave travels outwardly. Considering the time-dependent part of the solution, one finds that the radial component of the displacement goes as $\exp[i(\pm\kappa_i r - \omega t)]$, where $\kappa_i = (\omega^2/\omega_c^2 - 1)^{1/2}$. The outwardly travelling solution is thus obtained by taking the positive sign in Eq. (18).

Under this boundary condition, waves with $\omega > \omega_c$ will simply propagate away, loosing their energy through the boundary.[2] Here I am interested only in waves that are fully trapped inside the star, loosing no energy through the boundary, hence I will consider only the case when $\omega < \omega_c$. In the particular case when $\omega \ll \omega_c$ the expression for κ can be expanded and the boundary condition [Eq. (19)] approximated by

$$p' = \frac{p_0}{H}\left(1 - \Gamma_1 \frac{\omega^2}{4\omega_c^2}\right)\xi_r\,. \tag{20}$$

Finally, I note that as the frequency decreases, the second term inside the brackets on the right-hand side of Eq. (20) gets smaller and this boundary condition approaches the one defined in Eq. (17), justifying the adequacy of the latter in the case of sufficiently low frequencies.

2.5 *Eigenvalues*

The system of equations (13) and associated boundary conditions constitute an eigenvalue problem that needs to be solved numerically. The system admits non-trivial solutions only for discrete values of the eigenvalues ω. The discrete solutions

[2]In a real stellar atmosphere there can be partial reflection of the wave energy even when $\omega > \omega_c$. Accounting for that would require modifying the atmospheric model and, thus, the outer boundary condition accordingly.

can be associated to an integer number n, denominated by radial order. Once the depth-dependent amplitudes of the perturbations are computed, the full solutions can be derived from Eq. (11).

In summary, the eigenvalues, $\omega = \omega(n, l, m)$, of the 3-dimensional problem set by the pulsation equations (10) (after separation of time) are characterized by three quantum numbers, n, l, and m, where the absolute value of the first, $|n|$, is related to the number of nodes of the perturbation along the radial direction, while the other two, introduced in Sect. 2.3, are related to the angular dependence of the solutions, in particular to the horizontal scale of the perturbation and its orientation on the stellar surface. The discrete nature of the eigenvalues is clearly seen in the power density spectrum of the Sun shown in Fig. 1.

As discussed before, in the absence of physical agents that break the spherical symmetry of the problem, the solutions must be degenerate in m. Hence, in that case one has $\omega = \omega(n, l)$ and any linear combination of the $2l + 1$ independent solutions associated with the spherical harmonic of degree l, and different m values, is still an eigensolution for that eigenvalue.

A discussion of the full solutions obtained from numerical integration of the pulsation equations will be presented in Sect. 4. First, however, it is useful to analyse the second-order equation that is derived from the system of equations (13) under particular approximations. That will be discussed in the next section.

3 Trapping of the Oscillations

The full solutions of the linear, adiabatic pulsation equations must be computed numerically. Nevertheless, under the Cowling approximation, valid for large absolute values of the radial order, $|n|$, or for large degree l, the system of equations (13) reduces to second order on the variables p' and ξ_r, namely,

$$\frac{1}{r^2}\frac{\mathrm{d}}{\mathrm{d}r}\left(r^2\xi_r\right) - \frac{g_0}{c_0^2}\xi_r - \left(\frac{S_l^2}{\omega^2} - 1\right)\frac{1}{c_0^2\rho_0}p' = 0\,,$$

$$\frac{\mathrm{d}p'}{\mathrm{d}r} + \frac{g_0}{c_0^2}p' - \rho_0\left(\omega^2 - N_0^2\right)\xi_r = 0\,. \tag{21}$$

These equations can be combined to find a second-order wave equation for a single variable. To do so, I follow the work of Deubner and Gough (1984), which, in addition to the Cowling approximation, assumes that locally the oscillations can be treated as in a plane-parallel layer under constant gravity, hence neglecting[3] the derivatives of r and g_0. Let us introduce a new variable,

$$\Psi = \rho_0^{1/2} c_0^2 \nabla \cdot \boldsymbol{\xi}\,, \tag{22}$$

[3]For a more general case in which these assumptions are not made, see Gough (1993).

which through the adiabatic condition and the continuity equation can be seen to be directly related to the Lagrangian pressure perturbation ($\delta p = -\rho_0^{1/2}\Psi$). The second-order system (21), under the approximations mentioned above, can be manipulated to derive a wave equation for Ψ,

$$\frac{\mathrm{d}^2\Psi}{\mathrm{d}r^2} + \kappa_r^2\Psi = 0\,, \tag{23}$$

where κ_r is the local radial wavenumber given by

$$\kappa_r^2 = \frac{1}{c_0^2}\left[S_l^2\left(\frac{N_0^2}{\omega^2} - 1\right) + \omega^2 - \omega_c^2\right], \tag{24}$$

and the critical frequency is now given by

$$\omega_c^2 = \frac{c_0^2}{4H^2}\left(1 - 2\frac{\mathrm{d}H}{\mathrm{d}r}\right). \tag{25}$$

Notably, k_r depends critically on the three characteristic frequencies introduced before, namely, S_l, N_0 and ω_c. Examples of ω_c for a solar model and a red-giant model are shown in Figs. 4 and 5, respectively. It is small in the stellar interior, where the density varies on large scales, but it becomes large near the surface, where structural variations take place on a much shorter scale.

3.1 Mode Propagation Cavities

If κ_r^2 were constant and positive, the solution would be oscillatory everywhere. However, in a star κ_r^2 is a function of r, and one may generally expect it to be positive in some region(s) and negative in others. Where it is positive, the solution is locally wave-like (oscillatory in r) while where it is negative it is locally exponential. Because κ_r^2 depends on ω, the regions where wave-like solutions are found will depend on the mode under consideration. When there is only one region of the star where $\kappa_r^2 > 0$, the mode is said to be trapped there and that region is often called the mode propagation cavity. Away from that cavity, where the energy of the mode decreases exponentially, the mode is said to be evanescent. When κ_r^2 is positive in more that one region of the star, separated by regions where it is negative, the mode propagates in more than one cavity. However, often most of its energy is concentrated in one of these cavities whose structure, in turn, is the most determinant for the properties of the mode.

The radii at which $\kappa_r^2 = 0$ are called the turning points and define the edges of the propagation cavities. Setting the left-hand side of Eq. (24) to zero, one finds a second-order algebraic equation for ω^2, whose roots, $\omega_{l,\pm}^2$, are given by

$$\omega_{l,\pm}^2 = \frac{1}{2}\left(S_l^2 + \omega_c^2\right) \pm \frac{1}{2}\sqrt{\left(S_l^2 + \omega_c^2\right)^2 - 4S_l^2N_0^2}\,, \tag{26}$$

where the index l was used to recall that the roots depend on the mode degree through their dependence on the Lamb frequency. Rewriting Eq. (24) as

$$\kappa_r^2 = \frac{1}{c_0^2}\left[\omega^2 - \omega_{l,+}^2\right]\left[\omega^2 - \omega_{l,-}^2\right] , \tag{27}$$

one sees that modes propagate if

$$\omega > \omega_{l,+} \quad \text{or} \quad \omega < \omega_{l,-} , \tag{28}$$

and modes are evanescent if

$$\omega_{l,-} < \omega < \omega_{l,+} . \tag{29}$$

Figure 6 shows the frequencies $\omega_{l,\pm}$ for a model of the Sun in what is usually called a propagation diagram. Comparison with Fig. 4 shows the resemblance between N_0 and $\omega_{l,-}$. The frequency $\omega_{l,+}$, on the other hand, resembles S_l in the deeper layers of the stellar model, but it is clearly dominated by ω_c in the outer layers. Three modes, two at high frequency and one at low frequency, are illustrated by the horizontal lines in Fig. 6. For each of these modes, the propagation cavity

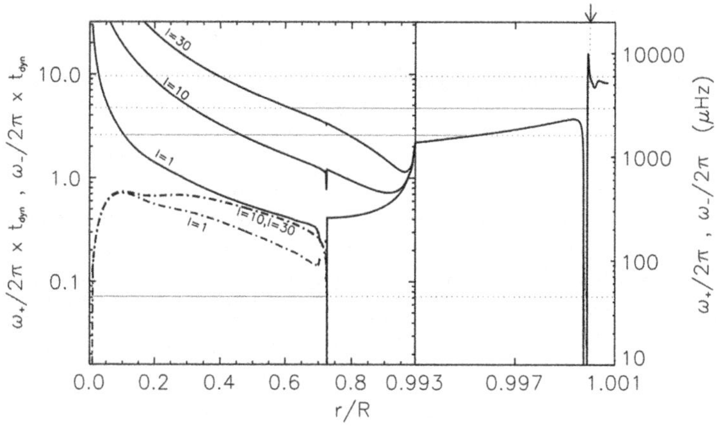

Fig. 6 Propagation diagram for model S of the Sun (Christensen-Dalsgaard et al. 1996). The frequencies ω_+ (*continuous thick lines*) and ω_- (*shaded-dotted thick lines*) are shown for three different values of mode degree, l. Note that in the outer layers (*right-hand section* of the plot), the horizontal axis varies much slower than in the inner layers (*left-hand section* of the plot). The *arrow* at the top marks $r/R = 1$. The *horizontal lines* mark four different frequencies and are continuous where the mode is trapped. At the lowest end, we have a characteristic g-mode frequency for this model, with $\nu \sim 45\,\mu$Hz. At the highest end we have a frequency $\nu \sim 6000\,\mu$Hz that is too high to be trapped inside the star. In between, we have two p-modes, one with $l = 1$, $\nu \approx 1613\,\mu$Hz, and one with $l = 30$, $\nu \approx 2936\,\mu$Hz. The *left-hand side vertical axis* shows the range of values taken by these frequencies when they are scaled according to their dependence on the dynamical timescale, while the *right-hand side axis* shows their true physical values

corresponds to the section in which the horizontal line is continuous. A fourth frequency is marked, with $\nu \sim 6000\,\mu$Hz, which is too high to correspond to a trapped mode.

3.2 *Acoustic Versus Internal Gravity Waves*

Inspection of the propagation diagram shown in Fig. 6 points towards the existence of two families of solutions, one at lower frequencies, where the mode cavity is essentially determined by the buoyancy frequency, and one at higher frequencies, where the propagation cavity is determined by the combination of the Lamb frequency, in the deeper regions, and the critical frequency, near the surface. This can also be seen by considering the lower and higher frequency limits of κ_r^2 in the propagation region, as discussed below.

3.2.1 Acoustic Waves

Let us consider first the higher frequency limit, namely, the case when $\omega^2 \gg N_0^2$, in regions where $\kappa_r^2 > 0$. Except near the surface, where ω_c becomes large, Eq. (24) then gives

$$\kappa_r^2 \approx \frac{\omega^2}{c_0^2} - \frac{l\,(l+1)}{r^2}\,. \tag{30}$$

Recalling that the last term on the right-hand side is κ_h^2 [cf. Eq. (12)], one finds that

$$|\boldsymbol{\kappa}|^2 \equiv \kappa_r^2 + \kappa_h^2 \approx \frac{\omega^2}{c_0^2}\,. \tag{31}$$

One thus has, for $\omega^2 \gg N_0^2$, a dispersion relation of the type:

$$\omega \approx c_0|\boldsymbol{\kappa}|\,, \tag{32}$$

which is characteristic of acoustic waves. These waves are maintained by the gradient of the pressure perturbation, i.e., the first term on the right-hand side of the perturbed momentum equation. Note that in this case the frequency of the mode increases as κ increases. Taking the radial order for this family of solutions as being positive integers, n, one thus finds that the frequency of acoustic modes increases both with increasing radial order, n, and with increasing degree, l.

Looking back at Eq. (30), one can further establish the lower turning point for the acoustic modes, that defines the lower boundary of their propagation cavity. Setting

κ_r^2 to zero implies $\omega^2 = S_l^2$, from which the lower turning point is found to be defined by

$$r_{1,l} = \frac{\sqrt{l(l+1)}c_0(r_{1,l})}{\omega} , \tag{33}$$

where the subscript l has been used to emphasize that the lower turning point depends on the mode degree and it has been explicitly indicated that the value of c_0 is to be taken at the turning point. From Eq. (33), we see that the lower turning point of acoustic modes depends strongly on the mode degree. As the mode degree increases, the lower turning point gets closer to the surface, implying that the propagation cavity of the mode becomes shallower. That is also evident in Fig. 6, where one can see that the depth at which a horizontal line in the high-frequency regime crosses lines of $\omega_{l,+}$ for different degrees, becomes smaller as the degree increases. In addition, we see from Eq. (33), and also form inspection of Fig. 6, that for a fixed mode degree, the lower turning point decreases as the mode frequency increases. That means that higher frequency acoustic modes propagate deeper, for fixed mode degree.

The upper turning point of the acoustic modes is, in turn, determined by comparing the oscillation frequency with the critical frequency, which near the surface is much greater than S_l. There, one may approximate,

$$\kappa_r^2 \approx \frac{\omega^2 - \omega_c^2}{c_0^2} . \tag{34}$$

Thus, one finds that $\kappa_r = 0$ in the outer layers if

$$\frac{c_0}{2H}\left[1 - 2\frac{\mathrm{d}H}{\mathrm{d}r}\right]^{1/2} \approx \omega , \tag{35}$$

which provides an implicit condition for the upper turning point of acoustic modes, r_2. Note that unlike the case of the lower turning point, to this approximation the upper turning point of acoustic modes is independent of the mode degree.

The upper turning point of acoustic modes in a solar-like model is best seen in the right-hand section of Fig. 6. For the lowest of the three frequencies in the high-frequency regime ($\nu \approx 1613\,\mu$Hz), the upper turning point is below the photosphere, while for the second lowest ($\nu \approx 2936\,\mu$Hz) it is at the photosphere. For that reason, the former is significantly less sensitive to the details of the outer layers than the latter. This is relevant because these layers are particularly difficult to model. The oscillation frequencies derived from models are therefore affected by the incomplete modelling of the outer layers and, as a result, show systematic differences when compared with the observations. This is less so for the lower frequency acoustic modes, which may, in that case, serve as an anchor with which to get a handle on the systematic errors (assuming some kind of frequency dependence of these errors). The other point to notice is that there is a frequency above which no trapping is possible. This has been discussed in Sect. 2.4.2, where the critical frequency for a plane-parallel, ideal, isothermal atmosphere has been introduced. Independently of taking the latter, or the more realistic critical frequency shown in Fig. 6, it is clear that above ~5.3 mHz full trapping of the modes no longer occurs. Since ω_c also scales with the inverse of the dynamical timescale (as seen when comparing Figs. 4 and 5), the maximum expected observed frequency is strongly dependent on the evolutionary state of the star.

3.2.2 Internal Gravity Waves

Let us now turn our attention to the low frequency limit of Eq. (24) in regions where $\kappa_r^2 > 0$. Let us consider that $\omega^2 \ll S_l^2$ throughout that propagation region and, in addition, that $\omega_c^2 \ll S_l^2$ there. In that case k_r^2 is given approximately by

$$\kappa_r^2 \approx \frac{S_l^2}{c_0^2\omega^2}\left[N_0^2 - \omega^2\right]. \tag{36}$$

Recalling that $\kappa_h^2 = l(l+1)/r^2 = S_l^2/c_0^2$, we then find the dispersion relation

$$\omega^2 \approx \frac{N_0^2}{1 + k_r^2/\kappa_h^2}, \tag{37}$$

which is characteristic of internal gravity waves. Internal gravity waves are maintained by the gravity acting on the perturbation to the density. If one considers a slow upwards displacement of an element of fluid whose pressure is kept in equilibrium with the surrounding, buoyancy will respond to restore the fluid towards the equilibrium position if its density is larger than that of the surroundings. That, in turn, can lead to the oscillatory motion associated to gravity waves. Because an element of fluid cannot move strictly vertically, there is always an horizontal component of the motion, which, in turn, means that gravity waves can never be associated to spherically symmetric perturbations, i.e., there are no gravity waves of degree $l = 0$. Moreover, since buoyancy will only oppose to the motion of the element of fluid where $N_0^2 > 0$, i.e., in convectively stable regions, the gravity waves will only propagate where there is no convection.

Looking back at the dispersion relation in Eq. (37), there are additional points that should be noted. First, the frequency of gravity waves is always smaller than N_0. This just confirms the role of buoyancy in maintaining the dynamics of gravity waves. The second point is that the frequency of gravity waves depends critically on the shape of the perturbation. In fact, when the perturbation is "needle-like", meaning that $K_r^2/K_h^2 \ll 1$, the frequency is higher, tending to N_0 as that ratio tends to zero. In the other limit, for wide perturbations with $K_r^2/K_h^2 \gg 1$, the oscillation frequency is smaller, tending to zero as the ratio becomes increasingly higher. This can be understood if we recall that the amount of material displaced horizontally is larger in the latter case than in the former. This horizontal displacement, and the horizontal pressure gradient that it originates, increases the effective inertia of the element of fluid on which the buoyancy force is acting. The result is a smaller acceleration of the element of fluid and, consequently, a smaller mode frequency.

It is worth emphasizing that the aspects discussed above are in striking contrast with what was previously found for acoustic waves. Indeed, for acoustic waves the frequency is found to depend essentially on the characteristic scale of the perturbation, determined by the total wavenumber $|\boldsymbol{\kappa}|$, while for gravity waves the frequency depends in addition, and very critically, on the relation between the horizontal and vertical scales. Moreover, considering modes of a given degree (hence fixing the horizontal scale), we find that the frequency of gravity waves decreases with increasing κ_r. Taking the radial orders n as negative integers, as is commonly done for gravity waves, we see that their frequencies decrease with increasing $|n|$, again in contrast with what was found earlier for acoustic waves.

Finally, from Eq. (36) we find that under the conditions assumed, gravity waves propagate between the radii at which $N_0 = \omega$. The latter thus provides an implicit condition for the lower and upper turning points of these modes which to this approximation are independent of the mode degree, l. This is, again, in contrast with the case of acoustic modes, for which the lower turning point, and, hence, the extent of the propagation cavity, was found to be strongly dependent on the mode degree. For the case of a star like the Sun, we see from Fig. 6 that the lower frequency gravity modes are essentially trapped between the centre of the star and the base of the convective region. For the highest frequency gravity modes the upper turning point gets smaller and the modes are trapped in deeper layers. For main-sequence stars more massive than the Sun, the innermost layers are convectively unstable and, thus, the cavity of gravity waves is bounded on the inner side by the edge of the convectively unstable core. On the other hand, since the convective envelope gets shallower for more massive stars, gravity waves there can propagate almost to the stellar surface. In more evolved stars, the trapping region again depends on the existence, or not, of convection in the core, as well as on the extent of the convective envelope, which can get very deep, as happens, for instance, along the red-giant branch. Moreover, the steep increase of the buoyancy frequency in the core of evolved stars, such as that shown in Fig. 5, can lead to several cavities of propagation for the same mode, in which case the analysis becomes more complex than in the cases discussed above. I will get back to that case in the next section, where the full numerical results of the pulsation equations shall be discussed. Table 1 provides a summary of the properties of acoustic and gravity waves.

3.3 The Surface Gravity Waves

For completeness, in this section I will consider the case of perturbations obeying $\delta p = 0$, hence, $\Psi = 0$. Going back to the system of equations (21), and considering,

Table 1 Summary of the properties of acoustic and gravity waves discussed in this chapter

Acoustic waves	Gravity waves
Maintained by the gradient of pressure fluctuation	Maintained by gravity acting on density fluctuation
Radial or non-radial	Always non-radial
Propagate in convectively stable or unstable regions	Propagate in convectively stable regions only
Propagation cavity strongly dependent on l	Propagation cavity largely independent of l
Frequency increases with increasing n and increasing l	Frequency always $< \max(N_0)$. It increases with increasing l, and decreases with increasing $\lvert n \rvert$

as before, that locally the oscillations can be treated as in a plane-parallel layer under constant gravity, we find:

$$\frac{\mathrm{d}\xi_r}{\mathrm{d}r} - \frac{g_0\kappa_h^2}{\omega^2}\xi_r + \frac{1}{\rho_0 c_0^2}\left(1 - \frac{c_0^2\kappa_h^2}{\omega^2}\right)\delta p = 0\,,$$

$$\frac{\mathrm{d}\delta p}{\mathrm{d}r} + \frac{g_0\kappa_h^2}{\omega^2}\delta p - \rho_0 g_0\left(\frac{\omega^2}{g_0} - \frac{g_0\kappa_h^2}{\omega^2}\right)\xi_r = 0\,. \tag{38}$$

For $\delta p = 0$, the system above is satisfied by an exponential solution of the type $\xi_r = \exp[\kappa_h (r - r_0)]$, where $\kappa_h = \omega^2/g_0$. Here, r_0 is a fiducial depth in the vicinity of which the plane-parallel approximation is being made and g_0 is the gravitational acceleration at $r = r_0$. Thus, the depth-dependent amplitude of the solution in this case decays exponentially with depth and the dispersion relation is $\omega = \sqrt{g_0\kappa_h}$. Moreover, under the approximations considered here, it is independent of the stratification of the star. Since the characteristic scale of the amplitude decay is κ_h^{-1}, the plane-parallel approximation is particularly adequate when κ_h, hence the degree l of the mode, is not too small and the mode is concentrated near the surface of the star.

This solution can be identified as a surface gravity wave, similar to a wave propagating at the surface of a deep ocean. Since $\delta p = 0$, $\nabla \cdot \boldsymbol{\xi} = 0$ and the fluid is not compressed during the perturbation.

4 Numerical Solutions to the Pulsation Equations

The full solutions to the pulsation equations and associated boundary conditions must be computed numerically. In this section, I briefly discuss the range of solutions obtained for a model of the Sun and discuss also a specific example for a star in a different evolutionary state.

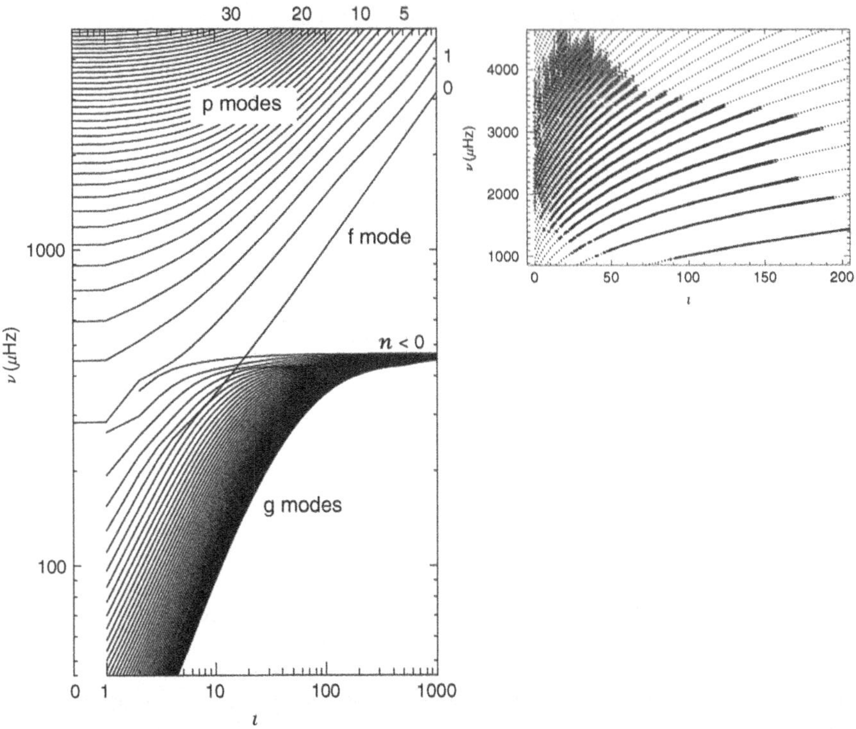

Fig. 7 *Left:* Cyclic frequencies computed for a model of the Sun, as function of mode degree, *l*. The discrete eigenvalues for each radial order have been joined by *continuous lines* with a few examples of the radial order identified on the right-hand side and top of the figure. *Right:* Frequencies of the Sun derived from 144 days of observations with the instrument MDI on board the *SOHO* spacecraft. The depicted error bars correspond to 1000σ. The *dotted lines* show the model results, for comparison. Figure adapted from Christensen-Dalsgaard (2008c) and Aerts et al. (2010)

The left panel of Fig. 7, adapted from Aerts et al. (2010), shows the cyclic frequencies ($=\omega/2\pi$) computed with the Aarhus adiabatic oscillation package (ADIPLS; Christensen-Dalsgaard 2008a) for a solar model obtained with the Aarhus STellar Evolution Code (ASTEC; Christensen-Dalsgaard 2008b). The discrete eigenvalues for each radial order have been joined by continuous lines with a few examples of the radial order identified on the right-hand side and top of the figure. Three families of solutions are identified in the figure: (1) The acoustic (or p-) modes, at higher frequencies; (2) the gravity (or g-) modes, at lower frequencies, and; (3) the f-mode, at intermediate frequencies. These correspond to the cases discussed in the previous section, based on the analysis under the Cowling approximation. For comparison, the eigenfrequencies of the real Sun, derived from data acquired with the instrument MDI on board of the *SOHO* spacecraft are shown on the right panel of the same figure, where the model results are overplotted as dotted curves. Only the p- and f-modes are seen in the real data. In fact, the observation of gravity modes in the Sun has been a long-standing goal, but despite all efforts and some claims of detection of signatures of g-modes and, possibly, individual g-modes in the Sun (García et al. 2007; García 2010), the matter is still not settled (Appourchaux et al. 2010).

A closer look at the different families of solutions displayed in Fig. 7 shows that the behaviour of the solutions with changing radial order and mode degree also follows what was found from the analysis in Sect. 3. For the p-modes, at fixed degree, l, we can see an increase of the frequency with radial order. Likewise, following a single line of fixed radial roder, n, we also see an increase of the frequency with increasing degree. Both of these dependencies were expected from the dispersion relation for acoustic waves derived earlier. Regarding the g-modes, we can identify an upper bound to the frequency, which corresponds to the maximum value of the buoyancy frequency in that model. In addition, I note that there are no results for spherically symmetric modes, $l = 0$, as expected given that the displacement associated with these waves can never be purely radial (always involves a horizontal displacement of the fluid). Moreover, it is seen that at fixed degree the frequency decreases with increasing absolute value of the radial order, $|n|$, while at fixed radial order, the frequency increases with increasing degree. I also note that the fact that the lines of g-mode solutions are very close to each other implies that at fixed frequency the radial order increases very rapidly with increasing mode degree. As for the case of p-modes, the numerical results discussed here are in agreement with the dispersion relation for g-modes derived in Sect. 3. Finally, the f-mode eigenfrequencies for moderate to high mode degrees (the ones for which the plane-parallel approximation used in Sect. 3 is adequate), are found between the p-mode and g-mode eigenfrequencies. Despite the resemblance of the high degree f-mode solutions and the p-mode solutions I wish to recall that these modes obey $\delta p = 0$. Thus, the perturbation takes place without compression or refraction of the fluid, reminding us that f-modes are not acoustic waves.

The trapping of the modes discussed in Sect. 3 can also be verified by inspection of the numerical eigenfunctions. In Fig. 8 three examples of p-mode (left panel) and g-mode (right panel) eigenfunctions are shown, for the solar model discussed above. The quantity plotted, $r^3 \delta p$, has the dimensions of energy and, in each case, it is normalized to its maximum value. Inspection of the p-mode solutions shows that the spherically symmetric pulsation ($l = 0$; top panel) propagates from the centre to the stellar surface. One may then expected it to carry average information about the entire star. In contrast, a p-mode of similar frequency but much higher degree ($l = 60$; bottom panel) has its energy concentrated in the outer layers of the star. This was expected from the analysis performed in Sect. 3, where it was found that the propagation cavity of p-modes becomes shallower as the mode degree increases. Looking now at the right panel we see that, in contrast with the p-modes, the energy of the g-modes is concentrated towards the innermost layers of the star. The modes are trapped below the convective envelope (marked by the vertical dotted line), in a cavity that is mostly independent of mode degree. The modes shown have similar frequency, so as anticipated from Fig. 7, even a small increase in mode degree results

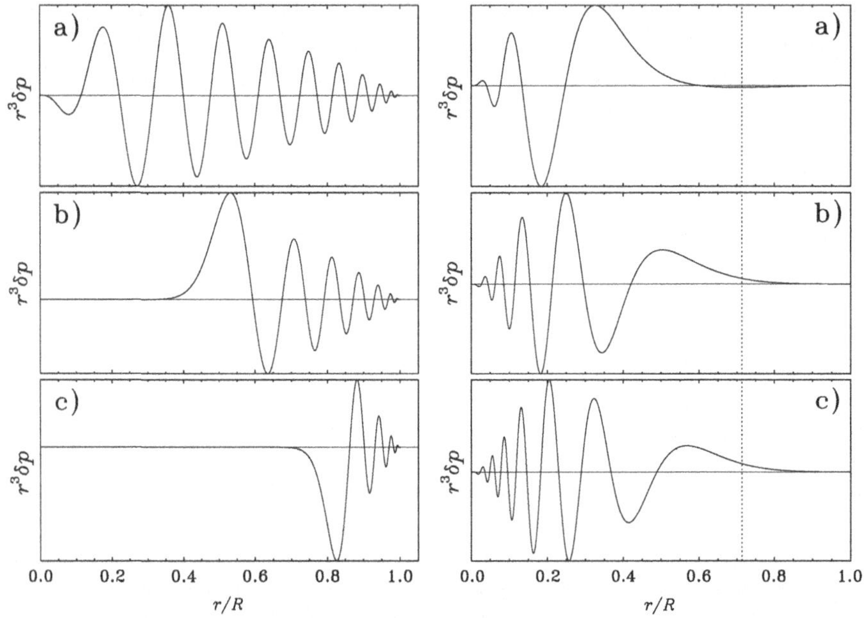

Fig. 8 Normalized eigenfunctions for a model of the Sun, as function of fractional radius. The chosen eigenfunction, $r^3\delta p$, has the dimensions of energy and is normalized to its maximum value. *Left:* Results for three p-modes: (*a*) ($l = 0, n = 21, \nu = 3038.0\,\mu\text{Hz}$), (*b*) ($l = 20, n = 14, \nu = 2939.2\,\mu\text{Hz}$), and (*c*) ($l = 60, n = 9, \nu = 3043.2\,\mu\text{Hz}$). *Right:* Results for three g-modes: (*a*) ($l = 1, n = -5, \nu = 109.2\,\mu\text{Hz}$), (*b*) ($l = 2, n = -10, \nu = 102.6\,\mu\text{Hz}$), and (*c*) ($l = 3, n = -14, \nu = 104.1\,\mu\text{Hz}$). The *dotted line* marks the base of the convective envelope (figure courtesy of J. Christensen-Dalsgaard)

is a significant increase in the radial order, as seen from the increase in the number of nodes when comparing the upper and lower panels on the right-hand side.

The regions where modes propagate depend directly on the stellar structure and, hence, are different for stars of different masses or different evolutionary states. In Sect. 2.3 I have pointed out that as the star evolves beyond the main sequence, and the core contracts, the buoyancy frequency increases significantly towards the centre. This results in the appearance of mixed modes, i.e., modes that are maintained by gravity acting on density perturbations in the deep interior, and by the gradient of the pressure perturbation in the outer layers. An example of such a mode computed for the model shown in Fig. 5 is shown in Fig. 9. Two mode cavities can be identified, in the inner and outer layers of the star, respectively, separated by an evanescent region where the solution is not oscillatory. Despite the latter, coupling does exist between the two cavities and in a general case the solution is different from what would be found if the two cavities were considered independently of each other.

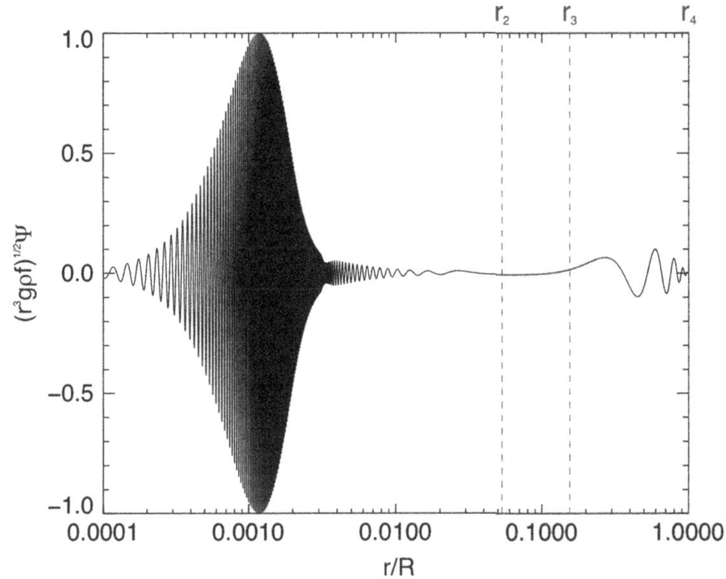

Fig. 9 Same as Fig. 8, for a dipole mode with frequency $\nu = 51.20\,\mu$Hz, in a model of a star in the red-giant branch. The *vertical, dashed lines* show the position of the two turning points bounding the evanescent region, namely, r_2 and r_3. The outermost turning point, r_4, is also shown, while the innermost turning point, r_1, is outside the plotted range. The g-mode cavity is located between the unseen r_1 and r_2, and the p-mode cavity is located between r_3 and r_4. Figure adapted from Cunha et al. (2015)

5 Discussion

As mentioned from the outset, a number of aspects of the theory of stellar pulsations had to be left out of these notes in the interest of space. In what follows, I identify issues that I find particularly important and that are discussed in detail in the books and lecture notes mentioned in the introduction.

To start with, it should be pointed out that it is possible to perform an asymptotic analysis of the second-order pulsation equations derived under the Cowling approximation to find approximate eigenvalues and eigenfunctions for modes of high radial orders and low degree. That has been performed in different ways by different authors (Vandakurov 1968; Tassoul 1980; Unno et al. 1989; Gough 1993). In the case of p-modes, the eigenfrequencies in this limit are found to be well approximated by

$$\nu_{nl} \simeq \left(n + \frac{l}{2} + \frac{1}{4} + \alpha\right) \Delta\nu_0 - [Al(l+1) - \delta]\frac{\Delta\nu_0^2}{\nu_{nl}}, \tag{39}$$

where

$$\Delta\nu_0 = \left(2\int_0^R \frac{\mathrm{d}r}{c}\right)^{-1} \tag{40}$$

is the inverse sound travel time across a stellar diameter, and

$$A = \frac{1}{4\pi^2 \Delta\nu_0}\left[\frac{c(R)}{R} - \int_0^R \frac{\mathrm{d}c}{\mathrm{d}r}\frac{\mathrm{d}r}{r}\right]. \tag{41}$$

Here, α is a slowly varying function of frequency determined by the reflection properties near the surface and δ is a small correction term predominantly related to the near-surface region. To leading order, Eq. (39) predicts the uniform spacing between frequencies of consecutive modes of the same degree, corresponding to the large frequency separation, $\Delta\nu$, observed in Fig. 1. Also, we see that modes of odd degree fall halfway between modes of even degree. The first term in Eq. (39) cancels out when subtracting the frequencies of modes of consecutive orders and degrees differing by two. That combination then gives the small frequency separation, $\delta\nu$, also seen in Fig. 1, and which is found to be

$$\delta\nu_{nl} = \nu_{n\,l} - \nu_{n-1\,l+2} \simeq -(4l+6)\frac{\Delta\nu_0}{4\pi^2\nu_{nl}}\int_0^R \frac{\mathrm{d}c}{\mathrm{d}r}\frac{\mathrm{d}r}{r}, \tag{42}$$

where the small term $c(R)$ in Eq. (41) has been neglected. From Eq. (42) we see that the small separation is particularly sensitive to the innermost layers, as a result of the r^{-1} dependence of the integrand. Other small separations and ratios of small to large separations are often considered in asteroseismology. In particular, the ratios (Roxburgh and Vorontsov 2003; Cunha and Metcalfe 2007) have the advantage that they are essentially independent of the inadequately modelled surface layers of the stars.

In the case of high radial order, low-degree g-modes, the first-order term of the asymptotic analysis predicts a uniform spacing in mode period, Π_{nl}, rather than in frequency. In this case, we have (e.g., Tassoul 1980; Smeyers and Moya 2007):

$$\Pi_{nl} \simeq \frac{\Pi_0}{\sqrt{l(l+1)}}\left(n + \alpha_{l,\mathrm{g}}\right), \tag{43}$$

where

$$\Pi_0 = 2\pi^2\left(\int_{r_1}^{r_2} N\frac{\mathrm{d}r}{r}\right)^{-1}. \tag{44}$$

The phase $\alpha_{l,\mathrm{g}}$ depends on whether the core of the star is radiative or convective, depending on mode degree in the first case but not in the latter. In the case of a radiative core one can write $\alpha_{l,\mathrm{g}} = l/2 + \alpha_\mathrm{g}$, where α_g is independent of mode degree. Note, however, that in both cases there is a strong dependence on mode degree of the period spacings between modes of the same degree and consecutive orders, due to the term $[l(l+1)]^{-1/2}$ in Eq. (43). This is in contrast with the case of p-modes for which the asymptotic large separation is, to first order, independent of mode degree.

Another important aspect that has been left out of this chapter is the impact on the oscillation spectrum of rotation and magnetic fields. A perturbative analysis of the impact of rotation on pulsations can be found, e.g., in Aerts et al. (2010), while the impact of an internal magnetic field can be found, e.g., in Gough (1993). Non-perturbative analyses of these phenomena, required in the cases of fast rotation or magnetic fields that permeate the stellar surface, are discussed by Lignières et al. (2006), Reese et al. (2006) and Cunha and Gough (2000), Cunha (2006), respectively.

The ultimate goal of stellar pulsations studies is to infer information about the physics and dynamics of stellar interiors from the asteroseismic data. That is commonly achieved through forward modelling, a method that is intrinsically model-dependent, or, in optimal cases, through inverse techniques, in which the solutions are not restricted to those of a set of models.

Finally, I note that no word has been said about the driving of the pulsations, with the exception of a brief reference, in Sect. 1, to the fact that modes can be intrinsically stable, as in solar-like pulsators, or unstable, as in classical pulsators. This topic is, however, well out of the scope of the present notes and I, thus, refer the interested reader to the literature listed in Sect. 1, and references therein.

Acknowledgements I am grateful to Ângela Santos for producing Fig. 1 and to Jørgen Christensen-Dalsgaard for producing Fig. 8. This work was supported by Fundação para a Ciência e a Tecnologia (FCT) through national funds (UID/FIS/04434/2013) and by FEDER through COMPETE2020 (POCI-01-0145-FEDER-007672), as well as through the Investigador FCT Contract No. IF/00894/2012/CP0150/CT0004. This work has received funding from EC, under FP7, through the grant agreement FP7-SPACE-2012-312844.

References

Aerts, C., Christensen-Dalsgaard, J., Kurtz, D.W.: Asteroseismology. Springer, Berlin (2010)
Appourchaux, T., Belkacem, K., Broomhall, A.-M., et al.: Astron. Astrophys. Rev. **18**, 197 (2010)
Baglin, A., Auvergne, M., Barge, P., et al.: In: Fridlund, M., Baglin, A., Lochard, J., Conroy, L. (eds.) The CoRoT Mission Pre-Launch Status - Stellar Seismology and Planet Finding, vol. 1306, p. 33. ESA Special Publication, Noordwijk (2006)
Basu, S.: Living Rev. Sol. Phys. **13**, 2 (2016)
Brown, T.M., Gilliland, R.L., Noyes, R.W., Ramsey, L.W.: Astrophys. J. **368**, 599 (1991)
Christensen-Dalsgaard, J.: Astrophys. Space Sci. **316**, 113 (2008a)
Christensen-Dalsgaard, J.: Astrophys. Space Sci. **316**, 13 (2008b)
Christensen-Dalsgaard, J.: The art of modeling stars in the 21st century. In: Deng, L., Chan, K.L. (eds.) IAU Symposium, vol. 252, pp. 135–147 (2008c)
Christensen-Dalsgaard, J., Dappen, W., Ajukov, S.V., et al.: Science **272**, 1286 (1996)
Cunha, M.S.: Mon. Not. R. Astron. Soc. **365**, 153 (2006)

Cunha, M.S., Gough, D.: Mon. Not. R. Astron. Soc. **319**, 1020 (2000)
Cunha, M.S., Metcalfe, T.S.: Astrophys. J. **666**, 413 (2007)
Cunha, M.S., Aerts, C., Christensen-Dalsgaard, J., et al.: Astron. Astrophys. Rev. **14**, 217 (2007)
Cunha, M.S., Stello, D., Avelino, P.P., Christensen-Dalsgaard, J., Townsend, R.H.D.: Astrophys. J. **805**, 127 (2015)
Deubner, F.-L., Gough, D.: Annu. Rev. Astron. Astrophys. **22**, 593 (1984)
Domingo, V., Fleck, B., Poland, A.I.: Sol. Phys. **162**, 1 (1995)
Fröhlich, C., Romero, J., Roth, H., et al.: Sol. Phys. **162**, 101 (1995)
García, R.A.: Highlights Astron. **15**, 345 (2010)
García, R.A., Turck-Chièze, S., Jiménez-Reyes, S.J., et al.: Science **316**, 1591 (2007)
Gilliland, R.L., Brown, T.M., Christensen-Dalsgaard, J., et al.: Publ. Astron. Soc. Pac. **122**, 131 (2010)
Gough, D.O.: In: Zahn, J.-P., Zinn-Justin, J. (eds.) Astrophysical Fluid Dynamics - Les Houches 1987, pp. 399–560 (1993)
Jiménez, A., Roca Cortés, T., Jiménez-Reyes, S.J.: Sol. Phys. **209**, 247 (2002)
Kjeldsen, H., Bedding, T.R.: Astron. Astrophys. **293**, 87 (1995)
Koch, D.G., Borucki, W.J., Basri, G., et al.: Astrophys. J. Lett. **713**, L79 (2010)
Lignières, F., Rieutord, M., Reese, D.: Astron. Astrophys. **455**, 607 (2006)
Reese, D., Lignières, F., Rieutord, M.: Astron. Astrophys. **455**, 621 (2006)
Roxburgh, I.W., Vorontsov, S.V.: Astron. Astrophys. **411**, 215 (2003)
Smeyers, P., Moya, A.: Astron. Astrophys. **465**, 509 (2007)
Takata, M.: Publ. Astron. Soc. Jpn. **57**, 375 (2005)
Takata, M.: Publ. Astron. Soc. Jpn. **68**, 109 (2016)
Tassoul, M.: Astrophys. J. Suppl. Ser. **43**, 469 (1980)
Unno, W., Osaki, Y., Ando, H., Saio, H., Shibahashi, H.: Nonradial Oscillations of Stars. University of Tokyo Press, Tokyo (1989)
Vandakurov, Y.V.: Sov. Astron. **11**, 630 (1968)
Verner, G.A., Elsworth, Y., Chaplin, W.J., et al.: Mon. Not. R. Astron. Soc. **415**, 3539 (2011)

An Introduction to Data Analysis in Asteroseismology

Tiago L. Campante

Abstract A practical guide is presented to some of the main data analysis concepts and techniques employed contemporarily in the asteroseismic study of stars exhibiting solar-like oscillations. The subjects of digital signal processing and spectral analysis are introduced first. These concern the acquisition of continuous physical signals to be subsequently digitally analyzed. A number of specific concepts and techniques relevant to asteroseismology are then presented as we follow the typical workflow of the data analysis process, namely, the extraction of global asteroseismic parameters and individual mode parameters (also known as peak-bagging) from the oscillation spectrum.

1 Introduction

Solar-like oscillations are excited by turbulent convection in the outer layers of stars (see, e.g., Christensen-Dalsgaard 2004, and references therein). Consequently, all stars cool enough to harbor an outer convective envelope may be expected to exhibit solar-like oscillations. Among several other classes of pulsating stars, solar-like oscillations are detectable in main-sequence core, and post-main-sequence shell, hydrogen-burning stars residing on the cool side of the Cepheid instability strip. The NASA *Kepler* mission (Borucki et al. 2010) has led to a revolution in the field of cool-star asteroseismology by allowing the detection of solar-like oscillations in several hundred solar-type stars (i.e., low-mass, main-sequence stars and cool subgiants) and in over ten thousand red giants (for a review, see Chaplin and Miglio 2013). Of all these stars displaying solar-like oscillations about one hundred

T.L. Campante (✉)
School of Physics and Astronomy, University of Birmingham, Birmingham B15 2TT, UK

Department of Physics and Astronomy, Stellar Astrophysics Centre (SAC), Aarhus University, Ny Munkegade 120, DK-8000 Aarhus C, Denmark

Institut für Astrophysik, Georg-August-Universität Göttingen, Friedrich-Hund-Platz 1, 37077 Göttingen, Germany
e-mail: campante@bison.ph.bham.ac.uk

T.L. Campante et al. (eds.), *Asteroseismology and Exoplanets: Listening to the Stars and Searching for New Worlds*, Astrophysics and Space Science Proceedings 49, DOI 10.1007/978-3-319-59315-9_3

are *Kepler* Objects of Interest (KOIs), i.e., candidate exoplanet-host stars (e.g., Campante et al. 2015; Lundkvist et al. 2016).

The present chapter is intended as a practical guide to some of the main data analysis concepts and techniques employed contemporarily in the asteroseismic study of stars exhibiting solar-like oscillations. The contents of this chapter strongly reflect the author's own experience as a data analyst. For that reason, special care has been taken to provide references to the work conducted by others, so that the reader can easily expand on the material presented herein.

Pre-processing of light curves, although an integrant part of the data analysis process, is beyond the scope of this contribution. I therefore start by introducing the subjects of digital signal processing and spectral analysis in Sect. 2. These concern the acquisition of continuous physical signals to be subsequently digitally analyzed. A number of specific concepts and techniques relevant to asteroseismology are then presented as we follow the typical workflow of the data analysis process (see Fig. 1). One must first establish whether signatures of solar-like oscillations

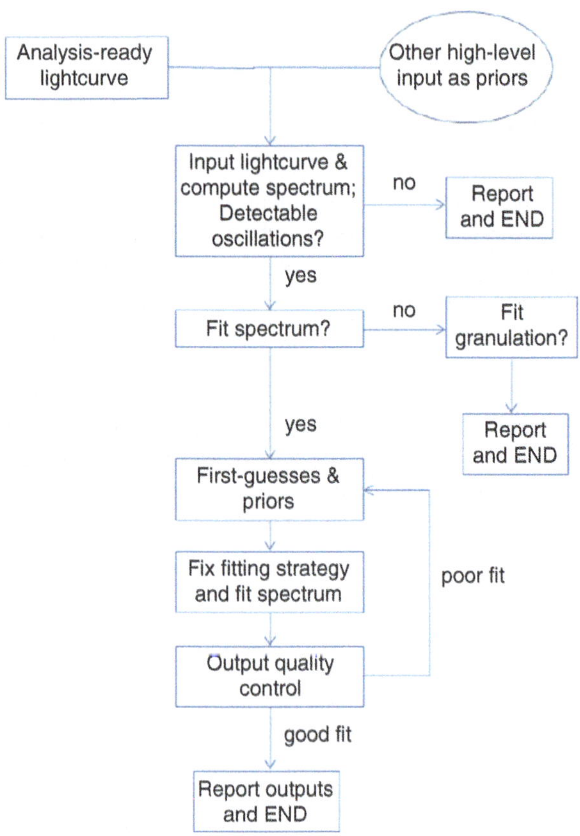

Fig. 1 Typical workflow of the data analysis process

are detectable in the power spectrum of the light curve. If they are, an attempt is made at extracting *global asteroseismic parameters* from the data (Sect. 3). One then establishes whether the oscillation spectrum is of sufficient quality to allow extraction of individual frequencies. If the answer is yes, *individual mode parameters* are then extracted by fitting a multiparameter model to the oscillation spectrum, i.e., by *peak-bagging* the oscillation spectrum (Sect. 4).

2 Digital Signal Processing and Spectral Analysis

Whereas some temporal phenomena can be understood through models in the time domain involving deterministic trends and/or stochastic autoregressive behavior, others are dominated by periodic behavior that is most effectively modeled in the frequency domain. The functional form of solar-like oscillations is that of a *stochastically-excited harmonic oscillator*. This being a periodic functional form, the Fourier transform becomes the obvious choice for performing data analysis.

2.1 Nyquist Sampling Theorem and Aliasing

Let us consider the idealized case of a continuous signal $x(t)$ sampled by a set of impulse functions regularly spaced by Δt. Since the Fourier transform of such a set of impulse functions is another set of impulse functions with separation $1/\Delta t$ in the frequency domain, one can use the convolution theorem to show that the transform of the sampled signal is periodic:

$$x(t) \sum_{n=-\infty}^{+\infty} \delta\,(t - n\,\Delta t) \Longleftrightarrow X(\nu) * \frac{1}{\Delta t} \sum_{n=-\infty}^{+\infty} \delta\left(\nu - \frac{n}{\Delta t}\right) , \qquad (1)$$

where $X(\nu)$ is the Fourier transform of $x(t)$, the symbol '$\Longleftrightarrow$' indicates a Fourier pair and the symbol '$*$' denotes convolution.

The *Nyquist sampling theorem* (Nyquist 1928; Shannon 1949) states that if the Fourier transform of a continuous signal is band-limited, i.e., is zero for all $|\nu| \geq \nu_{\rm lim}$, then $x(t)$ can be uniquely reconstructed from a knowledge of its sampled values at uniform intervals of $\Delta t \leq 1/(2\,\nu_{\rm lim})$. For a given uniform sampling interval Δt, the *Nyquist frequency* is defined as $\nu_{\rm Nyq} = 1/(2\Delta t)$. In case the continuous signal being sampled contains frequency components above the Nyquist frequency, these will give rise to an effect known as *aliasing*, whereby the transform of the continuous signal is distorted due to spectral leakage. The signal is then said to be undersampled and can no longer be uniquely recovered.

The Nyquist frequency can be thought of as the highest useful frequency to search for in the power spectrum. However, based on astrophysical arguments, one can also accept frequencies above $\nu_{\rm Nyq}$ (Murphy et al. 2013; Chaplin et al. 2014). Prospects for detecting solar-like oscillations in the *super-Nyquist* regime of *Kepler* long-cadence data, i.e., above the associated Nyquist frequency of $\sim 283\,\mu$Hz, are now being explored (Yu et al. 2016). Targets of interest are cool subgiants and stars lying at the base of the red-giant branch.

Regular gaps in the light curve due to diurnal interruptions and, for data sets spanning more than a year, caused by the annual motion of the Earth, are usually present in observations carried out from the ground, giving rise also to frequency aliasing. *Daily aliases*, appearing at splittings of ± 1 cycle/day (or, equivalently, $\pm 11.57\,\mu$Hz), are particularly problematic when observing solar-like oscillations, since frequency separations of that same magnitude are common (e.g., Arentoft et al. 2008; Bedding et al. 2010).

2.2 *Filtering*

Asteroseismic time series are often affected by low-frequency drifts, which can be either of instrumental origin or else intrinsic to the star. These low-frequency drifts introduce a background in the Fourier domain that ultimately leads to a degradation of the signal-to-noise ratio (SNR) in the oscillation modes. High-pass filters are widely used to reduce this effect while preserving the relevant signals.

Let us start by shedding some light on the process of *smoothing* of a time series. Smoothing consists in convolving a signal $x(t)$ with a weighting function $w(t)$:

$$x_{\rm low}(t) = x(t) * w(t) \Longleftrightarrow X_{\rm low}(\nu) = X(\nu)\, W(\nu)\,, \tag{2}$$

where $X(\nu)$ and $W(\nu)$ are the transforms of $x(t)$ and $w(t)$, respectively. Conversely, a *high-pass filter* can be implemented by simply computing $x_{\rm high}(t) = x(t) - x_{\rm low}(t)$:

$$x_{\rm high}(t) \Longleftrightarrow X_{\rm high}(\nu) = X(\nu)\,[1 - W(\nu)]\;. \tag{3}$$

A commonly used high-pass filter in helioseismology is the backwards-difference filter (García and Ballot 2008):

$$x_{\rm bd}(t) = x(t) - x(t - t_0) = x(t) - [x(t) * \delta(t - t_0)]\;, \tag{4}$$

where a time shift t_0 has been considered. It becomes immediately obvious that $w(t) = \delta(t - t_0)$ in Eq. (2). Using Eq. (3), one can then determine the transfer function of the backwards-difference filter:

$$|1 - W(\nu)|^2 = \left[2 \sin\left(\frac{\pi}{2}\frac{\nu}{\nu_{\rm c}}\right)\right]^2\,, \tag{5}$$

where the cut-off frequency, $\nu_{\rm c} = 1/(2\,t_0)$, has been introduced.

Typical examples of the weighting function $w(t)$ are a boxcar function, a triangular function (equivalent to the convolution of two boxcar functions) and a bell-shaped function (equivalent to the convolution of four boxcar functions or two triangular functions). The transform of the boxcar function is the sinc function and thus leads to an excessive ringing (or Gibbs-like) effect in the Fourier domain. Multiple-boxcar smoothing is therefore advisable.

2.3 *Power Spectral Density Estimation*

Attention is first drawn to the estimation of the Fourier transform of $x(t)$ based on a finite number of samples. Suppose there are N evenly spaced samples $x(t_n) = x(n\Delta t)$, with $n = 0, 1, \ldots, N-1$. The *Discrete Fourier Transform*[1] (DFT) is defined as:

$$X_{\rm DFT}(\nu_p) = \sum_{n=0}^{N-1} x(t_n)\, {\rm e}^{{\rm i}\, 2\pi\nu_p t_n} \;\; \text{for } \nu_p = p/(N\Delta t)\,,\; p = 0, 1, \ldots, N-1\,. \tag{6}$$

$X_{\rm DFT}(\nu_p)$ is the truncated transform of the sampled signal, which has periodicity $1/\Delta t$ or twice the Nyquist frequency. Then $p = 0$ corresponds to the transform at zero frequency and $p = N/2$ to the value at $\pm\nu_{\rm Nyq}$. Values of p between $N/2+1$ and $N-1$ correspond to the transform for negative frequencies.

Finally, I introduce the one-sided *power density spectrum* or *power spectrum*, $P(\nu_q)$, defined only for nonnegative frequencies (with $q = 0, 1, \ldots, N/2$):

$$\begin{aligned} P(\nu_0) &= \frac{\Delta t}{N}\, |X_{\rm DFT}(\nu_0)|^2\,, \\ P(\nu_q) &= \frac{\Delta t}{N}\left[|X_{\rm DFT}(\nu_p)|^2 + |X_{\rm DFT}(\nu_{N-p})|^2\right], \\ P(\nu_{N/2}) &= \frac{\Delta t}{N}\, |X_{\rm DFT}(\nu_{N/2})|^2\,, \end{aligned} \tag{7}$$

where $\nu_{N/2} = 1/(2\Delta t)$ (i.e., the Nyquist frequency). Based on *Parseval's theorem* (Parseval des Chênes 1806), we may then normalize $P(\nu_q)$ according to

$$\sum_{q=0}^{N/2} P(\nu_q)\, \Delta\nu = \frac{1}{N} \sum_{n=0}^{N-1} x^2(t_n)\,. \tag{8}$$

[1]Cooley and Tukey (1965) have introduced the *Fast Fourier Transform* (FFT), an efficient method of implementing the DFT.

According to the *Wiener–Khintchine theorem* (Wiener 1930; Khintchine 1934), the power spectrum and the autocorrelation function, $\phi(\tau)$, are a Fourier pair:

$$\phi(\tau) = \int_{-\infty}^{+\infty} P(\nu)\,\mathrm{e}^{-\mathrm{i}\,2\pi\nu\tau}\,\mathrm{d}\nu \Longleftrightarrow P(\nu) = \int_{-\infty}^{+\infty} \phi(\tau)\,\mathrm{e}^{\mathrm{i}\,2\pi\nu\tau}\,\mathrm{d}\tau\,, \tag{9}$$

where

$$\phi(\tau) = \lim_{T\to\infty} \frac{1}{T} \int_{-T/2}^{T/2} x(t)x(t+\tau)\,\mathrm{d}t\,. \tag{10}$$

The Wiener–Khintchine theorem is absolutely crucial to understanding the spectral analysis of random processes. It straightforwardly explains, for instance, why white noise, whose autocorrelation function is the Dirac delta function, has constant power spectral density.

2.4 Power Spectrum Statistics and Hypothesis Testing

In the following I consider the *statistics* of the power spectrum of a pure noise signal (see also Appourchaux 2013). Let $x(t)$ represent a random process from which a finite number of samples $x(t_n)$ are drawn. The samples are assumed to be independent and identically distributed (i.i.d.), and the process is further assumed to be stationary, with $\mathrm{E}\left[x(t_n)\right]=0$ and $\mathrm{E}\left[x^2(t_n)\right]=\sigma_0^2$ for all n. The DFT of the set $x(t_n)$ may be decomposed into its real and imaginary parts as:

$$\begin{aligned} X_{\mathrm{DFT}}(\nu_p) &= X_{\mathrm{DFT}}^{\mathrm{Re}}(\nu_p) + \mathrm{i}\,X_{\mathrm{DFT}}^{\mathrm{Im}}(\nu_p) \\ &= \sum_{n=0}^{N-1} x(t_n)\cos(2\pi\nu_p t_n) + \mathrm{i}\sum_{n=0}^{N-1} x(t_n)\sin(2\pi\nu_p t_n)\,. \end{aligned} \tag{11}$$

It follows from the Central Limit theorem that, for large N, both $X_{\mathrm{DFT}}^{\mathrm{Re}}$ and $X_{\mathrm{DFT}}^{\mathrm{Im}}$ are normally distributed with

$$\mathrm{E}\left[X_{\mathrm{DFT}}^{\mathrm{Re}}(\nu_p)\right] = \mathrm{E}\left[X_{\mathrm{DFT}}^{\mathrm{Im}}(\nu_p)\right] = 0\,, \tag{12}$$

$$\mathrm{E}\left[\left(X_{\mathrm{DFT}}^{\mathrm{Re}}(\nu_p)\right)^2\right] = \mathrm{E}\left[\left(X_{\mathrm{DFT}}^{\mathrm{Im}}(\nu_p)\right)^2\right] = \frac{N}{2}\sigma_0^2\,. \tag{13}$$

Finally, since $X_{\mathrm{DFT}}^{\mathrm{Re}}$ and $X_{\mathrm{DFT}}^{\mathrm{Im}}$ are independent and have the same normal distribution, the power spectrum, $|X_{\mathrm{DFT}}|^2$, then has by definition a *chi-squared distribution with 2 degrees of freedom* (i.e., χ_2^2).

Adopting $|X_{\mathrm{DFT}}|^2\,\Delta t/N$ as our normalization of the power spectrum yields a constant power spectral density for the noise of $\sigma_0^2\Delta t$ and variance $(\sigma_0^2\Delta t)^2$. Consequently, as N tends to infinity by sampling a longer stretch of data, the variance in the power spectrum remains unchanged. Furthermore, the probability density, $p(z)$, that the observed power spectrum takes a particular value z at a fixed frequency bin is given by

$$p(z) = \frac{1}{\langle z\rangle}\exp\left(-\frac{z}{\langle z\rangle}\right), \tag{14}$$

where $\langle z\rangle = \sigma_0^2\Delta t$. Equation (14) enables one to derive the probability that the power in one bin is greater than m times the mean level of the continuum, $\langle z\rangle$:

$$F(m) = \mathrm{e}^{-m}. \tag{15}$$

For instance, a *confidence level* of 99% or, equivalently, a *false alarm probability* of 1%, leads to $m \approx 4.6$. For a frequency band containing M bins, the probability that at least one bin has a normalized power greater than m is then:

$$F_M(m) = 1 - (1 - \mathrm{e}^{-m})^M, \tag{16}$$

which approximates to $F_M(m) = M\mathrm{e}^{-m}$ for $\mathrm{e}^{-m} \ll 1$.

In astrophysics it is very common to deal with unevenly sampled time series. In that event, an existing frequentist statistic known as the *Lomb–Scargle periodogram*[2] is widely used as an estimator of the power spectral density. The Lomb–Scargle periodogram can be formulated either as a modified Fourier analysis or as a least-squares regression of the data set to sine waves with a range of frequencies. It has the attractive property of retaining the χ_2^2 statistics.

2.5 Non-Fourier Periodograms

Astronomers have developed and extensively used a variety of *non-Fourier periodograms* for period searches in unevenly spaced data sets (e.g., Clarke 2002). The most common strategy involves folding the data modulo a trial period, computing a statistic on the folded time series (now a function of phase rather than time), and plotting the statistic for all independent periods. These methods measure the strength of signals that are strictly periodic, but not necessarily sinusoidal in shape. They are also relatively insensitive to the duration and uneven spacing of the data set, and

[2]Fast computation of the periodogram is achieved using the algorithm presented in Press and Rybicki (1989), whose trick is to carry out "extirpolation" of the data onto a regular mesh and subsequently employ the FFT.

some methods readily permit heteroscedastic weighting from measurement errors. An overview of the application of non-Fourier periodograms to asteroseismic data sets is given in Sect. 5.2 of the book by Aerts et al. (2010).

3 Extracting Global Asteroseismic Parameters

In order to fully characterize a star using asteroseismology, it is desirable that we obtain precise estimates of individual mode parameters (e.g., frequencies, amplitudes and linewidths). However, this is only possible for data above a certain SNR. *Global asteroseismic parameters*, indicative of the overall stellar structure, are on the other hand readily extractable using automated pipelines that are able to incorporate data with a lower SNR and for which a full peak-bagging analysis is not always possible. Furthermore, the automated nature of these pipelines is required if we are to efficiently exploit the large volumes of data made available by current and future space-based missions (Rauer et al. 2014; Campante et al. 2016b).

In this section I introduce an automated pipeline[3] which has been originally designed to extract global asteroseismic parameters of main-sequence and subgiant stars from *Kepler* power spectra (Campante et al. 2010; Campante 2012). This pipeline allows extracting the following information from the power spectrum (points 1–4 are covered below):

1. Frequency range of the oscillations;
2. Parameterization of the stellar background signal;
3. Average large frequency separation, $\Delta\nu$;
4. Frequency of maximum amplitude, $\nu_{\max}$;
5. Maximum mode amplitude, $A_{\max}$.

3.1 Detectability of Oscillations

We want to look for a frequency range in the power spectrum in which peaks appear at nearly regular intervals, one of the main signatures of the presence of solar-like oscillations. I note that the assumption of *quasi-regularity* may, however, be too strong in the case of evolved stars due to the presence of mixed modes. We start by partitioning the power spectrum into overlapping windows of variable width, w. The width w depends on the central frequency of the window, $\nu_{\rm central}$, used as a proxy for $\nu_{\max}$. We make use of the fact that the width of the p-mode bump approximately scales with $\nu_{\max}$ (e.g., Stello et al. 2007; Mosser et al. 2010), and so w is defined as $w = (\nu_{\rm central}/\nu_{\max,\odot})w_{\odot}$.

[3]A comparison of different pipelines used to extract global asteroseismic parameters is presented in Verner et al. (2011).

The next step consists in computing the *power spectrum of the power spectrum*, PS⊗PS, for each of these frequency windows. The presence of prominent features in the PS⊗PS around the predicted[4] values of $\Delta\nu/2$, $\Delta\nu/4$, and $\Delta\nu/6$ (the first, second, and third harmonics, respectively) is then examined. An hypothesis test is subsequently applied, whereby the presence of oscillations in a given window is established if the probability of the three above features being due to noise is less than 1%. Finally, the frequency range of the oscillations is determined based on the overall span of the windows with detected oscillations.

Figure 2 shows the detection of oscillations in the K2 power spectrum of a solar-type star vertical gray solid and dashed lines are separated by the estimated $\Delta\nu$, and mark the spacing on which we would expect to see modes. The inset shows the PS⊗PS, computed from the region around ν_{max}. The significant peak in the PS⊗PS lies at $\Delta\nu/2$ and is a signature of the near-regular spacing of solar-like oscillations.

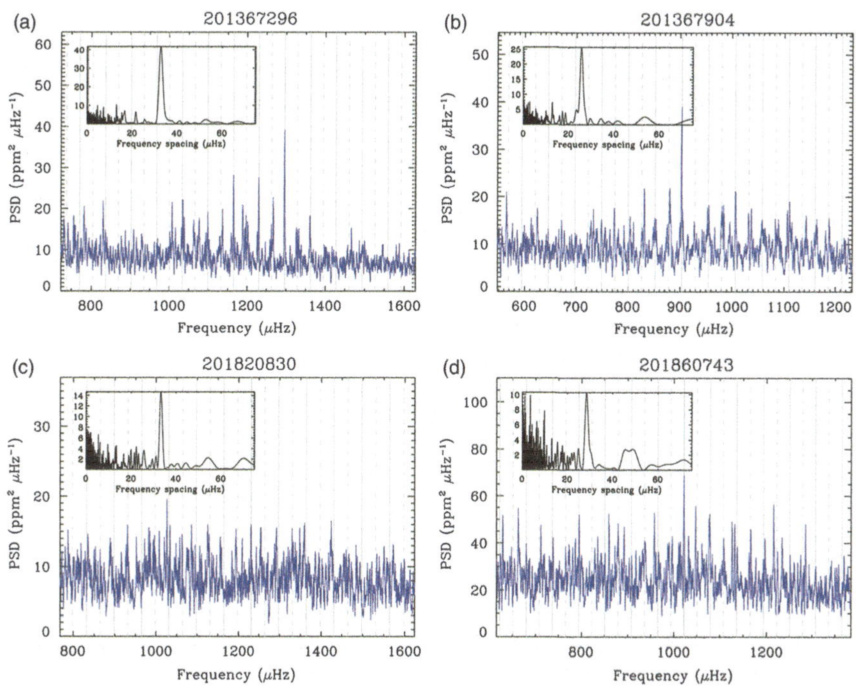

Fig. 2 K2 power spectrum (slightly smoothed) of a solar-type star with detected oscillations. *Inset* shows the PS⊗PS, computed from the region around ν_{max}. Adapted From Chaplin et al. (2015)

[4]The predicted value of $\Delta\nu$ is computed according to the relation $\Delta\nu \propto \nu_{central}^{0.77}$ (Stello et al. 2009).

3.2 Background Signal

The model of the *stellar background signal* is kept simple, merely containing a granulation component and photon shot noise. We fit this model to a smoothed version of the power spectrum employing a nonlinear least-squares fitting algorithm.

The frequency range of the oscillations (if detected) is excluded from the fitting window. The fitting window starts at 100 μHz to allow for the decay of any possible activity component, characterized by considerably longer timescales, and extends all the way up to the Nyquist frequency of *Kepler* short-cadence data (∼8300 μHz).

The granulation component is represented by a *Harvey-like profile* (e.g., Kallinger et al. 2014 and references therein) to which an offset is added to account for the *shot noise* component:

$$B(\nu) = B_0 + \eta^2(\nu)\left[\frac{B_{\rm gran}}{1+(2\pi\nu\,\tau_{\rm gran})^a}\right], \tag{17}$$

where $B_{\rm gran}$ is the height at $\nu = 0$ of the granulation component, $\tau_{\rm gran}$ is the characteristic turnover timescale and a calibrates the amount of memory in the process. Such a functional form is representative of a random non-harmonic field whose autocorrelation decays exponentially with time. The attenuation factor $\eta^2(\nu)$ takes into account the apodization of the oscillation signal due to the finite integration time.

The top panel of Fig. 3 displays the smoothed power spectrum of 16 Cyg A (dark red) overlaid on the original power spectrum (black). The fit to the background signal (red solid line) and both its components (red dashed lines) are also shown. The bottom panel displays the PS⊗PS over the frequency range of the oscillations. The features at $\Delta\nu/2$ (∼52 μHz), $\Delta\nu/4$ (∼26 μHz) and $\Delta\nu/6$ (∼17 μHz) are conspicuous.

3.3 Large Frequency Separation ($\Delta\nu$)

In order to estimate the *average large frequency separation*, $\Delta\nu$, we compute the PS⊗PS over the frequency range of the oscillations. The feature at $\Delta\nu/2$ (first harmonic) in the PS⊗PS is then located and its power-weighted centroid computed to provide an estimate of $\Delta\nu$. The standard deviation of grouped data, given by $\sqrt{[\sum hx^2 - (\sum hx)^2/\sum h]\;/\;(\sum h - 1)}$, is adopted as the error on $\Delta\nu$, meaning that the feature in the PS⊗PS is interpreted as an assembly of spectral heights (h) over a number of bins (with midpoint x).

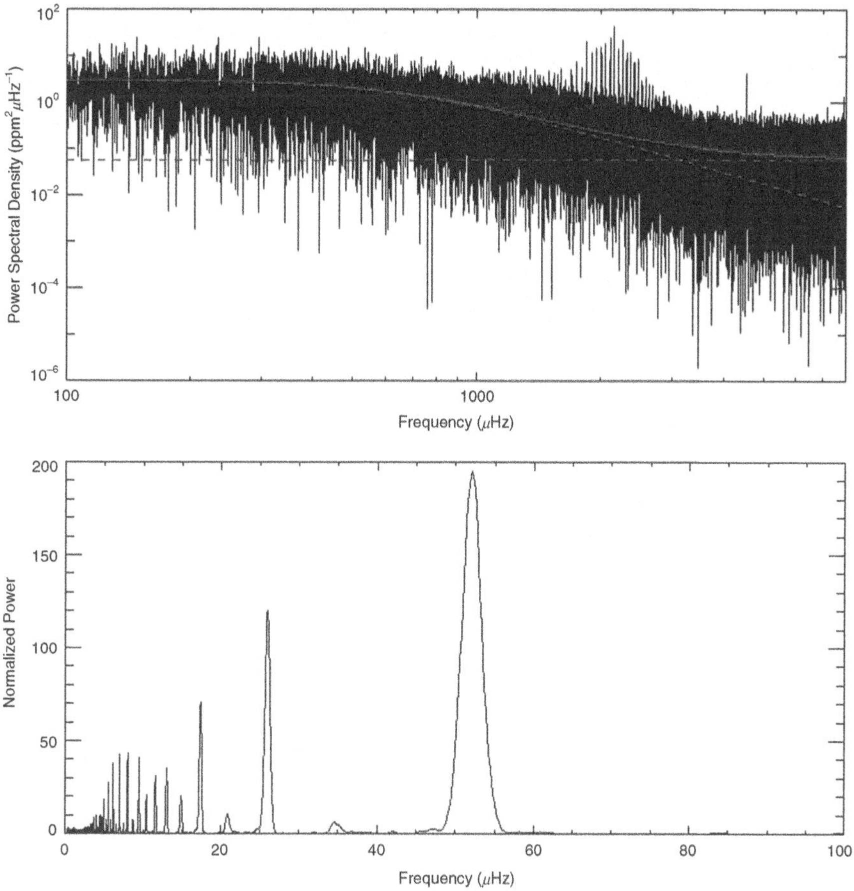

Fig. 3 Output from the analysis of the *Kepler* light curve of the bright G-type dwarf 16 Cyg A. *Top panel:* Modeling the stellar background signal. *Bottom panel:* Detection of oscillations in the PS⊗PS. From Campante (2012)

3.4 Frequency of Maximum Amplitude (ν_{max})

In order to estimate the *frequency of maximum amplitude*, ν_{max}, we average the p-mode power (after subtraction of the background fit) over contiguous rectangular windows of width $2\Delta\nu$ and convert to power per radial mode by multiplying by $\Delta\nu/c$, where c measures the effective number of modes per order (see Kjeldsen et al. 2008). An estimate of ν_{max} is then given by the power-weighted centroid, with the associated uncertainty derived from the standard deviation of grouped data (see Sect. 3.3).

4 Peak-Bagging

In this section I introduce a Bayesian peak-bagging tool that employs *Markov chain Monte Carlo* (MCMC) techniques (e.g., Campante et al. 2011; Handberg and Campante 2011; Campante 2012; Campante et al. 2016a). Besides making it possible to incorporate relevant prior information through Bayes' theorem, this tool also allows obtaining the marginal probability density function (pdf) for each of the model parameters. Such techniques are in many ways an extension of the *Maximum Likelihood Estimation* (MLE) methods originally introduced in helioseismology (Duvall and Harvey 1986; Anderson et al. 1990).

4.1 Power Spectrum of a Solar-Like Oscillator

Understanding the characteristics of the power spectrum of a solar-like oscillator is fundamental in order to extract information on the physics of the modes. The stochastic driving of a damped oscillator can be described by

$$\frac{\mathrm{d}^2}{\mathrm{d}t^2}y(t) + 2\eta\,\frac{\mathrm{d}}{\mathrm{d}t}y(t) + \omega_0^2\,y(t) = f(t)\,, \tag{18}$$

where $y(t)$ is the amplitude of the oscillator, η is the linear damping rate, ω_0 is the frequency of the undamped oscillator and $f(t)$ is a random forcing function. The Fourier transform of Eq. (18) is then expressed as

$$-\,\omega^2\,Y(\omega) - \mathrm{i}\,2\eta\omega\,Y(\omega) + \omega_0^2\,Y(\omega) = F(\omega). \tag{19}$$

When a realization of $y(t)$ is observed for a finite amount of time, an estimate of the power spectrum is then given by

$$P(\omega) = |Y(\omega)|^2 = \frac{|F(\omega)|^2}{(\omega_0^2 - \omega^2)^2 + 4\,\eta^2\omega^2}\,. \tag{20}$$

In the limit of taking the average of an infinite number of realizations, and assuming the damping rate to be very small compared to the frequency of oscillation, one obtains near the resonance the following expression for the *limit spectrum*:

$$\langle P(\omega)\rangle \simeq \frac{1}{4\,\omega_0^2}\frac{\langle P_f(\omega)\rangle}{(\omega-\omega_0)^2+\eta^2}\,. \tag{21}$$

The average power spectrum of the random forcing function, $\langle P_f(\omega)\rangle$, is a slowly-varying function of frequency. The result is thus a *Lorentzian profile*, characterized by the central frequency ω_0 and a width determined by the linear damping rate η.

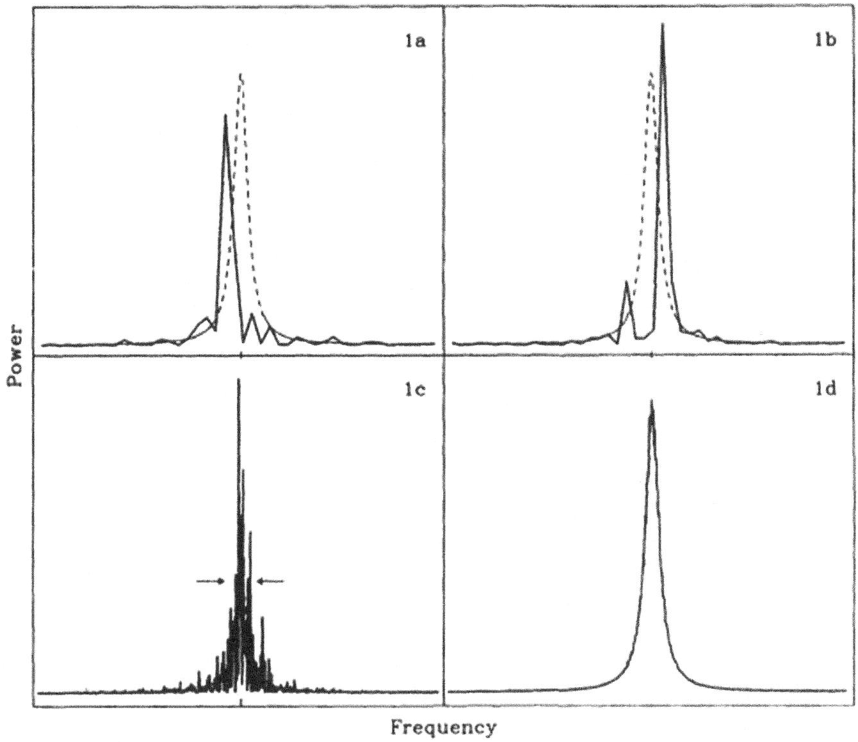

Fig. 4 Lorentzian profile (limit spectrum) and the erratic behavior of the power spectrum. From Anderson et al. (1990)

Panels (a) and (b) in Fig. 4 display two realizations of the same limit spectrum. Both power spectra appear as an erratic function concealing the underlying Lorentzian profile. Panel (c) displays a realization of the same limit spectrum, although with a resolution twenty times higher. Increasing the total observational span, hence the resolution, did nothing to reduce the variance in the power spectrum (cf. Sect. 2.4). Panel (d) displays the average of a large number of realizations with the same resolution as in (c), thus converging to the limit spectrum.

4.2 Modeling the Power Spectrum

We are primarily interested in performing a *global fit* to the power spectrum, whereby the observed modes are fitted simultaneously over a broad frequency range. We thus model the limit oscillation spectrum as a sum of standard Lorentzian profiles, $O(\nu)$, which sit atop a background signal described by $B(\nu)$:

$$\begin{aligned} P(\nu;\boldsymbol{\lambda}) &= O(\nu) + B(\nu) \\ &= \sum_{n',l} \sum_{m=-l}^{l} \frac{\mathscr{E}_{lm}(i_{\mathrm{s}}) H_{n'l}}{1 + \left[\frac{2(\nu - \nu_{n'l0} - m\nu_{\mathrm{s}})}{\Gamma_{n'lm}}\right]^2} + B(\nu)\,, \end{aligned} \tag{22}$$

where $\boldsymbol{\lambda}$ represents the set of model parameters. The inner sum in the above equation runs over the azimuthal components $\{m\}$ of each multiplet $\{n', l\}$, while the outer sum runs over the selection of observed modes. Figure 5 shows the power spectrum of HD 49933 (blue) based on 180 days of *CoRoT* photometry. The best-fitting model (red) is overlaid, with the shaded areas indicating the ranges of the uniform priors (see Sect. 4.3.1) on the mode frequencies.

At a given frequency bin j, the probability density, $f(P_j;\boldsymbol{\lambda})$, that the observed power spectrum takes a particular value P_j is related to the limit spectrum, $P(\nu_j;\boldsymbol{\lambda})$, by (cf. Eq. (14))

$$f(P_j;\boldsymbol{\lambda}) = \frac{1}{P(\nu_j;\boldsymbol{\lambda})} \exp\left[-\frac{P_j}{P(\nu_j;\boldsymbol{\lambda})}\right]. \tag{23}$$

We now want to specify the *likelihood function*, i.e., the joint pdf of the data sample $\{P_j\}$. Assuming the frequency bins to be uncorrelated, the joint pdf is simply given by the product of $f(P_j;\boldsymbol{\lambda})$ over some frequency interval of interest spanned by j:

$$L(\boldsymbol{\lambda}) = \prod_j f(P_j;\boldsymbol{\lambda})\,. \tag{24}$$

4.3 Bayesian Parameter Estimation Using MCMC

I now describe the formalism of a Bayesian approach to *parameter estimation* and *model comparison* that employs an MCMC algorithm. Let us consider a set of competing hypotheses, $\{H_i\}$, assumed to be mutually exclusive. One should be able to assign a probability, $p(H_i|D, I)$, to each hypothesis taking into account the observed data, D, and any available prior information, I. This is done through *Bayes' theorem* (Bayes and Price 1763):

$$p(H_i|D, I) = \frac{p(H_i|I)p(D|H_i, I)}{p(D|I)}\,. \tag{25}$$

The probability of the hypothesis H_i in the absence of D is called the *prior probability*, $p(H_i|I)$, whereas the probability including D is called the *posterior*

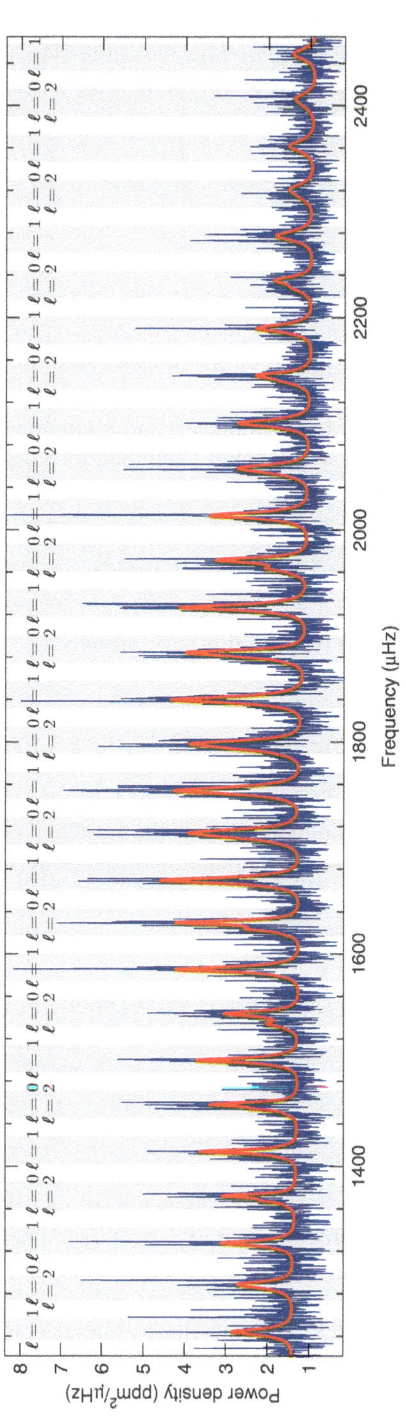

Fig. 5 Power spectrum of the F-type star HD 49933 (*blue*) based on 180 days of *CoRoT* photometry and best-fitting model (*red*). *Shaded areas* indicate the ranges of the uniform priors on the mode frequencies. From Handberg and Campante (2011)

probability, $p(H_i|D,I)$. The quantity $p(D|H_i,I)$ is called the *likelihood* of H_i. The denominator $p(D|I)$ is the *global likelihood* for the entire class of hypotheses. The sum of the posterior probabilities over the hypothesis space of interest is unity, hence one has:

$$p(D|I) = \sum_i p(H_i|I)p(D|H_i,I)\,. \tag{26}$$

4.3.1 Parameter Estimation

If a particular hypothesis, i.e., a given model M describing the physical process, is assumed true, then the hypothesis space of interest concerns the values taken by the model parameters, $\boldsymbol{\lambda}$. These parameters are continuous and one will be interested in obtaining their pdf. The global likelihood of model M is then given by the continuous counterpart of Eq. (26):

$$p(D|I) = \int p(\boldsymbol{\lambda}|I)p(D|\boldsymbol{\lambda},I)d\boldsymbol{\lambda}\,. \tag{27}$$

We restate Bayes' theorem to account for this new formalism:

$$p(\boldsymbol{\lambda}|D,I) = \frac{p(\boldsymbol{\lambda}|I)p(D|\boldsymbol{\lambda},I)}{p(D|I)}\,, \tag{28}$$

where $p(D|I)$ plays the role of a normalization constant. Ultimately, we are interested in using MCMC techniques to map the posterior pdf, $p(\boldsymbol{\lambda}|D,I)$. The procedure of *marginalization* allows computation of the posterior pdf for a subset of parameters $\boldsymbol{\lambda}_A$ by integrating over the remaining parameters (or *nuisance parameters*) $\boldsymbol{\lambda}_B$:

$$p(\boldsymbol{\lambda}_A|D,I) = \int p(\boldsymbol{\lambda}_A,\boldsymbol{\lambda}_B|D,I)d\boldsymbol{\lambda}_B\,. \tag{29}$$

4.3.2 Model Comparison

The problem of model comparison is analogous to that of parameter estimation. When facing a situation in which several parameterized models are available for describing the same physical process, one expects Bayes' theorem to allow for a statistical comparison between such models. Bayesian model comparison has a built-in *Occam's razor* by which a complex model is automatically penalized, unless the available data justify its additional complexity. Competing models may be either intrinsically different models or else similar but with varying number of parameters (i.e., nested models), or even the same model with different priors affecting its parameters.

Given two or more competing models and our prior information, I, being in the present context that one and only one of the models is true, we can assign individual probabilities similarly to what has been done in Eq. (25), after replacing H_i by M_i:

$$p(M_i|D,I) = \frac{p(M_i|I)p(D|M_i,I)}{p(D|I)}\,, \tag{30}$$

where the global likelihood of model M_i, $p(D|M_i,I)$, also called the *evidence* of the model, is given by Eq. (27). We are often interested in computing the ratio of the probabilities of two competing models:

$$O_{ij} \equiv \frac{p(M_i|D,I)}{p(M_j|D,I)} = \frac{p(M_i|I)p(D|M_i,I)}{p(M_j|I)p(D|M_j,I)} = \frac{p(M_i|I)}{p(M_j|I)}B_{ij}\,, \tag{31}$$

where O_{ij} is the *odds ratio* in favor of model M_i over model M_j, B_{ij} is the so-called *Bayes' factor* and $p(M_i|I)/p(M_j|I)$ is the *prior odds ratio*. The Bayesian odds ratio is the product of the ratio of the prior probabilities of the models and the ratio of their global likelihoods.

4.3.3 Markov Chain Monte Carlo

The need becomes clear for a mathematical tool that is able to efficiently evaluate the multidimensional integrals required in the computation of the marginal distributions. The aim is to draw samples from the *target distribution*, $p(\boldsymbol{\lambda}|D,I)$, by constructing a pseudo-random walk in parameter space such that the number of samples drawn from a particular region is proportional to its posterior density. This is achieved by generating a *Markov chain*, whereby a new sample, $\boldsymbol{\lambda}_{t+1}$, depends on the previous sample, $\boldsymbol{\lambda}_t$, according to a time-independent quantity called the *transition kernel*, $p(\boldsymbol{\lambda}_{t+1}|\boldsymbol{\lambda}_t)$. After a burn-in phase, $p(\boldsymbol{\lambda}_{t+1}|\boldsymbol{\lambda}_t)$ should be able to generate samples of $\boldsymbol{\lambda}$ with a probability density converging on the target distribution.

We generate a Markov chain by using the *Metropolis–Hastings algorithm* (Metropolis et al. 1953; Hastings 1970). Let us denote the current sample by $\boldsymbol{\lambda}_t$. We would like to steer the Markov chain toward the next sampling state, $\boldsymbol{\lambda}_{t+1}$, by first proposing a new sample, $\boldsymbol{\xi}$, to be drawn from a proposal distribution, $q(\boldsymbol{\xi}|\boldsymbol{\lambda}_t)$, which can have almost any form. The proposed sample is then accepted with a probability given by:

$$\alpha(\boldsymbol{\lambda}_t,\boldsymbol{\xi}) = \min(1,r) = \min\left[1, \frac{p(\boldsymbol{\xi}|D,I)}{p(\boldsymbol{\lambda}_t|D,I)}\frac{q(\boldsymbol{\lambda}_t|\boldsymbol{\xi})}{q(\boldsymbol{\xi}|\boldsymbol{\lambda}_t)}\right]\,, \tag{32}$$

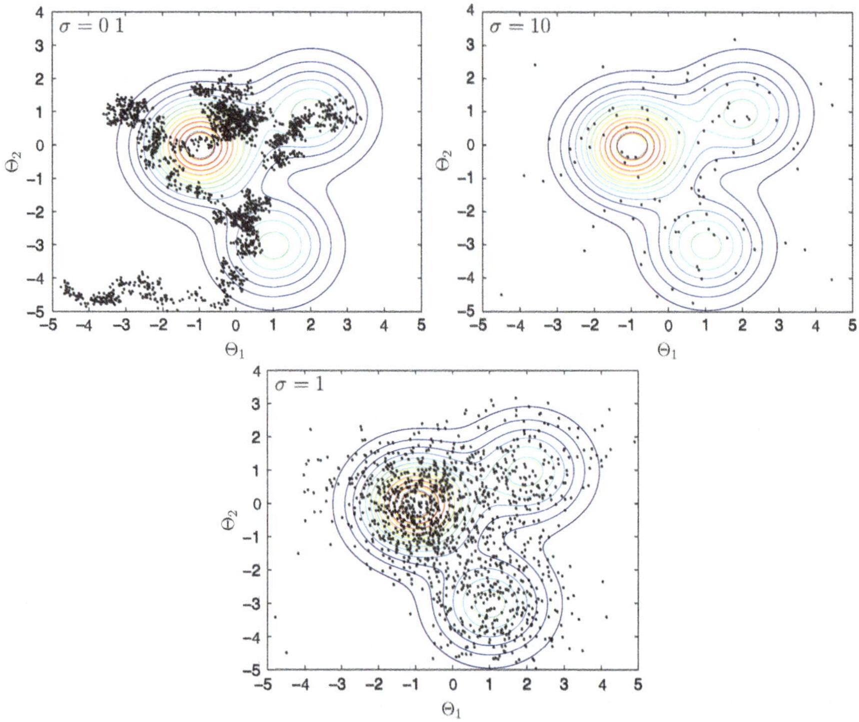

Fig. 6 Two-dimensional MCMC simulations of a triple-peaked posterior. The same target distribution is sampled by three chains, each characterized by a different set $\{\sigma\}$ of step sizes in parameter space. From Handberg and Campante (2011)

where $\alpha(\boldsymbol{\lambda}_t, \boldsymbol{\xi})$ is the *acceptance probability* and r is called the *Metropolis ratio*. If $\boldsymbol{\xi}$ is not accepted, then the chain will keep the current sampling state, i.e., $\boldsymbol{\lambda}_{t+1} = \boldsymbol{\lambda}_t$. Figure 6 shows the output from three two-dimensional MCMC simulations of the same triple-peaked posterior.

Once the posterior pdf, $p(\boldsymbol{\lambda}|D,I)$, has been mapped, the procedure of marginalization becomes trivial. The marginal posterior distribution of a given parameter λ, $p(\lambda|D,I)$, is then simply obtained by collecting its samples in a normalized histogram. An estimate of the k-th moment of λ about the origin is then given by

$$\langle \lambda^k \rangle \equiv \int \lambda^k p(\lambda|D,I) d\lambda \approx \frac{1}{N} \sum \lambda_t^k , \tag{33}$$

where N is the total number of samples.

The basic Metropolis–Hastings algorithm runs the risk of becoming stuck in a local mode of the target distribution. A way of overcoming this is to employ *parallel tempering*, whereby a discrete set of progressively flatter versions of the target distribution is created by introducing a *tempering parameter*, γ. We modify

Eq. (28) to generate the tempered distributions:

$$p(\boldsymbol{\lambda}|D,\gamma,I) \propto p(\boldsymbol{\lambda}|I)p(D|\boldsymbol{\lambda},I)^{\gamma}\,, \quad 0 < \gamma \leq 1\,. \tag{34}$$

For $\gamma = 1$, we retrieve the target distribution, while distributions with $\gamma < 1$ are effectively flatter versions of the target distribution. By running such a set of chains in parallel and allowing their parameter states to swap, we increase the mixing properties of the Markov chain.

Furthermore, the Metropolis–Hastings algorithm can be refined by implementing a statistical control system (e.g., Gregory 2005) allowing to automatically fine-tune the proposal distribution during the burn-in phase (see Fig. 6).

Acknowledgements The author acknowledges the support of the UK Science and Technology Facilities Council (STFC). Funding for the Stellar Astrophysics Centre is provided by The Danish National Research Foundation (Grant DNRF106).

References

Aerts, C., Christensen-Dalsgaard, J., Kurtz, D.W.: Asteroseismology. Springer, New York (2010)

Anderson, E.R., Duvall, T.L. Jr., Jefferies, S.M.: Astrophys. J. **364**, 699 (1990)

Appourchaux, T.: In: Pallé, P.L., Esteban, C. (eds.) Asteroseismology. Canary Islands Winter School of Astrophysics, vol. 123. Cambridge University Press, Cambridge (2013)

Arentoft, T., Kjeldsen, H., Bedding, T.R., et al.: Astrophys. J. **687**, 1180 (2008)

Bayes, T., Price, R.: Philos. Trans. **53**, 370 (1763)

Bedding, T.R., Kjeldsen, H., Campante, T.L., et al.: Astrophys. J. **713**, 935 (2010)

Borucki, W.J., Koch, D., Basri, G., et al.: Science **327**, 977 (2010)

Campante, T.L.: PhD thesis, Universidade do Porto (2012)

Campante, T.L., Karoff, C., Chaplin, W.J., et al.: Mon. Not. R. Astron. Soc. **408**, 542 (2010)

Campante, T.L., Handberg, R., Mathur, S., et al.: Astron. Astrophys. **534**, A6 (2011)

Campante, T.L., Barclay, T., Swift, J.J., et al.: Astrophys. J. **799**, 170 (2015)

Campante, T.L., Lund, M.N., Kuszlewicz, J.S., et al.: Astrophys. J. **819**, 85 (2016a)

Campante, T.L., Schofield, M., Kuszlewicz, J.S., et al.: Astrophys. J. **830**, 138 (2016b)

Chaplin, W.J., Miglio, A.: Annu. Rev. Astron. Astrophys. **51**, 353 (2013)

Chaplin, W.J., Elsworth, Y., Davies, G.R., et al.: Mon. Not. R. Astron. Soc. **445**, 946 (2014)

Chaplin, W.J., Lund, M.N., Handberg, R., et al.: Publ. Astron. Soc. Pac. **127**, 1038 (2015)

Christensen-Dalsgaard, J.: Sol. Phys. **220**, 137 (2004)

Clarke, D.: Astron. Astrophys. **386**, 763 (2002)

Cooley, J.W., Tukey, J.W.: Math. Comput. **19**, 297 (1965)

Duvall, T.L. Jr., Harvey, J.W.: In: Gough, D.O. (ed.) NATO ASIC Proceedings. Seismology of the Sun and the Distant Stars, vol. 169, p. 105. D. Reidel Publishing Co., Dordrecht (1986)

García, R.A., Ballot, J.: Astron. Astrophys. **477**, 611 (2008)

Gregory, P.C.: Bayesian Logical Data Analysis for the Physical Sciences: A Comparative Approach with 'Mathematica' Support. Cambridge University Press, Cambridge (2005)

Handberg, R., Campante, T.L.: Astron. Astrophys. **527**, A56 (2011)

Hastings, W.K.: Biometrika **57**, 97 (1970)

Kallinger, T., De Ridder, J., Hekker, S., et al.: Astron. Astrophys. **570**, A41 (2014)

Khintchine, A.: Math. Ann. **109**, 604 (1934)

Kjeldsen, H., Bedding, T.R., Arentoft, T., et al.: Astrophys. J. **682**, 1370 (2008)

Lundkvist, M.S., Kjeldsen, H., Albrecht, S., et al.: Nat. Commun. **7**, 11201 (2016)
Metropolis, N., Rosenbluth, A.W., Rosenbluth, M.N., Teller, A.H., Teller, E.: J. Chem. Phys. **21**, 1087 (1953)
Mosser, B., Belkacem, K., Goupil, M.-J., et al.: Astron. Astrophys. **517**, A22 (2010)
Murphy, S.J., Shibahashi, H., Kurtz, D.W.: Mon. Not. R. Astron. Soc. **430**, 2986 (2013)
Nyquist, H.: Trans. AIEE **47**, 617 (1928)
Parseval des Chênes, M.-A.: Mémoires présentés à l'Institut des Sciences, Lettres et Arts, pars divers savans, et lus dans ses assemblées. Sciences, mathématiques et physiques. Savans étrangers **1**, 638 (1806)
Press, W.H., Rybicki, G.B.: Astrophys. J. **338**, 277 (1989)
Rauer, H., Catala, C., Aerts, C., et al.: Exp. Astron. **38**, 249 (2014)
Shannon, C.E.: Proc. Inst. Radio Eng. (IRE) **37**, 10 (1949)
Stello, D., Bruntt, H., Kjeldsen, H., et al.: Mon. Not. R. Astron. Soc. **377**, 584 (2007)
Stello, D., Chaplin, W.J., Basu, S., Elsworth, Y., Bedding, T.R.: Mon. Not. R. Astron. Soc. **400**, L80 (2009)
Verner, G.A., Elsworth, Y., Chaplin, W.J., et al.: Mon. Not. R. Astron. Soc. **415**, 3539 (2011)
Wiener, N.: Acta Math. **55**, 117 (1930)
Yu, J., Huber, D., Bedding, T. R., et al.: Mon. Not. R. Astron. Soc. **463**, 1297 (2016)

Stellar Inversion Techniques

Daniel R. Reese

Abstract Stellar seismic inversions have proved to be a powerful technique for probing the internal structure of stars, and paving the way for a better understanding of the underlying physics by revealing some of the shortcomings in current stellar models. In this lecture, we provide an introduction to this topic by explaining kernel-based inversion techniques. Specifically, we explain how various kernels are obtained from the pulsation equations, and describe inversion techniques such as the Regularised Least-Squares (RLS) and Optimally Localised Averages (OLA) methods.

1 Introduction

Many of the problems which intervene in physics can be described in terms of forward and inverse problems. Generally speaking, a forward problem focuses on predicting the effects which result from a set of physical causes, such as deducing the gravitational field of an object from its distribution of matter. In an inverse problem, one typically tries to deduce the physical causes which led to a given set of results or effects (which are typically observations). Hence, trying to deduce the distribution of matter from the gravitational field of an object is an inverse problem.

The field of asteroseismology, i.e., the study of stellar pulsations, also fits this description. Trying to predict stellar pulsation frequencies for a given stellar model constitutes a forward problem. Likewise, trying to deduce the stellar structure which led to a given set of pulsation frequencies is an inverse problem. This inverse problem turns out to be quite difficult because, in general, the relation between stellar structure and oscillation frequencies is non-linear. Nonetheless, given the wealth of information on the internal structure of stars provided by pulsation

D.R. Reese (✉)
LESIA, Observatoire de Paris, PSL Research University, CNRS, Sorbonne Universités, UPMC Univ. Paris 06, Univ. Paris Diderot, Sorbonne Paris Citée, 92195 Meudon, France
e-mail: daniel.reese@obspm.fr

T.L. Campante et al. (eds.), *Asteroseismology and Exoplanets: Listening to the Stars and Searching for New Worlds*, Astrophysics and Space Science Proceedings 49, DOI 10.1007/978-3-319-59315-9_4

frequencies, a variety of approaches have been devised to tackle this problem, as expressed by Gough (1985) in the context of helioseismology:

> Inversions can conveniently be divided into three categories. The simplest consists of the execution of the forward problem using solar models with a few adjustable parameters, and the calibration of those parameters by fitting theory to observation. The second is the use of analytical methods. [...] Thirdly, there are the formal inversion techniques borrowed from geophysics that have been used on real and artificial solar data.

The first category of inversions is usually named "forward modelling" (not to be confused with the "forward problem") and corresponds to searching for an optimal model in a restricted parameter space. It typically includes methods such as grid searches (e.g., Silva Aguirre et al. 2015), MCMC methods (e.g., Bazot et al. 2012), or genetic algorithms (Metcalfe and Charbonneau 2003; Charpinet et al. 2005). The advantages of this approach is its obvious simplicity, and the fact that it produces physically coherent models. However, the parameter space is restricted and does not allow for hitherto unknown physical ingredients not included in the stellar models. Furthermore, such methods can be costly, especially if models are calculated on-the-fly. The second approach includes methods such as asymptotic methods or glitch fitting. These methods can provide a great deal of physical insight into stellar physics but are beyond the scope of the present lecture. Finally, formal inversion techniques typically consist in adjusting the structure of a reference stellar model so as to match a set of observed frequencies. The advantage of this approach is that it can potentially extract more information from the pulsation frequencies, and is therefore open to new physics. However, this method may lead to models which are not physically coherent, and can be more difficult to implement. These approaches are in fact complementary. Indeed, the forward approach typically provides a reference model, which can then be further refined via formal inversion techniques.

The present lecture focuses on the third category, i.e., formal inversion techniques. However, before tackling inversions, it is necessary to spend a bit of time on the forward problem in order to bring out some of the properties which apply in the context of inverse problems. This will be the subject of the next section. Then stellar inversion techniques will be described in Sect. 3. A short conclusion including a list of relevant references and available inversion codes will follow.

2 The Forward Problem

2.1 Adiabatic Pulsation Equations

Stellar pulsations, the periodic motion of gas or plasma within a star, are described by the Lagrangian displacement and the Eulerian perturbations to density, pressure, and gravitational potential, denoted $\boldsymbol{\xi}$, ρ', p' and ϕ', respectively. When applying the adiabatic approximation (i.e., when neglecting heat transfers during the periodic motions), these quantities are determined by Euler's equation, the continuity equation, and the adiabatic relation, which express the conservation of momentum, mass, and energy, as well as Poisson's equation. Through various analytical manipulations, and the use of Green's function and suitable boundary conditions for Poisson's equation, it is possible to express ρ', p' and ϕ' as a function of $\boldsymbol{\xi}$ alone. When inserted into Euler's equation, this leads to the following schematic equation:

$$\omega^2 \boldsymbol{\xi} = \mathscr{F}(\boldsymbol{\xi}), \tag{1}$$

where $\mathscr{F}$ is an integro-differential operator, and where we have assumed a time dependence[1] of the form $\exp(-\mathrm{i}\omega t)$. Equation (1), along with appropriate boundary conditions, is an eigenvalue problem, the solutions of which are known as "eigensolutions". Specifically, ω^2 is an eigenvalue and corresponds to the square of the pulsation frequency, whereas $\boldsymbol{\xi}$ is the eigenmode or eigenfunction, and specifies the geometric characteristics of the stellar pulsation.

The forward problem in this case, then corresponds to finding the above eigensolutions for a given stellar structure, i.e., for a given $\mathscr{F}$ operator. The inverse problem corresponds to finding the stellar structure (and hence $\mathscr{F}$) from a set of pulsation frequencies and some sort of mode identification, i.e., a partial characterisation of the structure of the pulsation modes. In the case of solar-like oscillators, a mode identification typically includes the harmonic degrees ℓ of the pulsations, and possibly the radial orders n (this is usually obtained from comparisons with models) and azimuthal orders m (only if frequency multiplets, typically caused by stellar rotation, can be resolved). Given that the forward problem is non-linear, the inverse problem will also be non-linear. However, to make the problem more tractable, one typically linearises it. Linearising Eq. (1) leads to the following equation[2]:

$$(\delta\omega^2)\boldsymbol{\xi} + \omega^2(\delta\boldsymbol{\xi}) = \delta\mathscr{F}(\boldsymbol{\xi}) + \mathscr{F}(\delta\boldsymbol{\xi}). \tag{2}$$

This equation simply expresses how a small modification to the stellar structure leads to small modifications of the pulsation modes, in particular frequency differences $\delta\omega$. Hence, in order to solve the inverse problem, one needs to find a reference stellar model (typically using some form of forward modelling) which is sufficiently close to the true stellar model so that the linear approximation applies, and *invert* the frequency differences, in order to find how to correct the

[1]If one assumes that modes are proportional to $\exp(\mathrm{i}m\varphi)$, φ being the longitude, such a time dependence will lead to $m > 0$ modes being prograde, where m is the azimuthal order. If one uses, instead, a time dependence of the form $\exp(\mathrm{i}\omega t)$, then $m > 0$ modes will be retrograde.

[2]Throughout these lectures, the δ notation will be used to indicate a modification of the equilibrium stellar structure and associated pulsations rather than a Lagrangian perturbation.

stellar model so that it more closely matches the actual star. However, Eq. (2) is not straightforward to use as it contains terms with $\delta\boldsymbol{\xi}$, the perturbation of the eigenmode. The next section shows how to remove these terms by exploiting an important property of the adiabatic pulsation equations, namely their symmetry.

2.2 Symmetry of the Adiabatic Pulsation Equations

Before explaining in what sense the pulsation equations are symmetric, it is necessary to introduce the following dot product:

$$\langle \boldsymbol{\eta}, \boldsymbol{\xi} \rangle = \int_V \rho_0 \boldsymbol{\eta}^* \cdot \boldsymbol{\xi} \mathrm{d}V\,, \tag{3}$$

where $\boldsymbol{\eta}^*$ is the complex conjugate of $\boldsymbol{\eta}$, and V the stellar volume. We note that this is a complex dot product, hence: $\langle \boldsymbol{\eta}, \boldsymbol{\xi} \rangle = \langle \boldsymbol{\xi}, \boldsymbol{\eta} \rangle^*$.

The adiabatic pulsation equations are symmetric with respect to the above dot product:

$$\langle \boldsymbol{\eta}, \mathscr{F}(\boldsymbol{\xi}) \rangle = \langle \mathscr{F}(\boldsymbol{\eta}), \boldsymbol{\xi} \rangle\,, \tag{4}$$

where $\boldsymbol{\xi}$ and $\boldsymbol{\eta}$ are any displacement fields, which need not necessarily be eigenfunctions at this point. In order to prove this symmetry, we start by introducing the associated pressure and gravitational potential perturbations as deduced from the relevant equations: $(\boldsymbol{\xi}, p', \phi')$ and $(\boldsymbol{\eta}, \pi', \psi')$. We then calculate the dot product between $\boldsymbol{\eta}$ and Euler's equation (applied to $\boldsymbol{\xi}$). After various manipulations (integration by parts etc.), this leads to the following formula (e.g., Unno et al. 1989):

$$\begin{aligned}\langle \boldsymbol{\eta}, \mathscr{F}(\boldsymbol{\xi}) \rangle &= \int_V \frac{(\pi')^* p'}{\rho_0 c_0^2} \mathrm{d}V + \int_V \rho_0 N_0^2 (\boldsymbol{\eta}^* \cdot \mathbf{e}_r)(\boldsymbol{\xi} \cdot \mathbf{e}_r) \mathrm{d}V \\ &+ \int_S \rho_0 g_0 (\boldsymbol{\eta}^* \cdot \mathbf{e}_r)(\boldsymbol{\xi} \cdot \mathbf{e}_r) \mathrm{d}S - \frac{1}{4\pi G} \int_{V_\infty} \nabla(\psi')^* \cdot \nabla \phi' \mathrm{d}V\,, \end{aligned} \tag{5}$$

where V is the star's volume, S its surface, V_∞ infinite space, $\mathbf{e}_r$ the unit vector in the radial direction, and N_0^2 the square of the Brunt–Väisälä frequency. In deriving the surface term, we assumed, as a boundary condition, that the Lagrangian pressure perturbation vanishes at the surface. Appendix C of Reese (2006) explains how to obtain the last term (integrated over V_∞). It is very clear from this explicit formulation that the pulsation equations are symmetric. More general forms of this equation have been derived, for instance, in the case of differentially rotating physical bodies (Lynden-Bell and Ostriker 1967).

This symmetry leads to a number of consequences. Firstly, the eigenvalues, ω^2, are real (meaning that the ω are either real or purely imaginary). Secondly, the eigenfunctions of distinct eigenvalues are orthogonal with respect to the above dot product. The third consequence is known as the "variational principle". According to this principle, the variational frequency, defined by $\omega_{\mathrm{var}}^2 = \langle \boldsymbol{\xi}, \mathscr{F}(\boldsymbol{\xi})\rangle / \langle \boldsymbol{\xi}, \boldsymbol{\xi}\rangle$, differs from the true eigenfrequency by an amount which is of second order or higher in terms of the error on the eigenfunction, i.e., $\omega^2 - \omega_{\mathrm{var}}^2 = \mathscr{O}\left(\|\Delta\boldsymbol{\xi}\|^2\right)$. This is useful as ω_{var}^2 provides an independent and potentially more accurate estimate of the eigenvalue than the numerical value and is therefore used as an accuracy test in various pulsation codes such as ADIPLS (Christensen-Dalsgaard 2008).

2.3 Kernels

We now return to our original problem, i.e., calculating the frequency variation caused by a small modification of the stellar structure. Taking the dot product between Eq. (2) and $\boldsymbol{\xi}$, and grouping terms with $\delta\boldsymbol{\xi}$ yields:

$$\delta\omega^2 \langle \boldsymbol{\xi}, \boldsymbol{\xi}\rangle - \langle \boldsymbol{\xi}, \delta\mathscr{F}(\boldsymbol{\xi})\rangle = \left\langle -\omega^2\boldsymbol{\xi} + \mathscr{F}(\boldsymbol{\xi}), \delta\boldsymbol{\xi}\right\rangle, \tag{6}$$

where we have made use of the symmetry of $\mathscr{F}$. The right-hand side vanishes because $\boldsymbol{\xi}$ is an eigenmode, and ω^2 the corresponding eigenvalue. Isolating $\delta\omega^2$ then yields:

$$\delta\omega^2 = 2\omega\delta\omega = \frac{\langle \boldsymbol{\xi}, \delta\mathscr{F}(\boldsymbol{\xi})\rangle}{\langle \boldsymbol{\xi}, \boldsymbol{\xi}\rangle}. \tag{7}$$

This last form is extremely useful because it relates modifications of the pulsation frequency directly to changes in the stellar model, *without needing* $\delta\boldsymbol{\xi}$.

The next obvious question is what types of perturbations can we expect in stars? A first type of perturbation, which in fact is ubiquitous, is stellar rotation. One can distinguish the 1D case, where the rotation profile, Ω, only depends on the radial coordinate r (also known as "shellular" rotation) from the 2D case where it depends on r and θ, the colatitude. A second type of perturbation is modifications to the stellar structure, as defined, for instance, by the ρ_0, $\Gamma_{1,0}$, c_0^2 etc., profiles. So far, structural modifications have only been envisaged in a 1D setting.

Rotation leads to two inertial accelerations: the centrifugal and the Coriolis acceleration. The former distorts the shape of the star but is a second order effect, so will be neglected. The latter intervenes in the oscillatory motions and leads to first order effects on the frequencies. To first order, Euler's equation takes on the form:

$$\omega^2\boldsymbol{\xi} = 2\omega m\Omega\boldsymbol{\xi} - 2i\omega\boldsymbol{\Omega}\times\boldsymbol{\xi} + \frac{\nabla p'}{\rho_0} - \frac{\rho'\mathbf{g}_0}{\rho_0} + \nabla\phi', \tag{8}$$

where Ω is the rotation profile, and $\boldsymbol{\Omega} = \Omega \mathbf{e}_z$. From this we deduce:

$$\delta \mathscr{F}(\boldsymbol{\xi}) = 2\omega m \Omega \boldsymbol{\xi} - 2i\omega \boldsymbol{\Omega} \times \boldsymbol{\xi} \,. \tag{9}$$

In the 1D case, the frequency shift is given by

$$\delta\omega_{n,\ell,m} = \omega_{n,\ell,m} - \omega_{n,\ell,0} = m \int_0^R K_\Omega^{n\ell}(r)\Omega(r)\mathrm{d}R \,, \tag{10}$$

where

$$K_\Omega^{n\ell} = \frac{\rho_0 r^2 \left(\xi_r^2 + \ell(\ell+1)\xi_h^2 - 2\xi_r\xi_h - \xi_h^2\right)}{\int_0^R \rho_0(r)\left(\xi_r^2 + \ell(\ell+1)\xi_h^2\right) r^2 \mathrm{d}r} \tag{11}$$

and where ξ_r and ξ_h are the radial and horizontal components of the Lagrangian displacement, respectively. $K_\Omega^{n\ell}$ is known as the "rotation kernel". As can be seen from this expression, frequencies with the same (n, ℓ) values are uniformly split as a function of m thanks to rotation. Figure 1 shows some examples of 1D rotation kernels.

If Ω is constant, then Eq. (10) simplifies to $\delta\omega = m(1 - \mathscr{C})\Omega$, where

$$\mathscr{C} = \frac{\int_0^R \rho_0 \left(2\xi_r\xi_h + \xi_h^2\right) r^2 \mathrm{d}r}{\int_0^R \rho_0(r)\left(\xi_r^2 + \ell(\ell+1)\xi_h^2\right) r^2 \mathrm{d}r} \,. \tag{12}$$

$\mathscr{C}$ is known as the Ledoux constant and represents the effects of the Coriolis force (see Ledoux 1951).

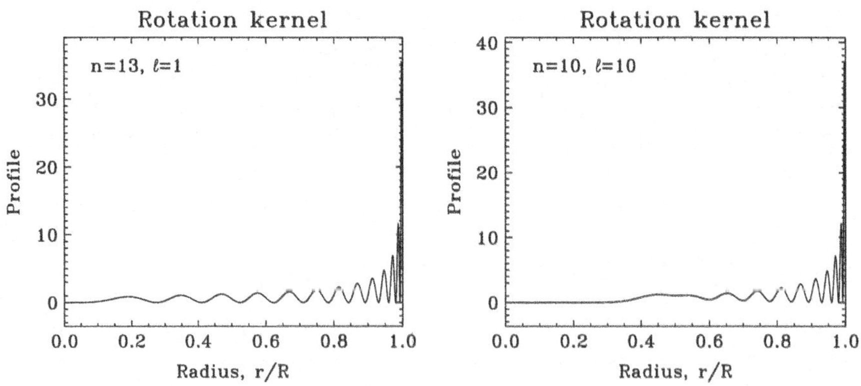

Fig. 1 Examples of 1D rotation kernels

In the 2D case, the rotational splitting is given by

$$\delta\omega_{n,\ell,m} = \int_0^R \int_0^\pi \mathscr{K}_{n,\ell,m}(r,\theta)\Omega(r,\theta) r \mathrm{d}r \mathrm{d}\theta\,, \tag{13}$$

where $\mathscr{K}_{n,\ell\, m}(r,\theta)$ is the 2D rotation kernel (expressions for such kernels may be found in Schou et al. 1994). This time, the splitting as a function of m may be non-uniform.

The acoustic structure of stars is typically determined by two variables, e.g., $(\rho_0, \Gamma_{1,0})$ and possibly some surface quantities such as the surface pressure. Accordingly, when modifying the structure of a star, the modifications to two structural quantities need to be specified, e.g., $(\delta\rho_0, \delta\Gamma_{1,0})$ (although in some cases, 3 functions need to be specified, e.g., Buldgen et al. 2017). As was the case for rotation, it is possible to relate changes in frequency to structural modifications of stars using kernels. The easiest structural kernels to derive are those for the variables (ρ, c^2). After a (very) lengthy derivation, one can show that

$$\frac{\delta\omega}{\omega} = \int_0^R \left[K_{c^2,\rho}(r)\frac{\delta c_0^2(r)}{c_0^2(r)} + K_{\rho,c^2}(r)\frac{\delta\rho_0(r)}{\rho_0(r)} \right] \mathrm{d}r\,, \tag{14}$$

where:

$$K_{c^2,\rho} = \frac{\rho_0 c_0^2 \chi^2 r^2}{2I\omega^2}\,, \tag{15}$$

$$\begin{aligned} K_{\rho,c^2} = \frac{\rho_0 r^2}{2I\omega^2} \Bigg\{ & c_0^2\chi^2 - \omega^2\left(\xi_r^2 + \ell(\ell+1)\xi_h^2\right) - 4\pi G \int_{s=r}^R \left(2\rho_0\xi_r\chi + \frac{\mathrm{d}\rho_0}{\mathrm{d}s}\xi_r^2 \right) \mathrm{d}s \\ & -2g_0\xi_r\chi + 2g_0\xi_r\frac{\mathrm{d}\xi_r}{\mathrm{d}r} + 4\pi G\rho_0\xi_r^2 + 2\left(\xi_r\frac{\mathrm{d}\phi'}{\mathrm{d}r} + \frac{\ell(\ell+1)\xi_h\phi'}{r} \right) \Bigg\}\,, \end{aligned} \tag{16}$$

$$I = \int_0^R \rho_0\left(\xi_r^2 + \ell(\ell+1)\xi_h^2\right) r^2 \mathrm{d}r\,, \quad \chi = \frac{\nabla\cdot\boldsymbol{\xi}}{Y_m^\ell} = \frac{\mathrm{d}\xi_r}{\mathrm{d}r} + \frac{2\xi_r}{r} - \frac{\ell(\ell+1)\xi_h}{r}\,.$$

Figure 2 gives an example of (ρ, c^2) kernels. We note that in deriving Eq. (14), we neglected various surface terms which result from integration by parts. Also, the modelling of surface layers in stars tends to be inaccurate. Accordingly, Eq. (14) typically includes an extra ad hoc adjustable surface term.

Besides these kernels, other structural kernels can also be obtained: (ρ, Γ_1), (P, Γ_1), $(u \equiv \frac{P}{\rho}, \Gamma_1)$, (g, Γ_1), (u, Y), (A, Γ_1), (N^2, c^2) etc. (see Masters 1979; Gough and Thompson 1991; Elliott 1996; Basu and Christensen-Dalsgaard 1997; Kosovichev 1999; Buldgen et al. 2017). Some of these require using the equation of state and its derivatives. Figure 2 shows an example of (ρ, Γ_1) kernels.

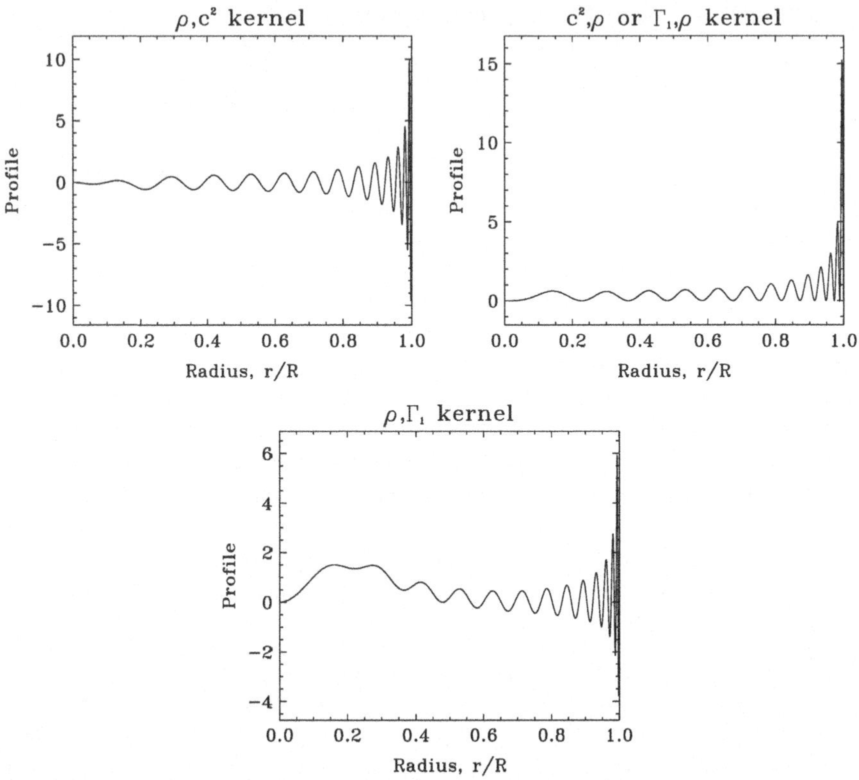

Fig. 2 Kernels for the $(n, \ell) = (13, 1)$ pulsation mode, for the structural pairs (ρ_0, c_0^2) and $(\rho_0, \Gamma_{1,0})$

3 The Inverse Problem

As described at the beginning of this lecture, the seismic inverse problem consists in deducing the stellar structure from a set of *identified* pulsation frequencies, i.e., with known quantum numbers. Inverse methods have proved to be a powerful way of solving such a problem. These typically involve correcting a reference stellar model so as to obtain a new model which reproduces the pulsation frequencies more accurately. Inverse methods come into two broad categories, namely linear and non-linear methods. The linear methods are further subdivided into the Regularised Least-Squares (RLS) and Optimally Localised Averages (OLA) methods. For the non-linear inversions, there are iterated versions of the RLS method, as well as a method which adjusts the internal phases of the eigenmodes. In what follows, we will focus on linear inverse methods, beginning with rotation inversions, as these provide a good starting point to illustrate the different methods.

3.1 Rotation Inversions

The rotation inverse problem can be expressed by the following set of equations:

$$S_{n_l,\ell_l} = \frac{\omega_{n_l,\ell_l,m_l} - \omega_{n_l,\ell_l,0}}{m_l} = \int_0^R K_{\Omega}^{n_l,\ell_l}(r)\Omega(r)\mathrm{d}r + \varepsilon_{n_l,\ell_l}, \quad 1 \leq l \leq L, \tag{17}$$

where the S_{n_l,ℓ_l} are the "rotational splittings" (i.e., the observations), $\Omega(r)$ the unknown rotation profile, and ε_{n_l,ℓ_l} the errors on the splittings, characterised by a standard deviation of $\sigma_{n_l,\ell_l} = \langle\varepsilon_{n_l,\ell_l}\rangle$. In what follows, we will use the index "l" as shorthand for (n_l, ℓ_l).

The goal of the inverse problem is to recover $\Omega(r)$ from the set of available rotational splittings. At first, this problem looks impossible. Indeed, the unknown is a function, whereas there is a finite number of observational constraints. Furthermore, the problem is ill-conditioned, i.e., it is highly sensitive to noise. In order to address these issues, it is necessary to inject a priori assumptions when solving the inverse problem. Accordingly, we should always bear in mind these assumptions when looking at and interpreting the results.

3.1.1 Regularised Least Squares (RLS)

A first approach to tackling this problem involves decomposing the rotation profile over a set of basis functions:

$$\Omega_{\mathrm{inv}}(r) = \sum_k a_k f_k(r), \tag{18}$$

where the a_k are unknown coefficients, and the f_k basis functions. In general, the number of unknown coefficients should be equal to or less than the number of observed splittings. Typical choices for the f_k include b-spline functions of various degrees. For instance, zeroth degree b-splines produce step-wise functions, whereas cubic splines produce functions with a continuous second derivative (which can be useful for regularisation terms).

When substituted into Eq. (17), this leads to the following theoretical rotational splittings, $\tilde{S}_l$, for the above rotation profile:

$$\tilde{S}_l = \int_0^R K_{\Omega}^l(r)\Omega_{\mathrm{inv}}(r)\mathrm{d}r. \tag{19}$$

An obvious way of choosing the a_k is by minimising (typically in a least-squares sense) the distance between the observed splittings, S_l, and the theoretical ones. However, a naive application of such a procedure leads to poor results as illustrated by the dotted grey curve in the top panel of Fig. 3. Indeed, the problem is ill-conditioned, and any errors in the observations will be strongly amplified.

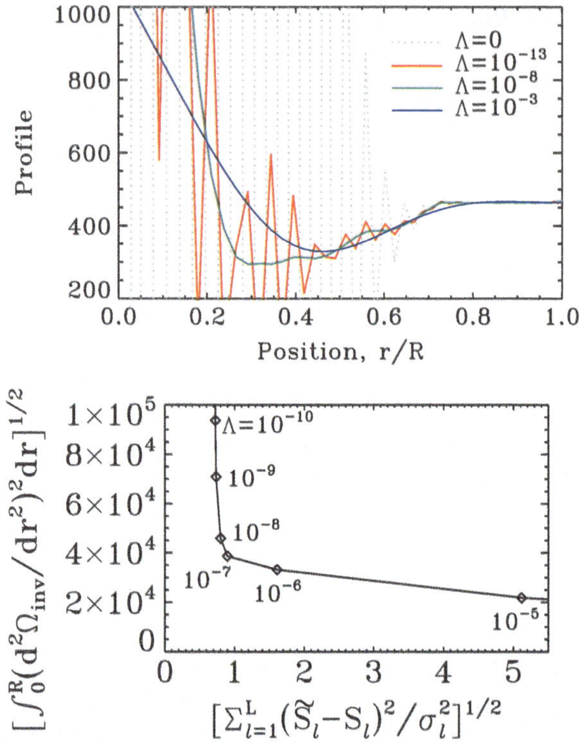

Fig. 3 *Top:* Inverted rotation profiles based on the RLS method for different values of the regularisation parameter. *Bottom:* L-curve which shows the two components of the RLS cost function as a function of Λ

A standard remedy to this problem is to include a supplementary regularisation term to obtain a smooth solution when carrying out the minimisation, hence the name "Regularised Least-Squares" (RLS) method. This leads to the following typical cost function:

$$J(a_k) = \sum_{l=1}^{L} \frac{(S_l - \tilde{S}_l)^2}{\sigma_l^2} + \Lambda \left\langle \frac{1}{\sigma^2} \right\rangle \int_0^R \left(\frac{\mathrm{d}^2 \Omega_{\mathrm{inv}}}{\mathrm{d}r^2} \right)^2 \mathrm{d}r\,, \tag{20}$$

where $\left\langle \frac{1}{\sigma^2} \right\rangle = \frac{1}{L} \sum_{l=1}^{L} \frac{1}{\sigma_l^2}$, and Λ is a regularisation parameter which can be adjusted. The cost function is minimised by numerically finding the a_k coefficients for which the gradient of J is zero.

Figure 3 shows various solutions obtained for the rotation inverse problem based on a set of rotational splittings from Christensen-Dalsgaard et al. (1990). As can be seen in the top panel, larger values of Λ lead to solutions that are smoother. The bottom panel shows that such solutions are a worse fit to the S_l. Hence, there is a trade-off between obtaining smooth solutions and fitting the data. The best solutions are obtained for intermediate values of Λ as can be seen by comparing the solutions in Fig. 3 to the true solution given in Figs. 3 and 11 of Christensen-Dalsgaard et al. (1990).

3.1.2 Various Error Measurements

It is also possible to calculate error bars around the inverted solution. To demonstrate this, we start with a given grid point, r_0. The relationship between $\Omega_{\text{inv}}(r_0)$ and the a_k coefficients is linear. Likewise, the relationship between the a_k and the S_l is also linear. Hence the relationship between $\Omega_{\text{inv}}(r_0)$ and the S_l is linear and can be expressed as follows:

$$\Omega_{\text{inv}}(r_0) = \sum_l c_l(r_0) S_l \,. \tag{21}$$

Assuming the errors on the splittings are uncorrelated, the 1σ error bar on the inverted value of the rotation rate will simply be

$$\sigma_{\Omega(r_0)} = \sqrt{\sum_l (c_l(r_0)\sigma_l)^2} \,. \tag{22}$$

In the specific case where the errors are uniform, the error is amplified by the quantity $\sqrt{\sum_l (c_l(r_0))^2}$ which is known as the "error magnification". It is important to bear in mind that these error bars only take into account how the observational errors propagate through the inversion. They do not actually measure the quality of the inversion, which could, for example, be poor due to over-regularisation.

In order to evaluate the quality of the inversion at a given point, it is useful to look at the "averaging kernel". If we replace the S_l in Eq. (21) by the expressions given in Eq. (17), then it is possible to establish a relationship between $\Omega(r)$ and $\Omega_{\text{inv}}(r_0)$:

$$\Omega_{\text{inv}}(r_0) = \int_0^R \underbrace{\sum_l c_l(r_0) K_\Omega^l(r)}_{\mathcal{K}_{\text{avg}}(r_0,r)} \, \Omega(r) \text{d}r + \sum_l c_l(r_0)\varepsilon_l \,. \tag{23}$$

This expression shows that $\Omega_{\text{inv}}(r_0)$ is in fact an average of the true rotation profile $\Omega(r)$. The corresponding weight function, $\mathcal{K}_{\text{avg}}(r_0, r)$, is the averaging kernel. Ideally, this function should have a strong amplitude at r_0 and be close to zero elsewhere. Figure 4 shows a few examples of averaging kernels for the RLS method.

3.1.3 Optimally Localised Averages (OLA)

The notion of averaging kernels naturally leads to the Optimally Localised Averages (OLA) methods. The basic idea in these methods is to optimise the coefficients c_l so as to obtain *optimal* averaging kernels. Two variants include the Multiplicative and the Subtractive OLA, abbreviated MOLA and SOLA, respectively.

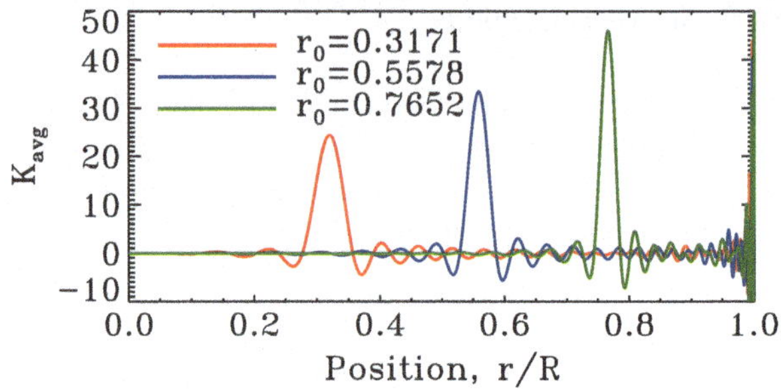

Fig. 4 Averaging kernels for the RLS method at various positions

The MOLA method comes from Backus and Gilbert (1968). In this method, the averaging kernel is *multiplied* by a penalty function that increases in amplitude as you move away from the target position r_0. Hence, the coefficients $c_l(r_0)$ are obtained by minimising the following cost function:

$$J(c_l) = \underbrace{\int_0^R P(r_0, r)\left[\mathcal{K}_{\text{avg}}(r_0, r)\right]^2 \mathrm{d}r}_{\text{fit data}} + \underbrace{\frac{\tan\theta}{\langle\sigma^2\rangle}\sum_{l=1}^{L}(c_l\sigma_l)^2}_{\text{regularisation}} + \underbrace{\lambda\left\{1 - \int_0^R \mathcal{K}_{\text{avg}}\right\}}_{\mathcal{K}_{\text{avg}}\,\text{unimodular}}, \quad (24)$$

where $\langle\sigma^2\rangle = \frac{1}{L}\sum_{l=1}^{L}\sigma_l^2$, θ is a trade-off parameter between fitting data and reducing error (i.e., a regularisation parameter), $P(r_0, r)$ the penalty function (usually $12(r - r_0)^2$), and λ a Lagrange multiplier used to ensure that the averaging kernel is "unimodular", i.e., $\int_0^R \mathcal{K}_{\text{avg}}(r_0, r)\mathrm{d}r = 1$. This last condition is important for ensuring that the inverted value, $\Omega_{\text{inv}}(r_0) = \sum_l c_l(r_0)S_l$ is a proper average of the underlying rotation profile.

The SOLA method was first described in Pijpers and Thompson (1992). In this method, the *difference* between the averaging kernel and a suitable target function is minimised. Hence, the coefficients $c_l(r_0)$ are obtained by minimising the following cost function:

$$J(c_l) = \int_0^R \left[\mathcal{T}(r_0, r) - \mathcal{K}_{\text{avg}}(r_0, r)\right]^2 \mathrm{d}r + \frac{\tan\theta}{\langle\sigma^2\rangle}\sum_{l=1}^{L}(c_l\sigma_l)^2 + \lambda\left\{1 - \int_0^R \mathcal{K}_{\text{avg}}\right\}, \quad (25)$$

where $\mathcal{T}(r_0, r)$ is the target function. Ideally, $\mathcal{T}$ should be a Dirac function centred on r_0. However, given the finite number of rotational splittings and hence rotation kernels to work with, trying to achieve such a target is impossible and would lead to poor numerical results. Generally, Gaussian or similar functions are used as targets:

$$\mathcal{T}(r_0, r) = \frac{1}{A}\exp\left(-\frac{(r - r_0)^2}{2\Delta(r_0)^2}\right), \quad (26)$$

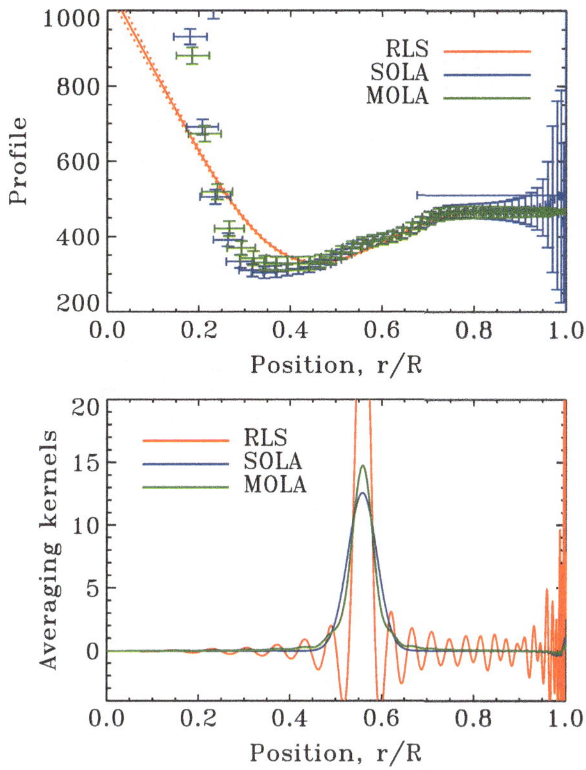

Fig. 5 *Top:* Inversion results and error bars for the RLS, MOLA and SOLA methods. *Bottom:* Averaging kernels at $r_0 = 0.5578\,R$ for these three methods

where A is a normalisation constant to ensure that $\int_0^R \mathcal{T}(r_0, r)\mathrm{d}r = 1$ (it is not simply $1/\sqrt{2\pi}\Delta(r_0)$ since the integration interval is not from $-\infty$ to ∞), and $\Delta(r_0)$ the width of the target function. A good choice for $\Delta(r_0)$ when dealing with acoustic modes is $\Delta(r_0) \propto c_0(r_0)$ (e.g., Thompson 1993).

Figure 5 shows inversion results for the RLS, MOLA and SOLA methods as well as some averaging kernels. The advantages of the MOLA method compared to the SOLA method is that it has fewer free parameters and tends to produce slightly better results. Conversely, the SOLA method has a much smaller computational cost. Indeed, minimising the SOLA cost function for different values of r_0 leads to systems of equations where only the right-hand side changes. Accordingly, only one matrix inversion (or factorisation) is needed for the entire inversion.

3.1.4 Applications

The first and most spectacular examples of rotation profile inversions are those done for the Sun. Indeed, the Sun's close proximity has enabled the detection of countless

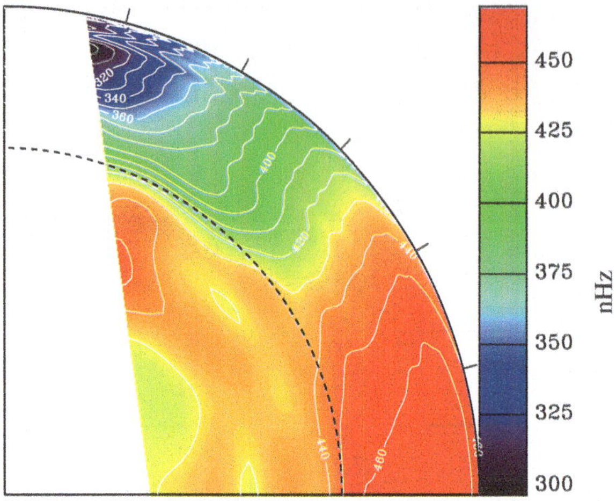

Fig. 6 2D solar rotation profile from Thompson et al. (2003) (see also Schou et al. 1998) based on a SOLA inversion technique. Figure courtesy of M.J. Thompson and J. Christensen-Dalsgaard

rotational splittings going to high ℓ values. This, in turn, has enabled 2D inversions of the solar rotation profile such as the one shown in Fig. 6, taken from Thompson et al. (2003) (see also Schou et al. 1998). Such profiles were not in agreement with the theoretical predictions at the time and have accordingly led to various theoretical investigations and numerical simulations to gain a better understanding of the Sun and its internal rotation (e.g., Thompson et al. 2003; Brun et al. 2004).

A more recent example of stellar rotation inversions are those in subgiants and red giants (Deheuvels et al. 2012, 2014). These results as well as results from ensemble asteroseismology have shown that although the core of these stars rotate much faster than the envelope, the difference in rotation speeds is orders of magnitude smaller than what is expected theoretically (Eggenberger et al. 2012; Ceillier et al. 2013; Marques et al. 2013). It is still an open question what transport mechanisms are involved in these stars and could solve this discrepancy.

3.2 *Structural Inversions*

We now turn our attention to structural inversions. In contrast to rotation inversions, there are *two* functions to invert simultaneously. As was derived in Sect. 2.3, the linearised relationship between modifications of the stellar structure and shifts in the frequency can be expressed as follows:

$$\underbrace{\frac{\delta\omega_l}{\omega_l}}_{\text{obs.}} = \int_0^R \underbrace{K^l_{a,b}(r)}_{\text{known}} \underbrace{\frac{\delta a}{a}}_{\text{unknown}} \mathrm{d}r + \int_0^R \underbrace{K^l_{b,a}(r)}_{\text{known}} \underbrace{\frac{\delta b}{b}}_{\text{unknown}} \mathrm{d}r + \frac{F_{\text{surf.}}(\omega_l)}{E_l} + \varepsilon_l, \tag{27}$$

where we have added an ad hoc surface correction term (i.e., the term with $F_{\text{surf.}}$) as well as the observational error, ε_l. The variables (a, b) represent two structural profiles (e.g., (ρ, Γ_1)). The structural inverse problem then consists in deducing the profiles $\delta a/a$ and $\delta b/b$ from the frequency shifts $\delta\omega_l/\omega_l$. The fact that there are two functions to invert leads to modifications of the RLS and OLA methods, as well as the introduction of "cross-term kernels", $\mathcal{K}_{\text{cross}}$.

3.2.1 Regularised Least Squares (RLS)

In the regularised least squares method, both functions ($\delta a/a$ and $\delta b/b$) are discretised over a set of basis functions, and the unknown coefficients are obtained by minimising a cost function of the form:

$$J\left(\frac{\delta a}{a}, \frac{\delta b}{b}\right) = \sum_l \frac{1}{\sigma_l^2}\left(\frac{\delta\omega_l}{\omega_l} - \int_0^R K_{a,b}^l \frac{\delta a}{a}\mathrm{d}r - \int_0^R K_{b,a}^l \frac{\delta b}{b}\mathrm{d}r\right)^2$$
$$+\Lambda\left\langle\frac{1}{\sigma^2}\right\rangle \int_0^R \left[\left(\frac{\mathrm{d}^2}{\mathrm{d}r^2}\frac{\delta a}{a}\right)^2 + \left(\frac{\mathrm{d}^2}{\mathrm{d}r^2}\frac{\delta b}{b}\right)^2\right]\mathrm{d}r\,. \tag{28}$$

Additional terms may be included to model surface effects.

In much the same way as for rotation inversions, the inverted functions are related in a linear way to the observables $(\delta\omega/\omega)_l$:

$$\left(\frac{\delta a}{a}\right)_{\text{inv}} = \sum_l c_l(r_0)\left(\frac{\delta\omega}{\omega}\right)_l, \quad \left(\frac{\delta b}{b}\right)_{\text{inv}} = \sum_l c_l'(r_0)\left(\frac{\delta\omega}{\omega}\right)_l. \tag{29}$$

These inversion coefficients can then be used to define the averaging and cross-term kernels:

$$\mathcal{K}_{\text{avg}}(r_0, r) = \sum_{l=1}^L c_l(r_0)K_{a,b}^l(r)\,, \quad \mathcal{K}_{\text{cross}}(r_0, r) = \sum_{l=1}^L c_l(r_0)K_{b,a}^l(r)\,, \tag{30}$$

$$\mathcal{K}_{\text{avg}}'(r_0, r) = \sum_{l=1}^L c_l'(r_0)K_{b,a}^l(r)\,, \quad \mathcal{K}_{\text{cross}}'(r_0, r) = \sum_{l=1}^L c_l'(r_0)K_{a,b}^l(r)\,, \tag{31}$$

which help to relate the inverted structural functions at r_0 to the true structural functions:

$$\left(\frac{\delta a}{a}\right)_{\text{inv}}(r_0) = \int_0^R \left[\mathcal{K}_{\text{avg}}(r_0, r)\frac{\delta a(r)}{a(r)} + \mathcal{K}_{\text{cross}}(r_0, r)\frac{\delta b(r)}{b(r)}\right]\mathrm{d}r\,, \tag{32}$$

$$\left(\frac{\delta b}{b}\right)_{\text{inv}}(r_0) = \int_0^R \left[\mathcal{K}_{\text{cross}}'(r_0, r)\frac{\delta a(r)}{a(r)} + \mathcal{K}_{\text{avg}}'(r_0, r)\frac{\delta b(r)}{b(r)}\right]\mathrm{d}r\,, \tag{33}$$

where we have neglected the contribution from surface effects and observational errors. As can be seen from these equations, the cross-term kernels help to quantify the amount of cross-talk between the two functions in the inversion.

In the particular case of solar inversions, where the mass is known through independent considerations, it is possible to constrain the inversion to preserve the mass by introducing a supplementary Lagrange multiplier, provided one of the structural variables being inverted is the density variation, $\delta\rho_0/\rho_0$. Indeed, if the mass is constant, then $\delta\rho_0/\rho_0$ obeys the following relation:

$$0 = 4\pi \int_0^R \rho_0(r)\frac{\delta\rho_0}{\rho_0} r^2 \mathrm{d}r\,. \tag{34}$$

3.2.2 Optimally Localised Averages (OLA)

The OLA methods will also be modified due to the presence of two functions which are being inverted. Given that the modifications to the MOLA and SOLA variants are similar, we will focus on the SOLA method in what follows. First of all, there will be two separate inversions, one for each of the functions being inverted. Secondly, not only do the averaging kernels need to be optimised, but the cross-term kernels need to be reduced as much as possible. These considerations lead to cost functions of the following form:

$$\begin{aligned} J(c_l(r_0)) = & \int_0^R \left\{\mathcal{T}(r_0, r) - \mathcal{K}_{\mathrm{avg}}(r_0, r)\right\}^2 \mathrm{d}r + \beta \int_0^R \left\{\mathcal{K}_{\mathrm{cross}}(r_0, r)\right\}^2 \mathrm{d}r \\ & + \frac{\tan\theta \sum_{l=1}^{L} (c_l(r_0)\sigma_l)^2}{\langle\sigma^2\rangle} + \lambda \left\{1 - \int_0^R \mathcal{K}_{\mathrm{avg}}(r_0, r)\mathrm{d}r\right\}\,. \end{aligned} \tag{35}$$

For each inversion, there is a regularisation parameter (θ), a supplementary parameter to adjust the trade-off between optimising the averaging kernel or minimising the cross-term kernel (β), a Lagrange multiplier to ensure the averaging kernel is unimodular (λ), and optionally some supplementary Lagrange multipliers used to suppress surface effects (Däppen et al. 1991). The target functions ($\mathcal{T}$) for each of the inverted functions can be adjusted independently.

In order to preserve the mass, for instance in the case of solar inversions, one can treat Eq. (34) as a supplementary observed relation. Specifically, 0 will play the role of $\delta\omega/\omega$ and the function $f(r) = 4\pi\rho r^2$ will be the kernel associated with the structural variable $\delta\rho_0/\rho_0$.

3.2.3 Applications

Up to now, structural inversions have been applied primarily to the Sun. Figure 7, which is based on the results of Basu et al. (2009), shows an example of such an

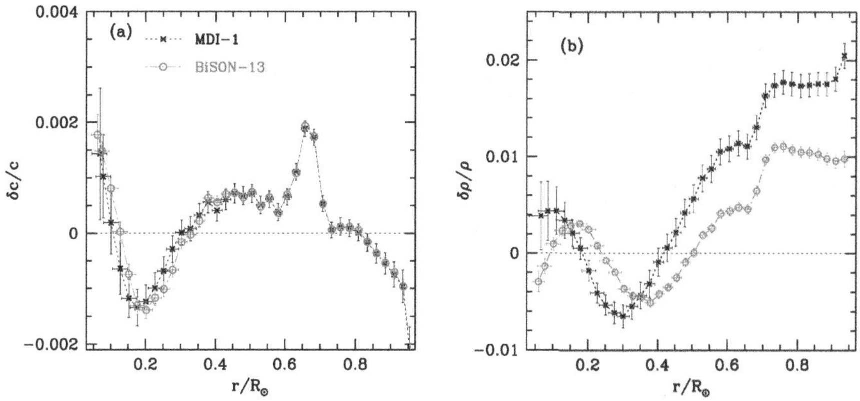

Fig. 7 (c, ρ) structural inversion for the Sun based on the results of Basu et al. (2009). Figure courtesy of S. Basu

inversion for the structural variables (c, ρ). In recent years, the downward revision of the solar metal abundances (e.g., Asplund et al. 2009) has led to a significant discrepancy between the results from solar structural inversions and models based on these new abundances (e.g., Basu et al. 2015). Indeed, helioseismic inversions led to a lower depth for the base of the convection zone compared to what is obtained from models with the revised abundances. Currently, it is not entirely clear how to solve this problem but different solutions are being investigated.

3.3 *Integrated Quantities*

In the case of stars other than the Sun, it is very difficult to carry out structural inversions due to the limited number of available modes (e.g., Basu et al. 2002). Indeed, because of cancellation effects in disc-integrated observations, only modes for which $\ell \leq 3$ are detected (Dziembowski 1977). One strategy in such a situation is to invert stellar parameters rather than structural profiles. Indeed, since structural inversions at a given grid point actually give a weighted average of the true underlying profile, one can use a SOLA inversion to directly target the appropriate weight function which yields the desired stellar parameter. The quantities which may be inverted by such a procedure include the total angular momentum (Pijpers 1998), the mean density (Reese et al. 2012), the acoustic radius and various core or internal mixing indicators (Buldgen et al. 2015b,a, 2017). Figure 8 shows inversions of the acoustic radius and mean density, as described in Buldgen et al. (2015b).

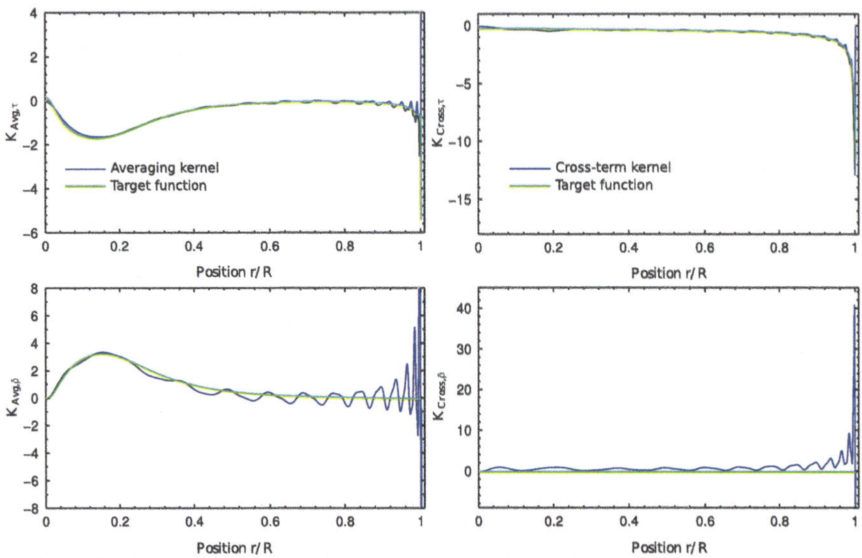

Fig. 8 Acoustic radius inversion (*top*) and mean density inversion (*bottom*) based on Buldgen et al. (2015b). Figure courtesy of G. Buldgen

4 Conclusion

As illustrated in this course, inversions can be used to probe stellar rotation profiles, probe the internal structure of stars, estimate various stellar parameters, and indirectly test new physics outside a given grid of stellar models. Nonetheless, one must not forget the limitations of seismic inversions, namely, the use of a priori assumptions about the smoothness of rotation or structural profiles and the linearisation of the relationship between frequencies and stellar structure (except in the case of non-linear inversions). Furthermore, it is important to keep in mind that inversions cannot yield more information than what is intrinsically contained in the observed pulsation modes.

In order to get a more in-depth understanding of inversions, we recommend the following articles or publications:

- Lynden-Bell and Ostriker (1967) and Christensen-Dalsgaard (2003): the variational principle
- Gough and Thompson (1991): structural kernels
- Christensen-Dalsgaard et al. (1990): error propagation and magnification, averaging kernels
- Rabello-Soares et al. (1999): adjusting the free parameters in inversions
- Reese et al. (2012) and Buldgen et al. (2015b): inversions of integrated quantities

We also note that the recent monograph by Pijpers (2006) contains several chapters on helioseismic and asteroseismic inversions.

Various seismic inversion software packages have also become freely available in recent years:

- InversionKit[3]: 1D inversions on individual stars
- InversionPipeline[4]: inversions of stellar parameters using a grid of models
- NonLinearKit[5]: non-linear 1D inversion tool still under development
- SOLA Pack[6]: 2D rotation inversions in the Sun

Acknowledgements I would like to thank the organisers of this Summer School for giving me the opportunity to give this lecture on inversions. Furthermore, I thank M.J. Thompson for introducing me to inversions, as well as S. Basu and G. Buldgen for many discussions on the topic.

References

Asplund, M., Grevesse, N., Sauval, A.J., Scott, P.: Annu. Rev. Astron. Astrophys. **47**, 481 (2009)
Backus, G., Gilbert, F.: Geophys. J. **16**, 169 (1968)
Basu, S., Christensen-Dalsgaard, J.: Astron. Astrophys. **322**, L5 (1997)
Basu, S., Christensen-Dalsgaard, J., Thompson, M.J.: In: Battrick, B., Favata, F., Roxburgh, I.W., Galadi, D. (eds.) Stellar Structure and Habitable Planet Finding, ESA Special Publication, vol. 485, pp. 249–252. European Space Agency, Noordwijk (2002)
Basu, S., Chaplin, W.J., Elsworth, Y., New, R., Serenelli, A.M.: Astrophys. J. **699**, 1403 (2009)
Basu, S., Grevesse, N., Mathis, S., Turck-Chièze, S.: Space Sci. Rev. **196**, 49 (2015)
Bazot, M., Bourguignon, S., Christensen-Dalsgaard, J.: Mon. Not. R. Astron. Soc. **427**, 1847 (2012)
Brun, A.S., Miesch, M.S., Toomre, J.: Astrophys. J. **614**, 1073 (2004)
Buldgen, G., Reese, D.R., Dupret, M.A.: Astron. Astrophys. **583**, A62 (2015a)
Buldgen, G., Reese, D.R., Dupret, M.A., Samadi, R.: Astron. Astrophys. **574**, A42 (2015b)
Buldgen, G., Reese, D.R., Dupret, M.A.: Astron. Astrophys. **598**, A21 (2017)
Ceillier, T., Eggenberger, P., García, R.A., Mathis, S.: Astron. Astrophys. **555**, A54 (2013)
Charpinet, S., Fontaine, G., Brassard, P., Green, E.M., Chayer, P.: Astron. Astrophys. **437**, 575 (2005)
Christensen-Dalsgaard, J.: Lecture Notes on Stellar Oscillations. http://astro.phys.au.dk/~jcd/oscilnotes/ (2003)
Christensen-Dalsgaard, J.: Astrophys. Space Sci. **316**, 113 (2008)
Christensen-Dalsgaard, J., Schou, J., Thompson, M.J.: Mon. Not. R. Astron. Soc. **242**, 353 (1990)
Däppen, W., Gough, D.O., Kosovichev, A.G., Thompson, M.J.: In: Gough, D., Toomre, J. (eds.) Challenges to Theories of the Structure of Moderate-Mass Stars. Lecture Notes in Physics, vol. 388, p. 111. Springer, Berlin (1991)
Deheuvels, S., García, R.A., Chaplin, W.J., et al.: Astrophys. J. **756**, 19 (2012)
Deheuvels, S., Doğan, G., Goupil, M.J., et al.: Astron. Astrophys. **564**, A27 (2014)
Dziembowski, W.: Acta Astron. **27**, 203 (1977)
Eggenberger, P., Montalbán, J., Miglio, A.: Astron. Astrophys. **544**, L4 (2012)
Elliott, J.R.: Mon. Not. R. Astron. Soc. **280**, 1244 (1996)

[3]See http://bison.ph.bham.ac.uk/spaceinn/inversionkit.
[4]See http://bison.ph.bham.ac.uk/spaceinn/inversionpipeline.
[5]See http://bison.ph.bham.ac.uk/spaceinn/nonlinearkit.
[6]See http://sun.stanford.edu/~rmunk/SOLApack/index.html.

Gough, D.: Sol. Phys. **100**, 65 (1985)
Gough, D.O., Thompson, M.J.: In: Cox, A.N., Livingston, W.C., Matthews, M.S. (eds.) Solar Interior and Atmosphere, pp. 519–561. University of Arizona Press, Tucson (1991)
Kosovichev, A.G.: J. Comput. Appl. Math. **109**, 1 (1999)
Ledoux, P.: Astrophys. J. **114**, 373 (1951)
Lynden-Bell, D., Ostriker, J.P.: Mon. Not. R. Astron. Soc. **136**, 293 (1967)
Marques, J.P., Goupil, M.J., Lebreton, Y., et al.: Astron. Astrophys. **549**, A74 (2013)
Masters, G.: Geophys. J. **57**, 507 (1979)
Metcalfe, T.S., Charbonneau, P.: J. Comput. Phys. **185**, 176 (2003)
Pijpers, F.P.: Mon. Not. R. Astron. Soc. **297**, L76 (1998)
Pijpers, F.P.: Methods in Helio- and Asteroseismology. Imperial College Press, London (2006)
Pijpers, F.P., Thompson, M.J.: Astron. Astrophys. **262**, L33 (1992)
Rabello-Soares, M.C., Basu, S., Christensen-Dalsgaard, J.: Mon. Not. R. Astron. Soc. **309**, 35 (1999)
Reese, D.: PhD thesis, Université Toulouse III - Paul Sabatier (2006). http://tel.archives-ouvertes.fr/tel-00120334
Reese, D.R., Marques, J.P., Goupil, M.J., Thompson, M.J., Deheuvels, S.: Astron. Astrophys. **539**, A63 (2012)
Schou, J., Christensen-Dalsgaard, J., Thompson, M.J.: Astrophys. J. **433**, 389 (1994)
Schou, J., Antia, H.M., Basu, S., et al.: Astrophys. J. **505**, 390 (1998)
Silva Aguirre, V., Davies, G.R., Basu, S., et al.: Mon. Not. R. Astron. Soc. **452**, 2127 (2015)
Thompson, M.J.: In: Brown, T.M. (ed.) Seismic Investigation of the Sun and Stars. Astronomical Society of the Pacific Conference Series, vol. 42, GONG 1992, p. 141 (1993)
Thompson, M.J., Christensen-Dalsgaard, J., Miesch, M.S., Toomre, J.: Annu. Rev. Astron. Astrophys. **41**, 599 (2003)
Unno, W., Osaki, Y., Ando, H., Saio, H., Shibahashi, H.: Nonradial Oscillations of Stars. University of Tokyo Press, Tokyo (1989)

Asteroseismology of Red Giants and Galactic Archaeology

Saskia Hekker

Abstract From the oscillations in red-giant stars measured in time-series data it is possible to derive more accurate stellar parameters (e.g., mass, radius and age) as can be done using only single-epoch spectroscopy or photometry. These stellar parameters combined with chemical composition and the position, distance and velocity of the stars play an important role in studying the formation and evolution of the Milky Way. In this chapter we discuss some key physical phenomena that are at play in (red-giant) stars as well as some important phases in red-giant evolution. Subsequently, oscillation characteristics that are of importance for the determination of stellar parameters (as indicated above) of red-giant stars are introduced followed by a description of the main components of the Milky Way. Finally, the role red giants can play in creating a detailed observational picture of the Milky Way and deciphering the formation and evolution of the Milky Way is discussed.

1 Introduction

Red-giant stars are low- to intermediate-mass ($M \lesssim 10\,\mathrm{M}_\odot$) stars that have exhausted hydrogen in the core. These extended, cool and hence red stars are key targets for stellar evolution studies as well as galactic studies for several reasons: (a) many stars go through a red-giant phase; (b) red giants are intrinsically bright; (c) large stellar internal structure changes as well as changes in surface chemical abundances take place over relatively short time; (d) red-giant stars exhibit global intrinsic oscillations.

Due to their large number and intrinsic brightness it is possible to observe many of these stars up to large distances. Furthermore, the global intrinsic oscillations provide a means to discern red-giant stars in the pre-helium core burning from the

S. Hekker (✉)
Max Planck Institute for Solar System Research, SAGE research group, Justus-von-Liebig Weg 3, 37077 Göttingen, Germany

Department of Physics and Astronomy, Stellar Astrophysics Centre (SAC), Aarhus University, Ny Munkegade 120, DK-8000 Aarhus C, Denmark
e-mail: hekker@mps.mpg.de

T.L. Campante et al. (eds.), *Asteroseismology and Exoplanets: Listening to the Stars and Searching for New Worlds*, Astrophysics and Space Science Proceedings 49, DOI 10.1007/978-3-319-59315-9_5

ones in the helium core burning phase and provide an estimate of stellar ages, a key ingredient for galactic studies.

In this lecture I will first discuss some physical phenomena that play a role in red-giant stars and several phases of red-giant evolution. Then, I will provide some details about asteroseismology—the study of the internal structure of stars through their intrinsic oscillations—of red-giant stars. I will conclude by discussing galactic archaeology—the study of the formation and evolution of the Milky Way by reconstructing its past from its current constituents—and the role red-giant stars can play in that.

The red-giant and asteroseismology parts of this lecture are based on the lecture notes by Onno Pols,[1] the book by Kippenhahn et al. (2012) and a review by Hekker and Christensen-Dalsgaard (2017). For more details I refer the reader to these sources.

2 Red-Giant Stars

In this section, I will provide a brief overview of some physical phenomena that are important in red-giant stars followed by a description of the different stages in stellar evolution of red-giant stars.

2.1 Physical Phenomena

2.1.1 Convection

There is a physical limit to the energy flux that can be transported by radiation through a specific medium. If the temperature gradient becomes too steep convection takes over as the primary means of energy transport (cf. Schwarzschild criterion; Schwarzschild 1906). The Schwarzschild criterion states that convection is activated once the radiative temperature gradient ($\nabla_{\rm rad}$) exceeds the adiabatic temperature gradient ($\nabla_{\rm ad}$), so the criterion for stability is:

$$\nabla_{\rm rad} < \nabla_{\rm ad}\,, \tag{1}$$

with

$$\nabla_{\rm rad} = \left(\frac{\partial \log T}{\partial \log P}\right)_{\rm rad} = \frac{3}{16\pi\, a\, c\, G}\frac{\kappa\, l\, P}{m\, T^4} \tag{2}$$

[1]https://www.astro.ru.nl/~onnop/education/stev_utrecht_notes/.

describing the logarithmic variation of temperature T with depth (expressed in pressure P) for a star in hydrostatic equilibrium in case energy is transported by radiation. Here, a is the radiation constant, c is the speed of light, G is the gravitational constant, κ is the opacity, l is the local luminosity and m is the mass coordinate, i.e., represents the mass contained inside a spherical shell of radius r. The adiabatic temperature gradient is defined as:

$$\nabla_{\rm ad} = \left(\frac{\partial \log T}{\partial \log P}\right)_{\rm ad} . \tag{3}$$

An alternative to the Schwarzschild criterion is the Ledoux criterion (Ledoux 1947), which, in addition to the temperature gradients, takes into account the spatial variation of the mean molecular weight, μ. For an ideal gas this takes the form:

$$\nabla_{\rm rad} < \nabla_{\rm ad} + \nabla_{\mu} , \tag{4}$$

with

$$\nabla_{\mu} = \left(\frac{\partial \log \mu}{\partial \log P}\right) . \tag{5}$$

Note that $\nabla_{\rm rad}$ and ∇_{μ} are spatial gradients, while $\nabla_{\rm ad}$ represents the temperature gradient in a gas element that undergoes a pressure variation.

Due to the nuclear reactions in the deep regions of a star, ∇_{μ} is generally positive throughout the star. Hence the right-hand side of Eq. (4) takes larger values than the right-hand side of Eq. (1) and thus the mean molecular weight gradient incorporated in the Ledoux criterion has a stabilising effect.

Semi-convection

In the regions that are convective according to the Schwarzschild criterion and stable according to the Ledoux criterion, the true behaviour of the material remains unclear (Gabriel et al. 2014). However, in stellar evolution codes some form of 'semi-convection' is often applied at the edge of a convective core. Semi-convection is a form of slow convection in which mixing takes place that is necessary to match observational constraints (e.g., Lattanzio 1983; Langer et al. 1985).

Mixing Length Parametrisation

Convection takes place over a large range of length scales which makes it complicated and expensive to model. To include convection in stellar modelling the mixing-length approximation is often used. The mixing length model was first proposed in 1925 by Ludwig Prandtl as a rough approximation of the distance or characteristic length a fluid parcel can travel before mixing with the surrounding fluid. In stellar structure the most commonly used implementation of the mixing length, l_{m}, is from Böhm-Vitense (1958):

$$l_{\mathrm{m}} = \alpha_{\mathrm{MLT}} H_P , \tag{6}$$

with H_P the pressure scale height, which is the radial distance over which the pressure changes by an e-folding factor,

$$H_P = \left| \frac{dr}{d \ln P} \right| = \frac{P}{\rho g} , \tag{7}$$

where ρ represents density and g represents gravity. Note that the mixing length parameter, α_{MLT}, is ordinarily calibrated to the Sun taking on a value between 1.2 and 2.2 depending on the stellar evolution code and choice of included physics. Although the solar value is often used as a fixed value in models of other stars there are indications that the value of α_{MLT} should change as a function of evolution (e.g., Trampedach et al. 2014). An alternative to the mixing-length formalism, described here, is the approach by, e.g., Canuto et al. (1996), whom devised a full spectrum of turbulence which considers convection on different length scales.

Convective Overshoot

The border between a radiative and convective layer may be soft in the sense that material on the convective side that approaches the boundary of stability with momentum penetrates into the radiative layer. This process, that is referred to as convective overshoot, extends the convective region. In case of a convective core, convective overshoot can bring fresh fuel into the core prolonging the ongoing core burning phase. The extent of convective overshoot, l_{ov}, can be expressed as a fraction, α_{ov}, of a local pressure scale height (Eq. (7)):

$$l_{\mathrm{ov}} = \alpha_{\mathrm{ov}} H_P , \tag{8}$$

where α_{ov} is a free parameter that can be calibrated against observations; typically $\alpha_{\mathrm{ov}} < 0.25$.

2.1.2 Electron Degeneracy

In very dense regions, such as the cores of low-mass (sub)giant stars ($M \lesssim 1.1\,\mathrm{M}_{\odot}$) or white dwarfs, the electron density is high enough to become degenerate. The density of electrons is described by Fermi–Dirac statistics as electrons are fermions with two spin states. Due to the Pauli exclusion principle that states that 'two identical fermions cannot occupy the same quantum state', fermions will be forced to higher momentum states when the fermion density increases above the number of quantum states. The maximum number density of electrons $n_{e,\max}$ with momentum p allowed by quantum mechanics is:

$$n_{e,\max}(p) = \frac{8\pi p^2}{h^3}, \tag{9}$$

with h the Planck constant. Hence, in very dense regions electrons have high momenta. The velocities of these electrons exerts a higher pressure than inferred from classical physics. This is called degeneracy pressure which is nearly independent of temperature.

The transition between the classical ideal gas situation and a state of strong degeneracy occurs smoothly, and is known as partial degeneracy. Partial degeneracy has to be taken into account when

$$n_e \gtrsim \frac{2(2\,\pi\, m_e\, k\, T)^{3/2}}{h^3}, \tag{10}$$

with m_e the electron mass and k the Boltzmann constant. The limit of strong (almost complete) degeneracy is reached when the electron density n_e is roughly a factor 10 higher than the right-hand side of Eq. (10).

2.1.3 Mirror Principle

At shell-burning regions, such as the hydrogen-shell burning region in red-giant stars, it is commonly seen that the region enclosed by the burning shell contracts, while at the same time the region outside the shell expands and vice versa. This is referred to as the mirror principle. This is not a physical law as such, but an empirical observation, supported by the results of numerical simulations.

The core contraction and envelope expansion of a star ascending the red-giant branch (i.e., evolve from D to G along the evolutionary track in Fig. 1) as well as the core expansion and envelope contraction while descending the red-giant branch (i.e., evolve from G to H along the evolutionary track in Fig. 1) are examples of the mirror principle.

2.2 Different Stages of Evolution of Red-Giant Stars

Evolutionary tracks of both a 1 and a 3 $M_\odot$ stellar model with solar composition are shown in Fig. 1. Red-giant stars burn hydrogen in a shell around an inert helium core during the subgiant phase (B–D in Fig. 1) and while ascending the red-giant branch (RGB; D–G in Fig. 1) along which their radius increases and their surface temperature decreases. After the onset of helium-core burning the stars reduce their

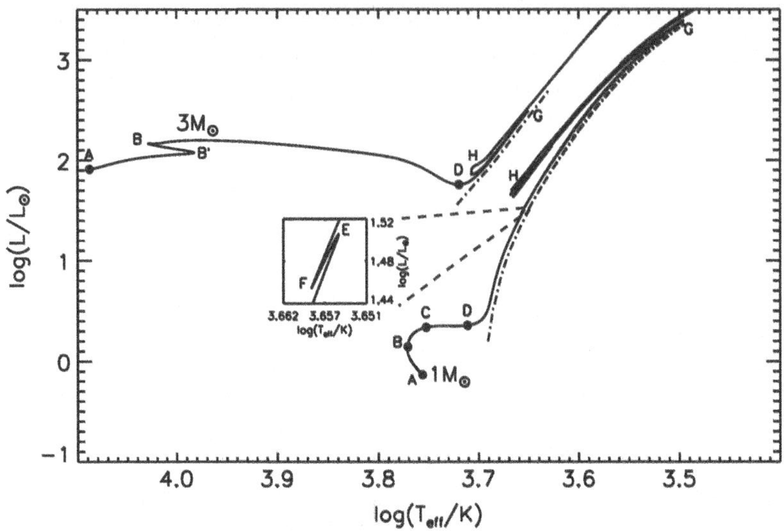

Fig. 1 Hertzsprung–Russell diagram of a 1 and a 3 $M_\odot$ evolutionary track. The *inset* shows the luminosity bump (see Sect. 2.2.4) of the 1 $M_\odot$ track. The stellar model tracks are computed using the MESA stellar evolution code (Paxton et al. 2011) with solar metallicity. The *letters* indicate different phases of evolution: A = zero-age main sequence; B′ = core hydrogen mass fraction ≈ 0.05, B = start of thick shell burning; C = maximum extent of thick shell burning (in mass); D = start of thin shell burning; E = maximum bump luminosity; F = minimum bump luminosity; G = tip of the red-giant branch; H = helium-core burning. The *dashed-dotted lines* provide a schematic indication of the location of the Hayashi lines for the two models. The phases B–D, D–G and G–H are referred to as subgiant, ascending red-giant and descending red-giant branches, respectively

size and increase their surface temperature (G–H in Fig. 1) while they reside in the so-called red-clump or secondary-clump phase where helium burning takes place in the core surrounded by a hydrogen-burning shell (CHeB stars; H in Fig. 1). Here, I discuss different stages of the evolution of red-giant stars indicated with different letters in Fig. 1 occurring in both low- and intermediate-mass stars.

Low- vs. Intermediate-Mass Stars

The distinction whether a star is a low-mass star or an intermediate-mass star is based on the onset of helium-core burning. For low-mass stars the inert helium core is degenerate on the RGB and helium ignition takes place under degenerate conditions (see Sect. 2.1.2 for more details about electron degeneracy). This occurs in stars with total masses between $\sim$0.48 and $\sim$2 $M_\odot$. These limits are defined by the lower limit of the critical mass needed to ignite helium and by the chemical composition of the star. The mass of the degenerate helium core is the same at ignition for these stars irrespective of the total mass (see Fig. 2).

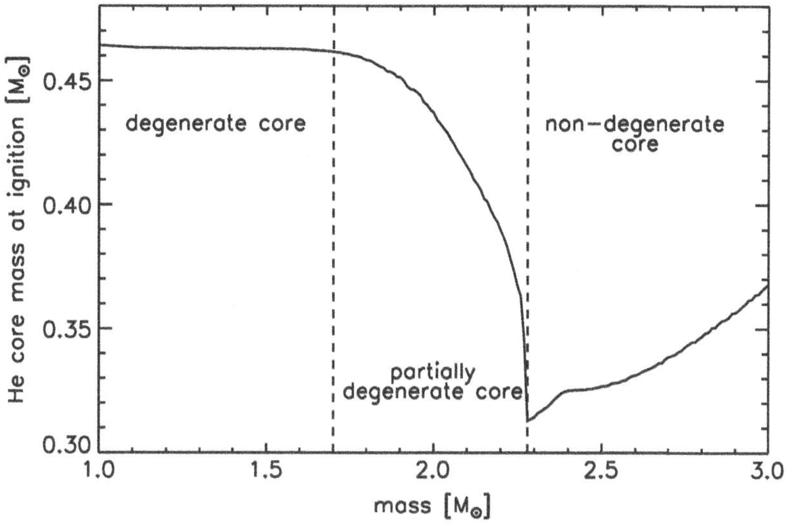

Fig. 2 Helium-core mass at ignition vs. stellar mass for stellar models of solar metallicity computed with the MESA stellar evolution code (Paxton et al. 2011). The *vertical dashed lines* are to guide the eye to the transitions in stellar mass of degenerate cores (*left*) to partially degenerate cores (*centre*) and non-degenerate cores (*right*). See Sect. 2.1.2 for a description of electron degeneracy

Intermediate-mass stars do not develop a degenerate core and have a more gentle onset of helium burning. Hence, for these stars the helium-core mass at ignition is a function of the total mass of the star. Figure 2 shows the helium-core mass at ignition as a function of total stellar mass.

2.2.1 Hook

In main-sequence stars (A–B in Fig. 1) above $\sim$1.1 $M_\odot$ the conditions in the core are such that a convective core is established. When hydrogen is nearly depleted ($X_{core} \approx 0.05$; B′ in Fig. 1) the star contracts to maintain the energy production. This contraction leads to an increase in effective temperature and luminosity until hydrogen is completely depleted in the centre. This results in a so-called 'hook' in the Hertzsprung–Russell diagram at the end of the main sequence indicated with B′–B in the 3 $M_\odot$ track in Fig. 1. Note that for stars with masses below $\sim$1.1 $M_\odot$ no convective core develops on the main sequence and no 'hook' is visible in the Hertzsprung–Russell diagram.

2.2.2 Bottom of Red-Giant Branch

During the subgiant phase (B–D in Fig. 1) the stellar envelope expands and cools while a star evolves from the main sequence towards the Hayashi line. The Hayashi line is the locus in the Hertzsprung–Russell diagram of fully convective stars, where a star cannot decrease its temperature further while maintaining hydrostatic equilibrium. The exact location of the Hayashi line is mass and metallicity dependent. The schematic locations of the Hayashi lines for a 1 and a 3 $M_\odot$ models with solar metallicity are indicated by the dashed-dotted lines in Fig. 1. When approaching the Hayashi line a further increase of the radius causes an increase in luminosity (bottom of red-giant branch; D in Fig. 1), such that the star stays on the hot side of the Hayashi line with an extended convective envelope (see Sect. 2.1.1 for a description of convection).

2.2.3 First Dredge-Up

While the outer layers of the star expand after the exhaustion of hydrogen in the core, the convective envelope penetrates deep into the star to regions where the chemical composition has been altered by nuclear processes earlier in its evolution. The convection transports the chemical elements of these deep layers to the surface. This so-called first dredge-up changes the surface chemical abundances, for instance the $^{12}C/^{13}C$ ratios are lowered. The use of 'first' refers to the fact that more dredge-up episodes take place at later stages of stellar evolution.

The first dredge-up occurs at the end of the subgiant phase and in the early red-giant phase, i.e., it starts on the 1 $M_\odot$ evolutionary track in Fig. 1 between C and D and ends before E. For an intermediate-mass star the evolution between B and D is very fast (Hertzsprung gap) and the first dredge-up takes place in the early phase of the RGB (D–G in Fig. 1). The end of the first dredge-up is when the convective region reaches a maximum depth in mass. Due to the advance of the hydrogen-burning shell the convection recedes, leaving behind a chemical (mean molecular weight) discontinuity. The mean molecular weight in the mixed region is due to the combination of the pristine stellar abundance and the products of partial hydrogen burning. Therefore the mean molecular weight in this region is lower than in the synthesised interior.

2.2.4 Luminosity Bump

Along the red-giant branch (D–G in Fig. 1) there is a clear zig-zag in the evolution path (E–F in Fig. 1): this is the so-called RGB luminosity bump. This temporal shift in luminosity and temperature happens when the hydrogen shell burning reaches the chemical discontinuity left behind by the first dredge-up at the deepest extent of the convective envelope. Naively, the decrease in the mean molecular weight at the chemical discontinuity causes the luminosity L to decrease following $L \propto \mu^7 M_{\rm core}^7$

(Refsdal and Weigert 1970), where M_{core} is the core mass. However, Christensen-Dalsgaard (2015) showed that the situation is more complex with the luminosity beginning to decrease prior to the shell burning through the discontinuity. This is most likely caused by the fact that the decrease in μ outside the discontinuity starts affecting the hydrostatic structure just below the discontinuity, and hence the temperature, within and above the hydrogen-burning shell before it reaches the discontinuity. The RGB luminosity bump is visible in stars up to a mass of about $2.2\,M_{\odot}$ as for more massive stars the hydrogen burning shell does not reach the chemical composition discontinuity before the onset of helium-core burning.

2.2.5 Onset of Helium-Core Burning

For low-mass stars with a degenerate core the onset of helium burning happens in a very short episode: the so-called He-flash (G in the $1\,M_{\odot}$ track in Fig. 1). In the highly degenerate core the pressure does not depend on the temperature and therefore there is no thermostatic control (see Sect. 2.1.2) to expand and cool the core. Therefore, at a temperature of about 10^8 K the onset of helium burning in degenerate conditions results in a thermal runaway process creating an enormous overproduction of nuclear energy during a very short time of order a few hours. This energy does not reach the stellar surface due to its absorption in non-degenerate layers. The onset of helium burning takes place at the location of maximum temperature that is generally not in the centre but in a concentric shell around the centre of the degenerate core. The temperature maximum is off centre because of extremely efficient neutrino losses in the core of electron degenerate material. Due to the off-centre ignition, it is predicted that the first main He-flash is followed by a series of subflashes till the degeneracy of the core is completely lifted and is back in equilibrium with helium burning in a convective core. Due to the degenerate state of the core prior to the He-flash the core mass at ignition does not depend on the total stellar mass (see Fig. 2).

For intermediate-mass stars with non-degenerate cores the pressure and temperature in the core are related. This thermostatic feedback allows for gentle ignition of helium in the core of these stars. In this case the luminosity at which helium ignites is a function of the stellar mass (see Fig. 2).

2.2.6 Helium-Core Burning

After the onset of helium burning the luminosity decreases and the surface temperature increases, i.e., the star descends the red-giant branch (G–H in Fig. 1). The drop in luminosity is due to the lower energy generation in the hydrogen shell burning layer due to its decreased density and temperature, while the increased surface temperature is caused by the mirror principle increasing the core radius and decreasing the total stellar radius. The descend of the red-giant branch is rapid, while returning to central fusion results in a quiescent long-lived phase (the helium-core-burning phase; H in Fig. 1). In this phase a star has two energy sources: helium burning in the core (producing carbon and oxygen) and hydrogen burning (producing helium) in a shell around the core.

Low-mass stars of different total masses have very similar helium-core masses upon ignition of helium in the core. Therefore, these stars occupy a small region in a Hertzsprung–Russell diagram and stay at this position for a relatively long time: the red clump, i.e., at this position in a Hertzsprung–Russell diagram stars 'clump' together when a whole population is observed. Within the red clump the luminosity is very tightly constraint by the helium-core mass, while there exists some temperature dependence on the total stellar mass and composition.

For intermediate-mass stars the helium-core mass at ignition is a function of stellar mass and therefore these stars do not reside in the red clump. Instead they form a secondary clump at lower luminosities and effective temperatures (Girardi 1999).

3 Asteroseismology of Red-Giant Stars

Asteroseismology is the study of the internal structure of stars through their intrinsic oscillations. In the outer layers of red-giant stars oscillations can be stochastically excited by the turbulent convection (e.g., Goldreich and Keeley 1977; Goldreich and Kumar 1988). Effectively, some of the convective energy is transferred into energy of global oscillations. These type of oscillations are referred to as solar-like oscillations as they also occur in the Sun. For an extensive overview of solar-like oscillators I refer the reader to the review by Chaplin and Miglio (2013). In this lecture I focus on asteroseismic inferences of red giants that are of importance for galactic archaeology. For more details on asteroseismic inferences and the underlying theory I refer the reader to Aerts et al. (2010) and the review by Hekker and Christensen-Dalsgaard (2017).

In stochastic oscillators essentially all modes are excited albeit with different amplitudes. This results in a clear oscillation power excess in the Fourier spectrum (see Fig. 3). Individual modes of oscillation in this power excess are described in terms of spherical harmonics. Hence the oscillations are described by their frequency ν and three quantum numbers: the radial order n indicating the number of nodes in the radial direction, the spherical degree l indicating the number of nodal lines on the surface, and the azimuthal order m indicating the number of nodal lines that pass through the rotation axis.

3.1 Oscillation Pattern in Fourier Space

From the Fourier power spectrum in Fig. 3 it is clear that oscillations reach observable amplitudes in a limited range in frequency with amplitudes that roughly

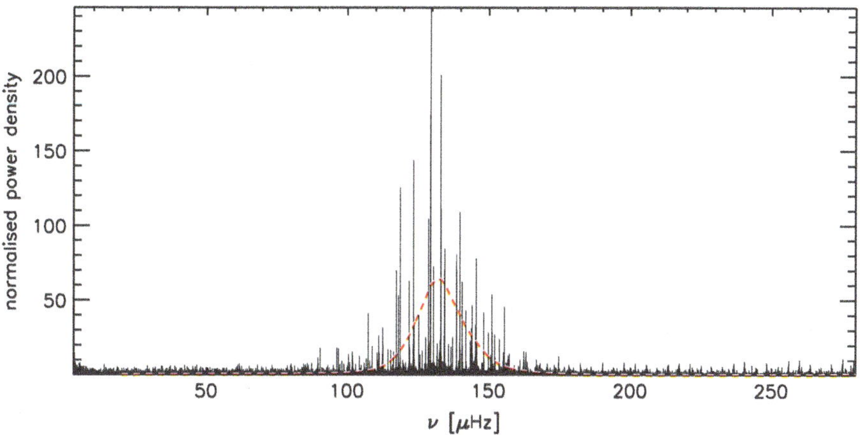

Fig. 3 Fourier spectrum normalised by the background signal of the red giant KIC 9145955. The *red dashed curve* is a heavily (triangular) smoothed power spectrum showing the power excess envelope of the oscillations. The amplitude of the smoothed power spectrum is enhanced for visual purposes

follow a Gaussian shape. The centre of the oscillation power excess is referred to as ν_{max}, which is empirically found to be tightly related to the acoustic cut-off frequency, ν_{ac} (Lamb 1932, using the approximation for an isothermal atmosphere),

$$\nu_{ac} = \frac{c}{4\,\pi\,H_P}\,, \tag{11}$$

with $\nu_{max} \approx 0.6\nu_{ac}$. The acoustic cut-off frequency is the maximum frequency of an acoustic eigenmode. At higher frequencies the waves are no longer trapped but travelling waves. It can be shown that ν_{max} provides a direct measure of the surface gravity (g) when the effective temperature (T_{eff}) is known (e.g., Brown et al. 1991; Kjeldsen and Bedding 1995):

$$\nu_{max} \propto \frac{g}{\sqrt{T_{eff}}} \propto \frac{M}{R^2\sqrt{T_{eff}}}\,, \tag{12}$$

where M and R indicate the stellar mass and radius, respectively. Belkacem et al. (2011) investigated the theoretical basis for this relation but a full explanation has not yet been found.

3.1.1 Acoustic Modes

Following asymptotic theory (Tassoul 1980), high-order acoustic oscillation modes of solar-like oscillators with pressure as restoring force follow a near-regular pattern

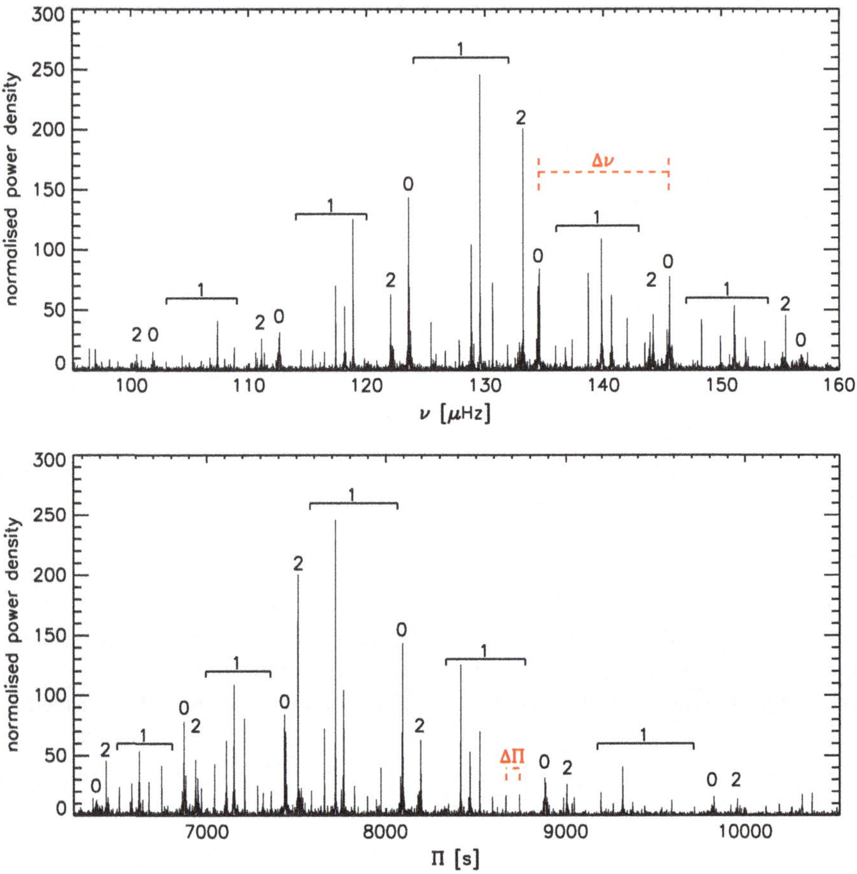

Fig. 4 *Top:* Fourier spectrum normalised by the background signal in the frequency range in which oscillations of the red giant KIC 9145955 are visible. The numbers indicate the degree l of the modes. The large frequency between two consecutive radial ($l = 0$) modes is indicated in *red*. *Bottom:* Same as top panel but now with period Π in seconds on the x-axis and the observed period spacing $\Delta\Pi$ between two mixed dipole ($l = 1$) modes indicated in *red*

in frequency:

$$\nu_{nl} \simeq \Delta\nu\left(n + \frac{l}{2} + \epsilon\right) - d_{nl}, \tag{13}$$

with $\Delta\nu$ the large frequency separation between modes of the same degree and consecutive radial order (see top panel of Fig. 4), ϵ an offset and d_{nl} a small correction to the leading order asymptotics, which is zero for $l = 0$.

$\Delta\nu$ is proportional to the inverse of the acoustic diameter, i.e., the sound travel time across a stellar diameter. Therefore, it can be shown that $\Delta\nu$ is a direct probe of the mean density ($\overline{\rho}$) of the star (Ulrich 1986):

$$\Delta\nu = \nu_{n\,l} - \nu_{n-1\,l} = \left(2\int_0^R \frac{dr}{c}\right)^{-1} \propto \sqrt{\overline{\rho}} \propto \sqrt{\frac{M}{R^3}}, \tag{14}$$

with r the distance to the centre of the star.

3.1.2 Gravity Modes

Following asymptotic theory (Tassoul 1980), high-order gravity modes with buoyancy as restoring force follow a near-regular pattern in period $\Pi_{n\,l}$,

$$\Pi_{n\,l} \simeq \Delta\Pi_l\,(n + \epsilon_g + 1/2)\,, \tag{15}$$

with ϵ_g a phase term and $\Delta\Pi_l$ the period spacing (see bottom panel of Fig. 4) defined as

$$\Delta\Pi_l = \frac{2\,\pi^2}{\sqrt{l(l+1)}}\left(\int_{r_1}^{r_2} N\frac{dr}{r}\right)^{-1}, \tag{16}$$

with r_1 and r_2 the turning points of the gravity mode cavity and N the Brunt–Väisälä frequency:

$$N^2 = g\left(\frac{1}{\Gamma_1}\frac{d\ln P}{dr} - \frac{d\ln\rho}{dr}\right), \tag{17}$$

where g is the local gravitational acceleration, P is pressure and $\Gamma_1 = (\partial\ln P/\partial\ln\rho)_{\rm ad}$. Note that N^2 is negative, and hence N imaginary, in convectively unstable regions.

In red-giant stars pure gravity modes cannot be observed due to the extended convective envelope. However, in red giants the frequencies of the gravity modes in the core and the frequencies of the acoustic modes in the envelope have similar values such that resonant interactions between the modes allow for mixed acoustic-gravity nature to occur. Or put differently: in red giants essentially all non-radial ($l > 0$) modes are mixed with different behaviour in different regions. The mixed mode frequencies are shifted by an amount depending on the coupling strength between the gravity and acoustic cavity. Due to these shifts the directly observed period spacing between consecutive mixed modes is smaller than the asymptotic value (Eq. (16)). The asymptotic $\Delta\Pi$ can be inferred from the observed period spacing (e.g., Mosser et al. 2015).

In practice mixed modes are mostly observed in dipole ($l = 1$) modes as for these modes the coupling between the pressure and gravity mode cavity is stronger and also because the period spacings are larger due to the dependence on $\sqrt{l(l+1)}$ (see Eq. (16)), and thus better resolved.

3.2 Evolutionary State

Red-giant stars on the red-giant branch and in the helium-core-burning phase can occupy the same region in the Hertzsprung–Russell diagram. Their surface properties can be very similar and hence it is difficult to distinguish between them based on classical observations such as surface temperature and brightness. However, mixed oscillation modes (Sect. 3.1.2) provide a means to probe the stellar cores where the difference between RGB and CHeB stars are significant. Stars on the RGB do not have convective cores while stars in the CHeB phase do have convective cores. Hence, the real part of the Brunt–Väisälä frequency (Eq. (17)) is zero in the convective core of CHeB stars and has a finite value in the core of RGB stars (see Fig. 5). According to Eq. (16) this results in a significantly larger value of $\Delta\Pi$ for CHeB stars compared to RGB stars (e.g., Bedding et al. 2011; Christensen-Dalsgaard 2014; Mosser et al. 2014, and references therein). Thus, asteroseismology provides a direct measure of the evolutionary phase of red-giant stars.

Alternatively it has been shown by Kallinger et al. (2012) and Christensen-Dalsgaard et al. (2014) that the differences in the core between RGB and CHeB stars also cause differences in the thermodynamic state of the envelope. This results in a different location of the second helium-ionisation zone for RGB and CHeB stars. This causes a difference in the 'local' phase term (ϵ in Eq. (13) when measured locally in a $2\Delta\nu$ interval around $\nu_{\rm max}$) for stars in different evolutionary phases.

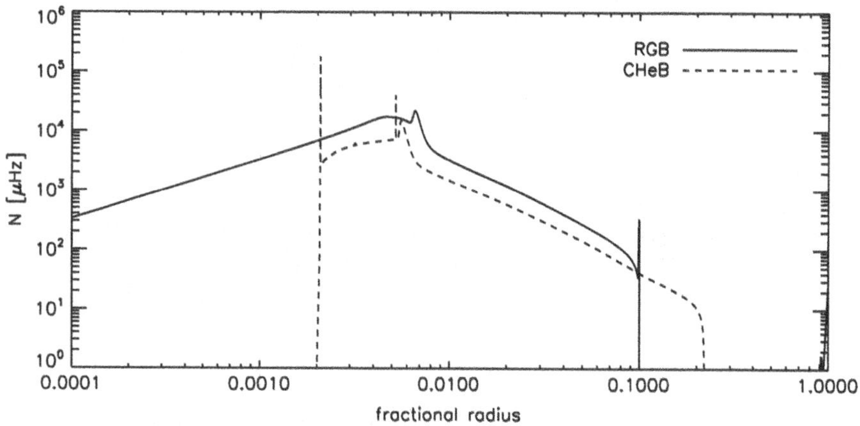

Fig. 5 Real part of the Brunt–Väisälä frequency N of a 1 $M_\odot$ RGB model (*solid line*; model 1 of Datta et al. 2015) and of a 1 $M_\odot$ CHeB model (*dashed line*; black model in Fig. 10 of Constantino et al. 2015)

3.3 Scaling Relations

Equations (12) and (14) are often referred to as asteroseismic scaling relations:

$$\nu_{\rm max} = \nu_{\rm max,ref} \frac{M}{R^2 \sqrt{T_{\rm eff}/T_{\rm eff,\,ref}}} \tag{18}$$

and

$$\Delta\nu = \Delta\nu_{\rm ref} \sqrt{\frac{M}{R^3}}, \tag{19}$$

with M and R expressed in solar values and 'ref' referring to reference values. These scaling relations can be used to derive the stellar mass and radius. To do so a scaling with solar values as references has commonly been applied. This implicitly assumes that the Sun and the observed star have the same internal structure and only vary in size. From knowledge of stellar evolution it is known that this is not the case. Indeed, the high-quality observations of the *CoRoT* (Baglin et al. 2009) and *Kepler* (Borucki et al. 2010) space telescopes as well as detailed analysis of models have shown that for Eq. (19) this assumption is not entirely correct for stars with different properties, such as a different metallicity, or stars in different evolution phases. White et al. (2011), Miglio et al. (2012), Hekker et al. (2013), Mosser et al. (2013), Guggenberger et al. (2016) and Sharma et al. (2016) have investigated the reference for the $\Delta\nu$ scaling relation (Eq. (19)) in detail and generally propose either a correction to solar reference values (White et al. 2011) or new reference values (Hekker et al. 2013; Mosser et al. 2013; Guggenberger et al. 2016; Sharma et al. 2016). The correction by Mosser et al. (2013) is based on the expanded asymptotic relation (Eq. (13)) known as the universal pattern (Mosser et al. 2011). Sharma et al. (2016) performed an interpolation in a model grid to find a reference value for each model from individual frequencies, while Guggenberger et al. (2016) used individual frequencies of models to derive a temperature- and metallicity-dependent reference function. Although there is consensus in the community that for red-giant stars scaling to reference values that take the evolutionary phase, mass, metallicity and effective temperature into account is preferred over scaling to solar values, the exact reference value or function to be used is still a matter of debate.

Note that for Eq. (12) a solar reference value is still commonly used as theoretical understanding of this relation is still insufficient to perform rigorous tests. Results for two red-giant binary stars show, however, that $\log(g)$ derived from the binary orbit and Eq. (12) are consistent (Themeßl et al. submitted), while stellar masses and radii derived using the orbits and asteroseismic scaling relations show discrepancies (e.g., Gaulme et al. 2016, Themeßl, Hekker, Southworth et al. submitted to MNRAS).

3.4 Comparison with Stellar Models

The scaling relations themselves do not account for any knowledge we have about stellar evolution. To include this knowledge, it is also possible to compare observables of individual stars, e.g., $\{\Delta\nu, \nu_{\max}, T_{\text{eff}}, [\text{Fe/H}]\}$ or individual frequencies with models. This can be done in a grid-based modelling approach in which the observables are compared with a grid of stellar evolution models for which $\Delta\nu$ and $\nu_{\max}$ are computed using scaling relations (or alternatively, in case of $\Delta\nu$, using individual frequencies). In this approach one does account for knowledge of stellar structure and evolution, as well as metallicity (e.g., Gai et al. 2011). An additional advantage is that it is possible in this way to obtain stellar ages in addition to stellar mass and radius, which is of importance for galactic archaeology. Note that the results of grid-based modelling for red giants are generally more accurate when the evolutionary state, i.e., RGB or CHeB (see Sect. 3.2), is known a priori.

In addition to grid-based modelling one can perform optimisation on a star-by-star basis (e.g., Creevey et al. 2017), although that is currently not feasible for large numbers of stars as required for galactic studies. Alternatively, one can use machine learning to obtain stellar parameters (including ages) in a fast and robust way (Bellinger et al. 2016; Angelou et al. 2017).

4 Galactic Archaeology

Galactic archaeology is the study of the formation and evolution of the Milky Way galaxy by reconstructing its past from its current constituents. To this end it is important to know and understand the properties of the current constituents of the Milky Way, in terms of position, kinematics, chemical composition and age. These quantities contain information on how and where the stars were formed and hence provide the possibility to reconstruct the past. As red-giant stars are common, intrinsically bright and show oscillations, these are prime targets to study the Milky Way.

I will first provide a brief overview of the main components of the Milky Way galaxy and their formation scenarios followed by a discussion of some recent insights on the Milky Way obtained from red-giant stars. For more details on the Galaxy structure and its formation I refer the reader to reviews by Freeman and Bland-Hawthorn (2002), Rix and Bovy (2013) and Bland-Hawthorn and Gerhard (2016) on which parts of this lecture are based.

4.1 Milky Way Galaxy

In this section I give brief descriptions of the main components of the Milky Way galaxy, i.e., the disc, the bulge, the halo and the dark matter halo. Star formation in the Milky Way has likely proceeded in phases, with limited overlap. Yet, the connection between the different components of the galaxy is far from understood (Allende Prieto 2010). However, N-body simulations such as the ones by, e.g., Athanassoula (2016) are very promising and are currently able to simulate galaxies that have quantitatively the same components and shape as observed for the Milky Way.

4.1.1 Disc

The disc is a flat rotating radially extended part of the galaxy containing spiral arms. Commonly the disc is thought to consist of a thin and thick component. Different definitions can be used to separate these components: by looking only at the dynamical properties of stars or only at the chemical properties of stars. The properties of the resulting populations are however not identical. Nevertheless, in general the thick disc is more extended and contains old (>10 Gyr), more metal-poor stars ([Fe/H] between −0.5 and −1 dex) with larger enhancement of α-elements compared to the thin disc (a chemically defined thick disc is therefore also commonly refer to as α-rich disc). The α-elements are elements for which the most abundant isotopes are integer multiples of four (the mass of the helium nucleus or α-particle): stable α-elements are carbon, oxygen, neon, magnesium, silicon, sulfur, argon and calcium. The enhancement in $[\alpha/\mathrm{Fe}]$ is usually interpreted as evidence that the thick disc formed predominantly at times that were dominated by type II supernovae (SNe), which have short-lived progenitors ($\sim$ few 10 Myr) and are the main source of α-elements. Additionally, thick disc stars show slow Galactic rotation and lag behind the thin disc rotation by roughly 50 km s^{-1} with a dependence on the distance from the plane (Allende Prieto 2010).

The thin disc stars are generally considered to be significantly younger than the thick disc stars and do not show enhancement in $[\alpha/\mathrm{Fe}]$. The thin disc is less extended than the thick disc and shows a dependence on age with younger stars more confined to the mid-plane. Additionally, the velocity dispersions (random motions in three dimensions) increase with age. This is often referred to as 'disk-heating'. This heating can occur due to gravitational scattering by objects (giant molecular clouds) or spiral density waves, or by collisions of satellite galaxies (Merrifield et al. 2001).

Several authors (e.g., Bovy et al. 2012; Kawata and Chiappini 2016) have argued that there is actually no thin disc/thick disc dichotomy and that the transition between thin and thick disc is rather a continuum of discs with (an abundance-dependent) scale length.

A relation between the stellar age and the mean metallicity or between stellar age and velocity dispersion, i.e., age-metallicity relation (AMR) or age-velocity relation

(AVR), would be fundamental observational input that could constrain the chemical and dynamical evolution of the galactic disc. However, there is much discussion about the existence of such relations (e.g., Freeman 2012; Bergemann et al. 2014; Kumamoto et al. 2017) and it may be that such relations can only be measured for young stars in the solar neighbourhood.

4.1.2 Bulge

The bulge is a densely populated part in the centre of the Milky Way with a boxy/peanut morphology and an X-shaped structure (Ness and Lang 2016, and Fig. 6), with its appearance depending on the angle it is viewed from and the stellar population that is looked at. A bulge can form naturally from the dynamics of a flat rotating disk of stars. If this is indeed happening the bulge formation takes 2–3 Gyr to act after the disc has been formed. Hence in this scenario the bulge structure is younger than the bulge stars that were originally part of the inner disc (Freeman 2012).

Another scenario could be that the bulge is formed out of mergers early in the formation of the galaxy. This is called a classical bulge and it is currently unclear whether the Milky Way contains such a bulge (Bland-Hawthorn and Gerhard 2016).

Fig. 6 *WISE* image of the Milky Way in W1 (3.4 μm) and W2 (4.6 μm) colour bands. An arcsinh stretch is used to allow the full dynamic range to be shown. Additionally, the median of each row of the image is subtracted to provide a better contrast which reveals the X-shaped structure in better detail. Figure taken from Ness and Lang (2016)

4.1.3 Halo

The halo is a sparsely populated extended (up to $\sim$100 kpc) part of the galaxy containing old metal-poor stars. The halo stars have typically [Fe/H] < -1 dex and ages greater than 12 Gyr. The stellar density distribution of the halo follows a power law: $\rho \propto r^{-3.5}$ (Freeman and Bland-Hawthorn 2002). The angular momentum of the halo is close to zero, while the velocity dispersion of the stars in three dimensions is large.

The galactic halo is formed at least partly through the accretion of small satellite galaxies that each carry their own signatures. Although accretion of dwarf galaxies is still taking place the most active accretion phase has probably ended about 12 Gyr ago before the disc formation. Some (dynamical) traces of past accretion events are still present in the structure of the halo (e.g., Helmi et al. 1999; Grillmair and Carlin 2016).

Globular clusters are collections of hundreds to thousands to millions of gravitationally bound stars that are an integral part of the stellar halo and orbit the galactic centre, both in the Milky Way and in other galaxies in the Local Group—the group of galaxies to which the Milky Way belongs. Globular clusters may contain some of the first stars produced in the galaxy, although the origin and role of globular clusters in galactic evolution are still unclear.

4.1.4 Dark Matter Halo

The dark matter halo is only detected by its gravitational field and its existence is inferred through the effects on the motions of the stars and the gas in the galaxy. The mass of the dark matter halo is generally larger than the total mass of the visible components of a galaxy and extends well beyond the edge of the visible galaxy with a density that decreases farther from the galactic centre. Furthermore, dark matter does not seem to interact with other matter present in the galaxy. I note here that Erik Verlinde proposed a theory that shows that dark energy can explain the emerging gravity and that dark matter does not exist as a particle, but that dark matter is a phenomenon that emerges from dark energy (Verlinde 2016).

4.2 Chemodynamic Model of the Milky Way

State of the art models of the Milky Way are computed in the cosmological context and take the kinematics, chemistry and ages of stars into account in a self-consistent way (e.g., Minchev et al. 2013, 2014). From these models it emerges naturally that stars can migrate in the radial direction of the disc. This causes difficulties for galactic archaeology in that stars can migrate significantly away from their place of birth (e.g., Loebman et al. 2016, and references therein). This migration has to be taken into account in retracing the history of the stars in the Milky Way.

4.3 Galactic Archaeology and Asteroseismology

A key ingredient for galactic archaeology is stellar age. Stellar ages are difficult to derive as there is no observable that is sensitive to age and age only (Soderblom 2010). The classical way to derive stellar ages is by isochrone fitting using effective temperature ($T_{\rm eff}$), logarithmic surface gravity ($\log(g)$) and metallicity ([Fe/H]). This technique is very powerful for stars that belong to a cluster, but is limited in use for single field stars, due to degeneracies.

To determine stellar ages, spectroscopic information (mostly $T_{\rm eff}$ and [Fe/H]) can be combined with asteroseismic information ($\Delta\nu$ and $\nu_{\rm max}$; see Sect. 3.1) and stellar models in a grid-based modelling approach (see Sect. 3.4). The asteroseismic input (including the evolutionary phase information for red giants; see Sect. 3.2) provides further constraints on the models and therefore more precise age estimates. Further improvements could be made when individual oscillation frequencies are considered, but this is currently not feasible for a large number of stars as would be required for studies of the Milky Way.

The number of stars for which spectroscopic data and future parallaxes from *Gaia* (Perryman et al. 2001; Lindegren et al. 2016) are and will be available is much larger than the number of stars with asteroseismic parameters as the latter require time-series data. Ideally, the asteroseismic subsample can serve as a calibration set to determine ages of the larger set of stars. Indeed this seems possible as discussed in Sect. 4.3.1.

4.3.1 Asteroseismic Calibration for Spectroscopic Samples

In case there are features in stellar spectra that change as a function of age it is possible to transfer knowledge of stellar spectra of stars that have asteroseismic ages (or other information known accurately) to spectra that do not have asteroseismic information. Such a data-driven transfer is made by 'The Cannon' (Ness et al. 2015) and allows to improve the scientific information obtained from spectra even if we do not understand the underlying reason of the correlation.

An example of age-dependent spectroscopic features are the carbon (C) and nitrogen (N) abundances. The first dredge-up (Sect. 2.2.3) causes a change in surface abundance as the stellar surface becomes mixed with material enriched in nitrogen and depleted in carbon, which causes a change in the ratio [C/N]. The value of the [C/N] abundance ratio depends on the CNO-processed material in the core at the end of the main sequence, and on the depth reached by the base of the convection zone, both depending on the stellar mass: higher-mass stars are comparatively richer in N and poorer in C with respect to lower-mass stars. Additional mixing processes can subsequently act to change the [C/N] ratio at the upper red-giant branch.

Martig et al. (2016) and Ness et al. (2016) have used the chemical abundances and ratio as mentioned above together with asteroseismic masses and ages for a subset of stars to obtain an empirical link between mass and age on the one hand and C, N and [C/N] and [Fe/H] ratios on the other hand. Such an approach can potentially be used to calibrate relations that can be used on stars for which spectroscopic, but no asteroseismic data are available.

4.3.2 α-Rich Young Stars

Here, I provide an example where stellar ages derived using asteroseismology played an important role in studies of populations of stars in the Milky Way.

As described in Sect. 4.1.1 generally stars rich in α-elements are formed at early times when the Milky way was dominated by type II SNe. Hence, α-rich stars are expected to be old. Chiappini et al. (2015) and Martig et al. (2015) did however detect a subsample of stars that are rich in α-elements and at the same time are young based on asteroseismic age determinations.

First studies by Jofré et al. (2016) and Yong et al. (2016) show that these stars could be evolved blue stragglers, suggesting that the apparent young age is a consequence of a merger or mass transfer.

5 Summary

In this lecture I have focussed on red-giant stars and the role they can play in studying the Milky Way galaxy. These common, intrinsically bright oscillating stars allow us to probe the Milky Way up to relatively large distances, and provide in specific fields improved precision on stellar parameters (e.g., ages) via asteroseismology. The lack of accurate stellar age determinations for large numbers of stars is currently a limiting factor in galactic archaeology.

As it is technically feasible to take a spectrum for more stars than it is to take a time series as required for asteroseismology, techniques have been and are being developed to use the more accurate and precise asteroseismic information and calibrate larger sets with these. One example of this is ‘The Cannon’ (Ness et al. 2015).

Additionally, the K2 mission (Howell et al. 2014) is currently still taking time-series data for thousands of stars for the purpose of galactic archaeology (Stello et al. 2017). In the near future the *TESS* (Ricker et al. 2014) and *PLATO* (Rauer et al. 2014) missions will also be launched. Although the prime aim of these missions is exo-planets and asteroseismology, the all-sky view of *TESS* and the step-and-stare mode of *PLATO* provide excellent prospects for galactic archaeology. These data combined with parallaxes from *Gaia* (Perryman et al. 2001; Lindegren et al. 2016) and data from ground-based spectrographs will allow to improve our current picture of the Milky Way and from that provide clues about the formation and evolution of the Milky Way.

Acknowledgements I would like to thank Alexey Mints, George Angelou and Maarten Mooij for useful discussions and comments on earlier versions of the manuscript. Furthermore, I acknowledge Earl Bellinger, Abishek Datta and Thomas Constantino for providing stellar evolution models used for figures. I acknowledge funding from the European Research Council under the European Community's Seventh Framework Programme (FP7/2007–2013)/ERC grant agreement no 338251 (StellarAges).

References

Aerts, C., Christensen-Dalsgaard, J., Kurtz, D.W.: Asteroseismology. Springer, Dordrecht (2010)
Allende Prieto, C.: In: Cunha, K., Spite, M., Barbuy, B. (eds.) Chemical Abundances in the Universe: Connecting First Stars to Planets. IAU Symposium, vol. 265, pp. 304–312 (2010)
Athanassoula, E.: Galactic Bulges, vol. 418, p. 391. Springer, Cham (2016)
Angelou, G.C., Bellinger, E.P., Hekker, S., Basu, S.: Astrophys. J. **839**, 116 (2017)
Baglin, A., Auvergne, M., Barge, P., et al.: In: Pont, F., Sasselov, D., Holman, M.J. (eds.) Transiting Planets. IAU Symposium, vol. 253, pp. 71–81 (2009)
Bedding, T.R., Mosser, B., Huber, D., et al.: Nature **471**, 608 (2011)
Belkacem, K., Goupil, M.J., Dupret, M.A., et al.: Astron. Astrophys. **530**, A142 (2011)
Bellinger, E.P., Angelou, G.C., Hekker, S., et al.: Astrophys. J., **830**, 31 (2016)
Bergemann, M., Ruchti, G.R., Serenelli, A., et al.: Astron. Astrophys. **565**, A89 (2014)
Bland-Hawthorn, J., Gerhard, O.: Annu. Rev. Astron. Astrophys. **54**, 529 (2016)
Böhm-Vitense, E.: Zentrum Arbeit Politik **46**, 108 (1958)
Borucki, W.J., Koch, D., Basri, G., et al.: Science **327**, 977 (2010)
Bovy, J., Rix, H.-W., Hogg, D.W.: Astrophys. J. **751**, 131 (2012)
Brown, T.M., Gilliland, R.L., Noyes, R.W., Ramsey, L.W.: Astrophys. J. **368**, 599 (1991)
Canuto, V.M., Goldman, I., Mazzitelli, I.: Astrophys. J. **473**, 550 (1996)
Chaplin, W.J., Miglio, A.: Annu. Rev. Astron. Astrophys. **51**, 353 (2013)
Chiappini, C., Anders, F., Rodrigues, T.S., et al.: Astron. Astrophys. **576**, L12 (2015)
Christensen-Dalsgaard, J.: In: Pallé, P.L., Esteban, C. (ed.) Asteroseismology, p. 194. Cambridge University Press, Cambridge (2014)
Christensen-Dalsgaard, J.: Mon. Not. R. Astron. Soc. **453**, 666 (2015)
Christensen-Dalsgaard, J., Silva Aguirre, V., Elsworth, Y., Hekker, S.: Mon. Not. R. Astron. Soc. **445**, 3685 (2014)
Constantino, T., Campbell, S.W., Christensen-Dalsgaard, J., Lattanzio, J.C., Stello, D.: Mon. Not. R. Astron. Soc. **452**, 123 (2015)
Creevey, O., Metcalfe, T.S., Schultheis, M.: Astron. Astrophys. **601**, A67 (2017)
Datta, A., Mazumdar, A., Gupta, U., Hekker, S.: Mon. Not. R. Astron. Soc. **447**, 1935 (2015)
Freeman, K.: Astrophysics and Space Science Proceedings, vol. 26, p. 137. Springer, New York (2012)
Freeman, K., Bland-Hawthorn, J.: Annu. Rev. Astron. Astrophys. **40**, 487 (2002)
Gabriel, M., Noels, A., Montalbán, J., Miglio, A.: Astron. Astrophys. **569**, A63 (2014)
Gai, N., Basu, S., Chaplin, W.J., Elsworth, Y.: Astrophys. J. **730**, 63 (2011)
Gaulme, P., McKeever, J., Jackiewicz, J., et al.: Astrophys. J. **832**, 121 (2016)
Girardi, L.: Mon. Not. R. Astron. Soc. **308**, 818 (1999)
Goldreich, P., Keeley, D.A.: Astrophys. J. **212**, 243 (1977)
Goldreich, P., Kumar, P.: Astrophys. J. **326**, 462 (1988)
Grillmair, C.J., Carlin, J.L.: In: Newberg, H.J., Carlin, J.L. (eds.) Astrophysics and Space Science Library, vol. 420, p. 87. Springer, New York (2016)
Guggenberger, E., Hekker, S., Basu, S., Bellinger, E.: Mon. Not. R. Astron. Soc. **460**, 4277 (2016)
Hekker, S., Christensen-Dalsgaard, J.: Astron. Astrophys. Rev. **25**, 1 (2017)
Hekker, S., Elsworth, Y., Basu, S., et al.: Mon. Not. R. Astron. Soc. **434**, 1668 (2013)

Helmi, A., White, S.D.M., de Zeeuw, P.T., Zhao, H.: Nature **402**, 53 (1999)
Howell, S.B., Sobeck, C., Haas, M., et al.: Publ. Astron. Soc. Pac. **126**, 398 (2014)
Jofré, P., Jorissen, A., Van Eck, S., et al.: Astron. Astrophys. **595**, A60 (2016)
Kallinger, T., Hekker, S., Mosser, B., et al.: Astron. Astrophys. **541**, A51 (2012)
Kawata, D., Chiappini, C.: Astron. Nachr. **337**, 976 (2016)
Kippenhahn, R., Weigert, A., Weiss, A.: Stellar Structure and Evolution. Springer, Berlin (2012). doi:10.1007/978-3-642-30304-3
Kjeldsen, H., Bedding, T.R.: Astron. Astrophys. **293**, 87 (1995)
Kumamoto, J., Baba, J., Saioh, T.R.: Publ. Astron. Soc. Japan **69**, 32 (2017)
Lamb, H.: Hydrodynamics. Dover, New York (1932)
Langer, N., El Eid, M.F., Fricke, K.J.: Astron. Astrophys. **145**, 179 (1985)
Lattanzio, J.: Proc. Astron. Soc. Aust. **5**, 232 (1983)
Ledoux, P.: Astron. J. **52**, 155 (1947)
Lindegren, L., Lammers, U., Bastian, U., et al.: Astron. Astrophys. **595**, A4 (2016)
Loebman, S.R., Debattista, V.P., Nidever, D.L., et al.: Astrophys. J. **818**, L6 (2016)
Martig, M., Rix, H.-W., Silva Aguirre, V., et al.: Mon. Not. R. Astron. Soc. **451**, 2230 (2015)
Martig, M., Fouesneau, M., Rix, H.-W., et al.: Mon. Not. R. Astron. Soc. **456**, 3655 (2016)
Merrifield, M.R., Gerssen, J., Kuijken, K.: In: Funes, J.G., Corsini, E.M. (eds.) Galaxy Disks and Disk Galaxies. Astronomical Society of the Pacific Conference Series, vol. 230, pp. 221–224 (2001)
Miglio, A., Brogaard, K., Stello, D., et al.: Mon. Not. R. Astron. Soc. **419**, 2077 (2012)
Minchev, I., Chiappini, C., Martig, M.: Astron. Astrophys. **558**, A9 (2013)
Minchev, I., Chiappini, C., Martig, M.: Astron. Astrophys. **572**, A92 (2014)
Mosser, B., Belkacem, K., Goupil, M.J., et al.: Astron. Astrophys. **525**, L9 (2011)
Mosser, B., Michel, E., Belkacem, K., et al., Astron. Astrophys. **550**, A126 (2013)
Mosser, B., Benomar, O., Belkacem, K., et al.: Astron. Astrophys. **572**, L5 (2014)
Mosser, B., Vrard, M., Belkacem, K., Deheuvels, S., Goupil, M.J.: Astron. Astrophys. **584**, A50 (2015)
Ness, M., Lang, D.: Astron. J. **152**, 14 (2016)
Ness, M., Hogg, D.W., Rix, H.-W., Ho, A.Y.Q., Zasowski, G.: Astrophys. J. **808**, 16 (2015)
Ness, M., Hogg, D.W., Rix, H.-W., et al.: Astrophys. J. **823**, 114 (2016)
Paxton, B., Bildsten, L., Dotter, A., et al.: Astrophys. J. Suppl. Ser. **192**, 3 (2011)
Perryman, M.A.C., de Boer, K.S., Gilmore, G., et al.: Astron. Astrophys. **369**, 339 (2001)
Rauer, H., Catala, C., Aerts, C., et al.: Exp. Astron. **38**, 249 (2014)
Refsdal, S., Weigert, A.: Astron. Astrophys. **6**, 426 (1970)
Ricker, G.R., Winn, J.N., Vanderspek, R., et al.: Space Telescopes and Instrumentation 2014: Optical, Infrared, and Millimeter Wave. Proceeding of SPIE, vol. 9143, p. 914320 (2014)
Rix, H.-W., Bovy, J.: Astron. Astrophys. Rev. **21**, 61 (2013)
Schwarzschild, K.: Astron. Nachr. **172**, 65 (1906)
Sharma, S., Stello, D., Bland-Hawthorn, J., Huber, D., Bedding, T.R.: Astrophys. J. **822**, 15 (2016)
Soderblom, D.R.: Annu. Rev. Astron. Astrophys. **48**, 581 (2010)
Stello, D., Zinn, J., Elsworth, Y., et al.: Astrophys. J. **835**, 83 (2017)
Tassoul, M.: Astrophys. J. Suppl. Ser. **43**, 469 (1980)
Trampedach, R., Stein, R.F., Christensen-Dalsgaard, J., Nordlund, Å., Asplund, M.: Mon. Not. R. Astron. Soc. **445**, 4366 (2014)
Ulrich, R.K.: Astrophys. J. **306**, L37 (1986)
Verlinde, E.P.: ArXiv e-prints. arXiv:1611.02269 (2016)
White, T.R., Bedding, T.R., Stello, D., et al.: Astrophys. J. **743**, 161 (2011)
Yong, D., Casagrande, L., Venn, K.A., et al.: Mon. Not. R. Astron. Soc. **459**, 487 (2016)

Synergies Between Asteroseismology and Exoplanetary Science

Daniel Huber

Abstract Over the past decade asteroseismology has become a powerful method to systematically characterize host stars and dynamical architectures of exoplanet systems. In this contribution I review current key synergies between asteroseismology and exoplanetary science such as the precise determination of planet radii and ages, the measurement of orbital eccentricities, stellar obliquities and their impact on hot Jupiter formation theories, and the importance of asteroseismology on spectroscopic analyses of exoplanet hosts. I also give an outlook on future synergies such as the characterization of sub-Neptune-size planets orbiting solar-type stars, the study of planet populations orbiting evolved stars, and the determination of ages of intermediate-mass stars hosting directly imaged planets.

1 Introduction: Know the Star, Know the Planet

Exoplanetary science has undergone a revolution over the past two decades, driven by ground-based Doppler surveys and high-precision, space-based photometry from missions such as *CoRoT* (Baglin et al. 2009) and *Kepler* (Borucki et al. 2010). At the time of writing nearly 3500 confirmed exoplanets are known, and future space-based missions such as *TESS* (Ricker et al. 2014) in combination with ground-based efforts are expected to continue this revolution over the coming decades.

The wealth of exoplanet discoveries has uncovered several important questions: How did gas-giant planets in close-in orbits (hot Jupiters) form? What are the origin and compositions of sub-Neptune-size planets, for which we have no equivalent in the solar system? What are the occurrence rates of exoplanets as a function of

D. Huber (✉)
Institute for Astronomy, University of Hawaii, 2680 Woodlawn Drive, Honolulu, HI 96822, USA

Sydney Institute for Astronomy (SIfA), School of Physics, University of Sydney, NSW 2006, Australia

Department of Physics and Astronomy, Stellar Astrophysics Centre (SAC), Aarhus University, Ny Munkegade 120, DK-8000 Aarhus C, Denmark
e-mail: huberd@hawaii.edu

T.L. Campante et al. (eds.), *Asteroseismology and Exoplanets: Listening to the Stars and Searching for New Worlds*, Astrophysics and Space Science Proceedings 49, DOI 10.1007/978-3-319-59315-9_6

their size, mass, orbital architecture, as well as their host star spectral type and evolutionary state? Do habitable planets exist outside our solar system?

Our ability to answer these questions depends strongly on our understanding of the host stars. This is primarily due to the fact that the majority of exoplanet detections are indirect—more than 98% of all exoplanets known to date were discovered using transits, the Doppler method, or microlensing, all of which measure properties of planets relative to the host star. Thus, constraining the physical properties of planets is often limited by the characterization of stars. *Indeed, for 99% of all planet candidates detected by Kepler the uncertainty in the planet radius is currently dominated by the uncertainty in the radius of the host star.* In addition to placing planet properties on an absolute scale, host star characteristics are also crucial to understand the planetary environments such as the extent of the habitable zone (Kane 2014).

The requirement for continuous high-precision monitoring has enabled a fortuitous synergy between asteroseismology and exoplanetary science, since the data can be simultaneously used to detect exoplanets and study stellar oscillations (see Fig. 1). In this review I will discuss some key synergies between both fields, and conclude with an outlook of what future synergies we can expect from current and future ground- and space-based facilities such as SONG, K2, *TESS*, *PLATO* and *WFIRST*.

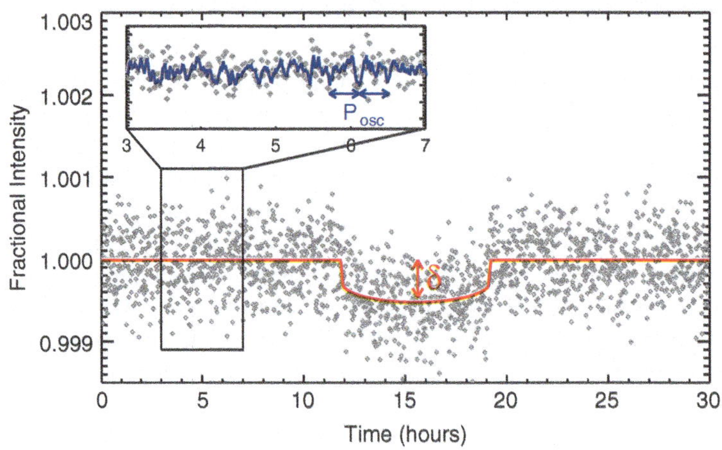

Fig. 1 *Kepler* short-cadence light curve showing a single transit of Kepler-36c. The *red solid line* is the transit model from Carter et al. (2012), and the *inset* shows the oscillations of the host star. The transit depth δ yields the size of the planet relative to the star, and the oscillation periods (P_{osc}) can be used to independently measure the size of the star

2 Characterization of Exoplanets

2.1 *The Connection Between Transits and Mean Stellar Density*

The primary observable for exoplanet transits is the transit depth,[1] ΔF, which for the simplified case of a uniformly bright stellar disk is related to the size of the planet (R_P) and the size of the star ($R_\star$) as:

$$\Delta F = \left(\frac{R_P}{R_\star}\right)^2 . \tag{1}$$

Accurate measurements of $R_P/R_\star$, however, are typically complicated by degeneracies between the transit depth, transit duration, impact parameter, limb darkening, and the size of the star (see Fig. 2). For example, for fixed $R_P/R_\star$ a larger impact parameter will lead to a shallower transit (due to limb darkening) with shorter duration. The same transit duration and depth, however, could likewise be caused by a smaller planet orbiting a smaller star with a lower impact parameter.

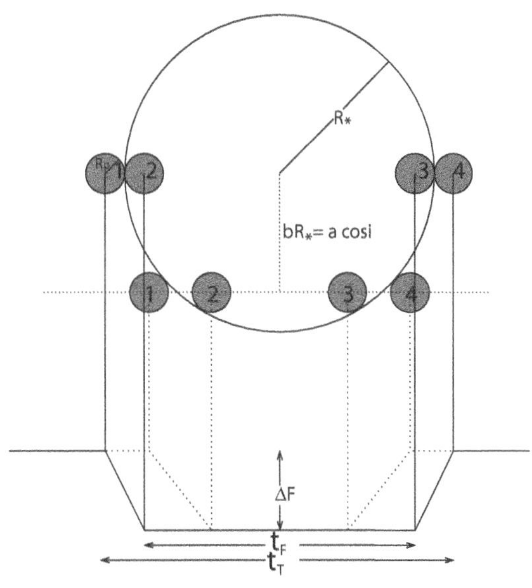

Fig. 2 Schematic transit light curves (*solid and dotted lines on the bottom*) and corresponding star-planet geometry (*top*). The four transit contact points are shown for both transits. The transit depth, ΔF, total transit duration, t_T, and transit duration between ingress and egress, t_F, are shown for the solid transit light curve. Also defined is the impact parameter, b. From Seager and Mallén-Ornelas (2003)

[1]The notation ΔF will be used hereafter to denote transit depth.

This degeneracy can be broken with independent knowledge of the host star density. Assuming $R_P \ll R_\star \ll a$ (where a is the orbit semi-major axis) and circular orbits, it can be shown that (Winn et al. 2010):

$$\frac{a}{R_\star} = \frac{2\,\Delta F^{1/4}}{\pi} \frac{P}{\sqrt{t_T^2 - t_F^2}} . \quad (2)$$

Here, P is the orbital period, t_T is the total transit duration and t_F is the transit duration between ingress and egress, as illustrated in Fig. 2. Using Kepler's third law, $a/R_\star$ can be expressed as

$$\frac{a}{R_\star} = \left(\frac{P^2\,G}{4\pi^2} \frac{M_\star}{R_\star^3}\right)^{1/3} , \quad (3)$$

and hence:

$$\rho_\star = \frac{3\pi}{G P^2} \left(\frac{a}{R_\star}\right)^3 . \quad (4)$$

The mean stellar density is therefore directly related to quantities which can be measured from a transit light curve (Seager and Mallén-Ornelas 2003).

Equation (4) is of key importance for the synergy between asteroseismology and exoplanetary science. Since asteroseismology measures the mean stellar density with a typical precision of a few percent or less, the combination of stellar oscillations and transits can be used to remove degeneracies when fitting exoplanet transits and accurately measure transit parameters. This is particularly important for small planets with low-*SNR* transits, for which ingress and egress durations often cannot be accurately measured and hence constraining $a/R_\star$ independently of $\rho_\star$ is difficult.

2.2 The Importance of Precise Exoplanet Radii

Precise host star radius measurements are important for understanding the composition of planets. Composition models depend sensitively on radius, especially in the regime of sub-Neptune-size planets, and density measurements from transit and Doppler surveys have indicated a threshold between mostly rocky and gaseous planet compositions of $\approx 1.6\,R_\oplus$ (Weiss and Marcy 2014; Rogers 2015). Uncertainties in planet radii due to indirect stellar characterization methods, however, have often led to ambiguities when interpreting exoplanet detections. For example, the $\approx 20\%$ radius uncertainty for Kepler-452b prevented firm conclusions about whether the planet, which orbits a G-type host star within the habitable zone, is indeed rocky (Jenkins et al. 2015).

Asteroseismology has provided some of the most precise characterizations of exoplanets to date. The first asteroseismic studies of exoplanet-host stars were performed using ground-based, radial-velocity observations of μ Ara (Bazot et al. 2005; Bouchy et al. 2005), space-based photometry using the *Hubble Space Telescope* of HD 17156 (Gilliland et al. 2011) and *CoRoT* photometry of HD 52265 (Ballot et al. 2011; Lebreton and Goupil 2014). The launch of *Kepler* led to a revolution in the synergy between asteroseismology and exoplanetary science, with over 70 confirmed *Kepler* exoplanet-host stars (see Fig. 3). This large sample allowed the first systematic precise characterization of planets in the *Kepler* sample (Huber et al. 2013a), including planet radii measured to $\approx 1\%$ (Ballard et al. 2014), as well as investigations of the effects of stellar incident flux on the radius distribution of close-in planets (Lundkvist et al. 2016).

More recent studies have focused not only on measuring global asteroseismic quantities (which are sensitive to densities, masses and radii) but also systematic modeling of individual oscillation frequencies, which allows precise constraints on stellar ages (Silva Aguirre et al. 2015; Davies et al. 2016). One of the most remarkable discoveries so far is Kepler-444, which consists of a K dwarf of age 11.2 ± 1.0 Gyr hosting five sub-Earth-size planets with orbital periods of less than 10 days (Campante et al. 2015). Kepler-444 demonstrated that sub-Earth-size planets have existed for most of the history of our Universe, and the discovery of a pair of low-mass companions in a highly eccentric orbit furthermore showed that the formation of small planets appears to be robust against early truncation of the protoplanetary disk (Dupuy et al. 2016).

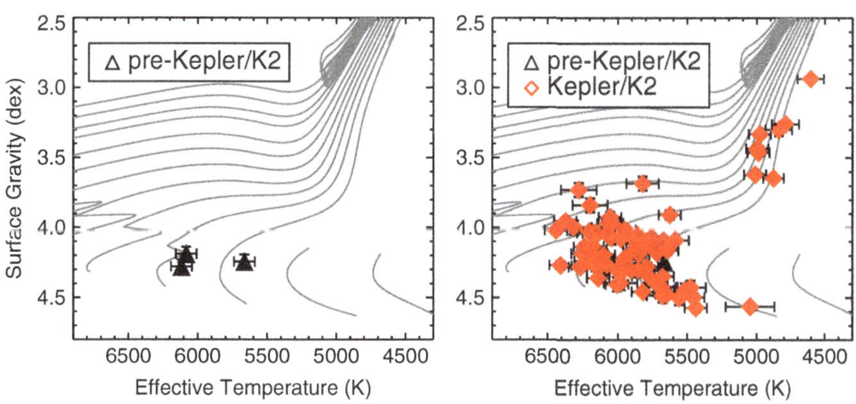

Fig. 3 Surface gravity versus effective temperature for exoplanet-host stars with asteroseismic detections before (*left panel*) and after (*right panel*) the launch of *Kepler*/K2. *Gray lines* show solar-metallicity evolutionary tracks from the BASTI database (Pietrinferni et al. 2004)

3 Orbital Eccentricities of Exoplanets

Orbital eccentricities play a key role in many areas of exoplanetary science, ranging from studies of the dynamics of multiplanet systems to the determination of the fraction of time a planet spends within the habitable zone. Traditionally, eccentricities can be measured through Doppler velocities, secondary transits, or transit-timing variations. However, these methods are either only applicable for relatively large gas-giant planets, or a small subset of multiplanet systems for which effects of eccentricity and mass can be successfully disentangled (e.g., Lithwick et al. 2012; Hadden and Lithwick 2014).

The combination of transit photometry and asteroseismology has opened up a powerful method to systematically measure orbital eccentricities of transiting planets. Since the eccentricity and orientation of the orbit to the observer control the transit duration (see Fig. 4), the ratio of the mean stellar density assuming a circular orbit [Eq. (4)] and true mean stellar density are related as (e.g., Kipping 2010):

$$\frac{\rho_\star}{\rho_{\star,\mathrm{transit}}} = \frac{(1-e^2)^{3/2}}{(1+e\sin\omega)^3}\,. \tag{5}$$

Here, e is the eccentricity and ω is the argument of periastron. Equations (4) and (5) demonstrate that if an independent measurement of $\rho_\star$ is available (for example, from asteroseismology), transits can be used to directly constrain the eccentricity of a planet without radial-velocity observations. Importantly, an accurate measurement of $\rho_{\star,\mathrm{transit}}$ requires an accurate estimate of the ingress and egress times.

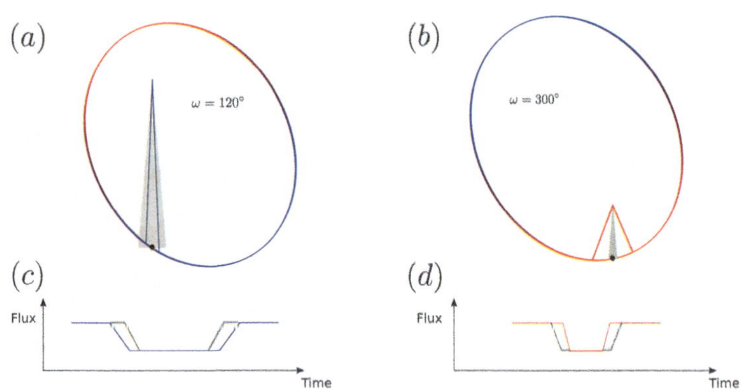

Fig. 4 Planetary orbit of eccentricity 0.6 with two different angles of periastron (*top panels*) and the corresponding observed transits (*bottom panels*). *Red and blue colors* correspond to the fast and slow part of the orbit, respectively. Observed transit durations are longer (*left panels*) or shorter (*right panels*) compared to circular orbits (*gray*) depending on the eccentricity and argument of periastron of the orbit. From Van Eylen and Albrecht (2015)

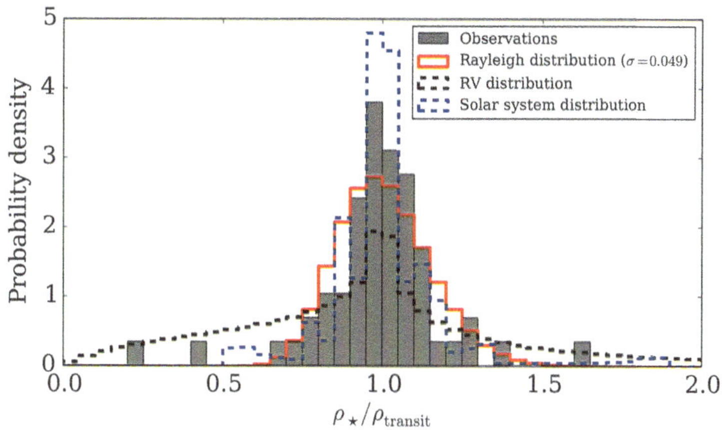

Fig. 5 Ratio of the asteroseismic mean stellar density and the density measured from transits assuming a circular orbit for 28 *Kepler* multiplanet host stars (*gray*). Corresponding distributions of solar system planets and planets detected by radial-velocity surveys are shown as *blue and black dashed histograms*, respectively. From Van Eylen and Albrecht (2015)

The first systematic study of eccentricities using asteroseismic densities concentrated on the identification of false positives in the *Kepler* planet candidate sample by comparing $\rho_\star$ and $\rho_{\star,\mathrm{transit}}$, yielding a significantly higher false-positive rate for red-giant-host stars (Sliski and Kipping 2014). A subsequent study by Van Eylen and Albrecht (2015) focused on 28 multiplanet systems, which are expected to have a small false-positive rate (Lissauer et al. 2012). Figure 5 shows a histogram of the derived ratios between transit and seismic density [left hand side of Eq. (5)] for their sample compared to the solar system and a sample of planets with eccentricities from radial-velocity surveys. The asteroseismic sample (red solid line) is consistent with circular orbits, similar to the solar system (blue dashed histogram), but in stark contrast to the radial-velocity sample (black dashed line). Since *Kepler* multiplanet systems include mostly small, low-mass planets compared to the more massive planets probed by Doppler surveys, this indicates that low-mass planets are preferentially on circular orbits. This conclusion is of great importance since circular orbits are frequently assumed when modeling exoplanets in the habitable zone (Barclay et al. 2013; Borucki et al. 2013; Quintana et al. 2014; Jenkins et al. 2015) or when estimating the detection completeness for planet occurrence studies (e.g., Howard et al. 2012; Dong and Zhu 2013; Petigura et al. 2013; Burke et al. 2015).

Expanding such studies holds promise to further constrain the dynamics of exoplanet systems using asteroseismology. For example, Xie et al. (2016) recently used stellar densities derived from spectroscopy to show that while multiplanet systems are indeed preferentially circular, single systems appear to show significantly higher eccentricities even for small (sub-Neptune-size) planets. Asteroseismic studies of

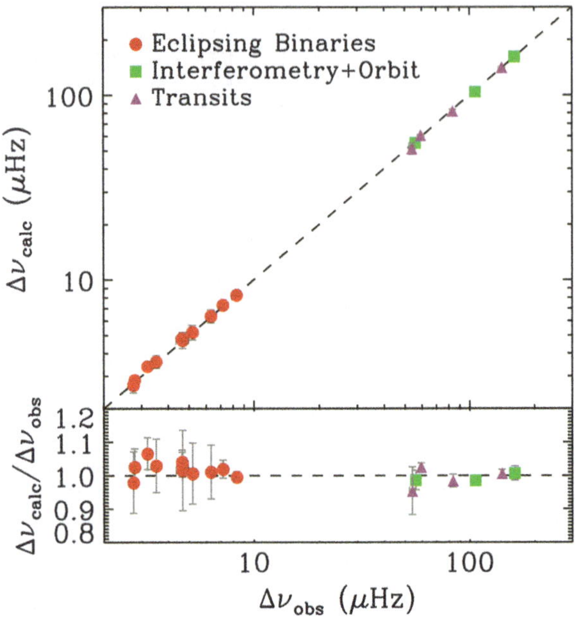

Fig. 6 Comparison of the mean stellar density from independent methods and as calculated from the asteroseismic scaling relation for the large frequency separation ($\Delta\nu$). Transit-derived densities account for more than half of the comparison values for subgiant and main-sequence stars

systems with single planets would be valuable to independently confirm this result with a smaller, but higher precision sample.

Transiting exoplanets for which the eccentricity can be measured independently (for example through radial velocities) can also be used as an independent test of asteroseismic densities calculated from the scaling relation for the large frequency separation ($\Delta\nu$). This is particularly valuable since the $\Delta\nu$ scaling relation has found widespread use for calculating stellar properties for thousands of stars in the era of "ensemble asteroseismology" (Kallinger et al. 2010; Chaplin et al. 2014). Figure 6 shows a comparison of the mean stellar density calculated from the $\Delta\nu$ scaling relation and from dynamically measured densities from double-lined eclipsing binaries (Frandsen et al. 2013; Gaulme et al. 2016), interferometric orbits (Procyon and α Cen A+B; see Bruntt et al. 2010; and references therein), as well as transiting exoplanets with known eccentricities: HD 17156 (Gilliland et al. 2011; Nutzman et al. 2011), TrES-2 (Southworth 2011; Barclay et al. 2012), HAT-P-7 (Christensen-Dalsgaard et al. 2010; Southworth 2011), and Kepler-14 (Southworth 2012; Huber et al. 2013a). Transit-derived densities account for more than half of the currently available comparison values for subgiant and main-sequence stars, and empirically demonstrate that the $\Delta\nu$ scaling relation is accurate to about $\approx 3\%$ (see also Huber 2015a).

4 Obliquities of Exoplanet Systems

The obliquity ψ is the angle between the host star rotation axis and the planetary orbital axis, and can be calculated as (Fabrycky and Winn 2009):

$$\cos\psi = \sin i_\star \cos\lambda \sin i_\mathrm{p} + \cos i_\star \cos i_\mathrm{p} \,. \tag{6}$$

Here, λ is the sky-projected spin-orbit angle, i_p is the angle between the line of sight and the orbital axis of the planet, and $i_\star$ is the inclination of the rotation axis to the line of sight of the observer. Figure 7 shows a graphical illustration of these angles for the HAT-P-7 system following Lund et al. (2014).

For transiting exoplanets, i_p can be typically constrained from the transit light curve, while λ can be measured through spectroscopic in-transit observations (the Rossiter–McLaughlin effect). Asteroseismic observations of the relative heights of rotationally split multiplets can be used to provide the measurement of the line-of-sight inclination of the stellar rotation $i_\star$ (Gizon and Solanki 2003). Thus, the combination of transits, Doppler velocities, and asteroseismology allow to uniquely measure the obliquity of exoplanet systems (Benomar et al. 2014; Lund et al. 2014). Importantly, the measurement of $i_\star$ using asteroseismology is independent of planet size, and hence can be used to constrain the obliquity even for systems with small planets in which Rossiter–McLaughlin measurements are typically not feasible. For transiting planets, a low stellar inclination in most cases automatically yields a misalignment of the orbital plane and the stellar equatorial plane (a high obliquity), while a value of near 90° for $i_\star$ implies that the star and the planets are likely (but not necessarily) aligned.

Obliquities have played a key role in constraining the formation mechanism for hot Jupiters, one of the longest standing problems in exoplanetary science. Hot Jupiters are typically thought to form at large orbital distances beyond the snow line, and subsequently migrate to the close-in orbits where they are currently observed (although in-situ formation has also been suggested; see Batygin et al. 2016). Two possible mechanisms have been proposed: migration of the planet through the protoplanetary disk (Lin et al. 1996) or dynamical perturbations such as planet-planet scattering (Chatterjee et al. 2008) or Kozai–Lidov oscillations (Fabrycky and Tremaine 2007) which cause the planet to attain a high orbital eccentricity, followed by shrinking and circularization of the orbit through tidal interactions (often referred to as high-eccentricity migration).

The observation that hot Jupiters show a wide range of obliquities (Johnson et al. 2009; Winn et al. 2010) has been interpreted as evidence for a dynamically violent formation scenario, thus favoring high-eccentricity migration as the dominant formation mechanism. However, this conclusion relies on the assumption that the stellar equator and the protoplanetary disk are initially aligned, and thus that the high obliquity observed today is indeed a consequence of dynamical interactions during the migration process. Key tests for this assumption are multiplanet systems which, if primordial alignments are common, should predominantly show low obliquities.

Fig. 7 Graphical illustration of the obliquity ψ, the sky-projected spin-orbit angle λ, the line-of-sight stellar inclination $i_{\star}$, and the line-of-sight orbit inclination $i_{\rm p}$. Note that the *top panel* shows $\lambda = 155°$, while the *middle panel* shows $\lambda = 180°$. From Lund et al. (2014)

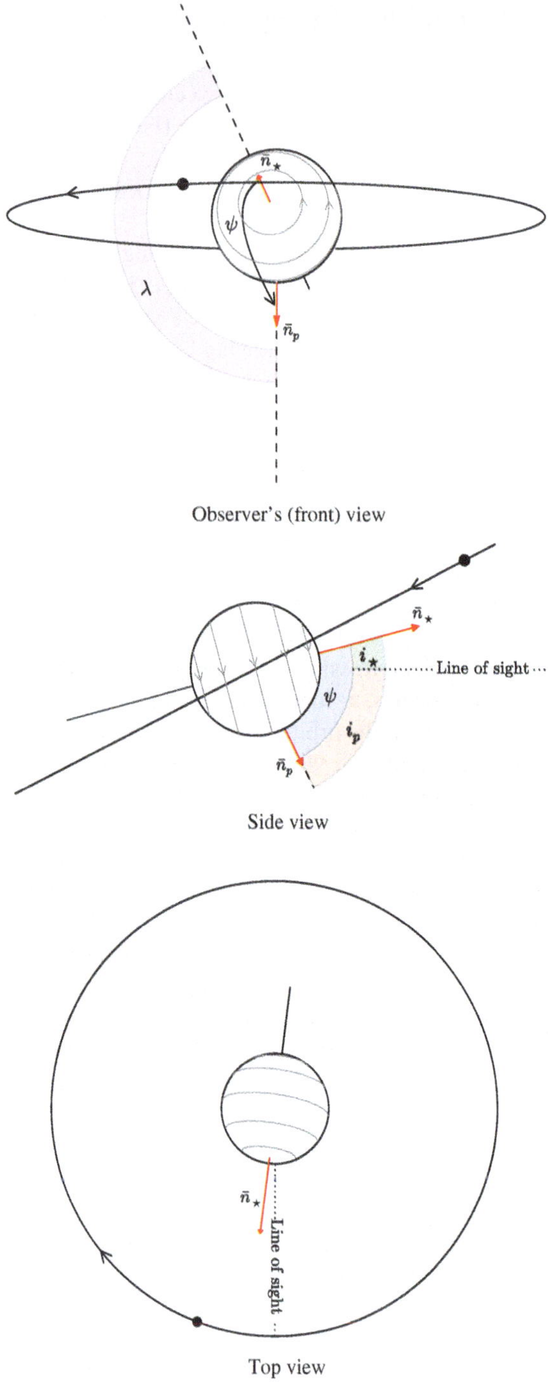

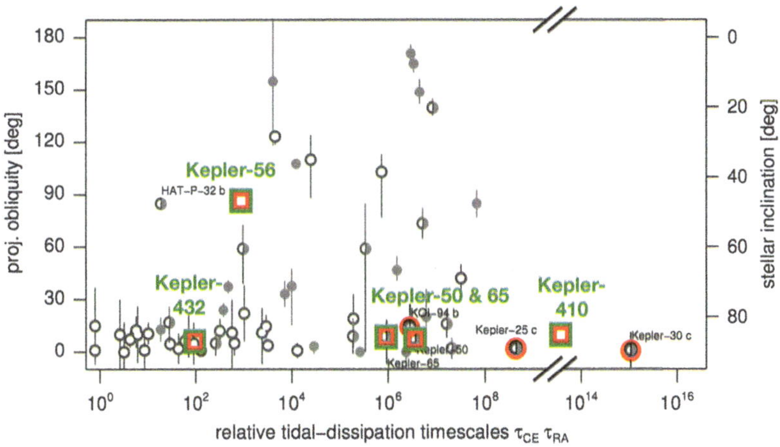

Fig. 8 Projected obliquity (*left ordinate*) and stellar inclination (*right ordinate*) versus relative tidal dissipation timescale for exoplanet systems. Systems with short dissipation timescales are expected to have been realigned even if they were misaligned by the formation process, while systems with long dissipation timescales are expected to preserve their configuration. Multiplanet systems without hot Jupiters are highlighted by *red circles*, and systems with inclinations measured using asteroseismology are highlighted with *green squares*. Positions for Kepler-56, Kepler-410 and Kepler-432 are approximate only. Adapted from Albrecht et al. (2013)

Asteroseismology has played an important role for testing this assumption since seismic inclination measurements are independent of planet size, and hence can be applied to multiplanet systems with small planets. Figure 8 shows the projected obliquity or stellar inclination for exoplanet systems as a function of relative tidal dissipation timescale, which is a proxy for how quickly a system can be realigned by tidal interactions if it was initially misaligned by the formation process (Albrecht et al. 2012). In line with expectations from high-eccentricity migration, hot-Jupiter systems with intermediate dissipation timescales are frequently observed to have high obliquities, while coplanar multiplanet systems without hot Jupiters have mostly low obliquities (e.g., Sanchis-Ojeda et al. 2013) despite long tidal dissipation timescales. Over half of the constraints for multiplanet systems come from asteroseismology (Chaplin et al. 2013; Van Eylen et al. 2014; Quinn et al. 2015).

Asteroseismology has also yielded the first intriguing counterexample for the observed trend of well-aligned multiplanet systems. Kepler-56, a red giant hosting two transiting planets confirmed through transit-timing variations (Steffen et al. 2012), revealed an inclination of $i_{\star} = 47° \pm 6°$, demonstrating the first spin-orbit misalignment in a multiplanet system (Huber et al. 2013b). Subsequent follow-up studies have confirmed that the misalignment is likely caused by the torque of a third planet on a wide orbit (Li et al. 2014; Otor et al. 2016; Gratia and Fabrycky 2017), and that such a configuration could be consistent with a primordial misalignment (Matsakos and Königl 2017). Future asteroseismic inclination measurements will be needed to determine whether spin-orbit misalignments in multiplanet systems are common, and whether high obliquities are indeed tracers of dynamical formation history of hot Jupiters.

5 Chemical Abundances of Exoplanet-Host Stars

Chemical abundances of exoplanet-host stars are tracers of the primordial composition of protoplanetary disks, and hence provide valuable clues about which conditions favor planet formation. For example, it is well established that gas-giant planets predominantly form around metal-rich stars (Gonzalez 1997; Fischer and Valenti 2005), whereas small planets form independently of host star metallicity (Buchhave et al. 2012). Going beyond metallicities, intriguing abundance differences in volatile and refractory elements between the Sun and solar twins with and without planets have been observed (Meléndez et al. 2009; Ramírez et al. 2009), although the link of these patterns to terrestrial planet formation is still being debated (e.g., Adibekyan et al. 2014).

Asteroseismology does not directly probe atmospheric abundances, but the combination of asteroseismology and spectroscopy can significantly improve our understanding of host star compositions. This is mainly due to the fact that bulk atmospheric parameters ($T_{\rm eff}$, $\log g$, [Fe/H], microturbulence) are often heavily correlated, which can lead to systematic errors in particular for spectral synthesis methods (Torres et al. 2012; Huber et al. 2013a), while spectral line analysis methods are typically less affected (Mortier et al. 2014). Recent efforts have shown that using $\log g$ from asteroseismology to inform spectroscopic modeling methods can significantly increase the accuracy of spectroscopic surface gravities without external constraints (see Fig. 9), thus also leading to more accurate abundances. Using stars with asteroseismology (ideally in combination with interferometry, which also yields an external constraint on $T_{\rm eff}$) as spectroscopic benchmarks promises to extend high-precision abundance work from solar twins to stars in different evolutionary states.

6 Future Prospects

The asteroseismology revolution initiated by *CoRoT* and *Kepler* is set to continue over the coming decades with the launch of *TESS* (Ricker et al. 2014), *PLATO* (Rauer et al. 2014) and *WFIRST* (Spergel et al. 2013). Each of these missions will provide high-precision, space-based photometry suitable for asteroseismology, with the expected number of detections of solar-like oscillations exceeding several million stars (see Fig. 10). Combined with ground-based efforts such as the SONG

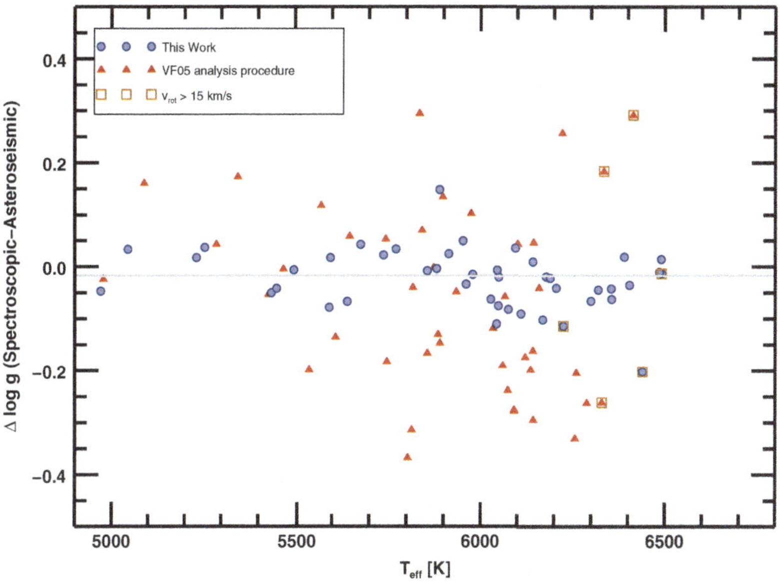

Fig. 9 Difference between asteroseismic and spectroscopic surface gravities before (*red*) and after (*blue*) improving the spectroscopic modeling procedure based on asteroseismic constraints on surface gravity. From Brewer et al. (2015)

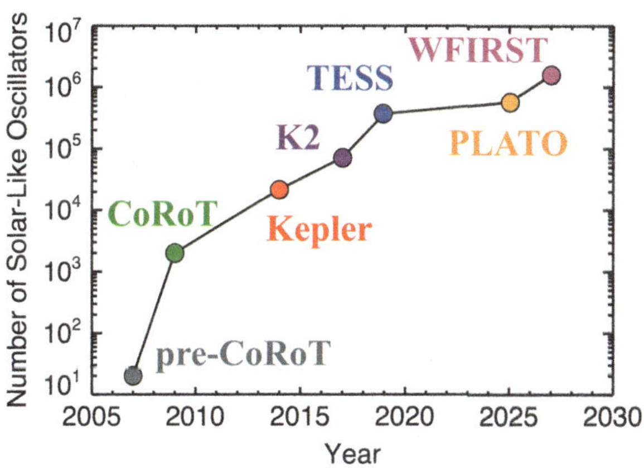

Fig. 10 Number of stars with detected solar-like oscillations as a function of time. The approximate projected yield for current and future missions is 5×10^4 for K2 (based on extrapolating classifications by Huber et al. 2016), 3×10^5 for *TESS* (assuming detections in all red-clump stars down to $I \approx 10$ mag with 27 days of data), 2×10^5 for *PLATO* (assuming a similar red-giant fraction to *Kepler*), and 10^6 for *WFIRST* (Gould et al. 2015). Note that > 90% of all detections are expected to be evolved stars, and *PLATO* will by far contribute the most detections for dwarfs and subgiants (≈ 80,000 stars; Rauer et al. 2014)

network (Grundahl et al. 2008), the key synergies between asteroseismology and exoplanet science are expected to be:

- **Densities and ages of sub-Neptune-size planets transiting asteroseismic solar-type stars:** In addition to *TESS*, ground-based radial velocities obtained by SONG will play an important role in characterizing stars hosting small exoplanets using asteroseismology. Such systems will provide the best opportunity to precisely study the composition diversity of sub-Neptunes by constraining host star radii and masses to a few percent. Importantly, *Gaia* parallaxes alone will not reach comparable precision due to model-dependent uncertainties such as bolometric corrections and reddening. The asteroseismology-exoplanet synergy is a core component of the *PLATO* mission, which will extend the reach of asteroseismology to characterize radii, masses and ages of solar-type stars with small, transiting planets in the habitable zone.
- **Gas-giant planets orbiting asteroseismic evolved stars:** Evolved stars provide an evolutionary "sweet spot" in which light curves with moderate cadence (such as the 30-min sampling provided by *Kepler*/K2 long-cadence data) can be used to detect transits and stellar oscillations simultaneously. Detections by *Kepler* and the "Giants Orbiting Giants Program" with the K2 Mission (see Fig. 11; Huber 2015b) have demonstrated that these planets can be used to address key questions in exoplanetary science such as the effects of host star evolution on the

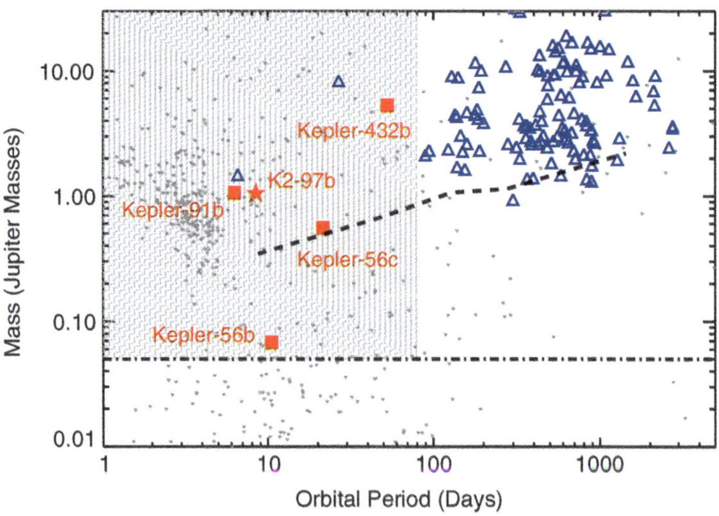

Fig. 11 Mass versus orbital period for exoplanets orbiting evolved stars ($R_\star > 3.5\,R_\odot$, $T_{\rm eff} < 5500$ K) detected with transits (*red symbols*) and radial velocities (*blue triangles*). Transiting systems are restricted to stars which are precisely characterized using asteroseismology. The *dashed line* shows the median RV detection limit for mean masses given by Bowler et al. (2010). The *dashed-dotted line* marks the mass of Neptune as an approximate K2 detection limit ($R_{\rm P} \gtrsim 0.5\,R_{\rm J}$), and the *gray shaded area* is probed by the K2 "Giants Orbiting Giants Program". Adapted from Huber (2015b)

radius inflation of hot Jupiters (Grunblatt et al. 2016). A particularly promising possibility to extend this synergy are full-frame images obtained by the *TESS* mission, which are expected to yield several hundred asteroseismic exoplanet-host stars (Campante et al. 2016). Preliminary simulations have shown that low-luminosity RGB stars in the ecliptic poles with 1 year coverage can also be used to measure rotational splittings, and hence extend the study of exoplanet obliquities of systems similar to Kepler-56 (see Sect. 4).
- **Planets orbiting pulsating A stars:** Near-diffraction-limited, infrared adaptive-optics imaging instruments such as GPI (Macintosh et al. 2008), SPHERE (Beuzit et al. 2008) and SCExAO (Guyon et al. 2010) will soon provide an increasing number of directly imaged planets orbiting young stars, including pulsating A stars such as HR 8799 (Zerbi et al. 1999; Marois et al. 2008). A common limitation for interpreting these discoveries is their unknown age, which is needed to determine whether the detected substellar companions are indeed planets. While mode identification in δ Scuti and γ Doradus stars is still challenging, the extension of the asteroseismology-exoplanet synergy to these systems will undoubtably become more important over the coming decades. Extended photometric monitoring provided over several years by *PLATO* may also provide future opportunities to detect planets in wide orbits around A stars using pulsation frequency shifts induced by the planet (Murphy et al. 2016).

The above list is by no means complete, and further synergies beyond those discussed here will certainly be explored. With the wealth of data from ground-based and space-based facilities there is little doubt that the exciting and fruitful synergy between asteroseismology and exoplanetary science will continue to grow over the coming decades.

Acknowledgements Many thanks to Tiago Campante, Mário Monteiro and all other organizers for a fantastic Summer School in Faial; and to Simon Albrecht, John Brewer, Vincent Van Eylen, Mikkel Lund, and Josh Winn for providing figures for this review. Financial support was provided by NASA grant NNX14AB92G and the Australian Research Council's Discovery Projects funding scheme (project number DE140101364).

References

Adibekyan, V.Z., González Hernández, J.I., Delgado Mena, E., et al.: Astron. Astrophys. **564**, L15 (2014)
Albrecht, S., Winn, J.N., Johnson, J.A., et al.: Astrophys. J. **757**, 18 (2012)
Albrecht, S., Winn, J.N., Marcy, G.W., et al.: Astrophys. J. **771**, 11 (2013)
Baglin, A., Auvergne, M., Barge, P., et al.: In: IAU Symposium, vol. 253. pp. 71–81 (2009)
Ballard, S., Chaplin, W.J., Charbonneau, D., et al.: Astrophys. J. **790**, 12 (2014)
Ballot, J., Gizon, L., Samadi, R., et al.: Astron. Astrophys. **530**, A97 (2011)
Barclay, T., Huber, D., Rowe, J.F., et al.: Astrophys. J. **761**, 53 (2012)
Barclay, T., Burke, C.J., Howell, S.B., et al.: Astrophys. J. **768**, 101 (2013)
Batygin, K., Bodenheimer, P.H., Laughlin, G.P.: Astrophys. J. **829**, 114 (2016)
Bazot, M., Vauclair, S., Bouchy, F., Santos, N.C.: Astron. Astrophys. **440**, 615 (2005)

Benomar, O., Masuda, K., Shibahashi, H., Suto, Y.: Publ. Astron. Soc. Jpn. (2014). arXiv:1407.7332
Beuzit, J.-L., Feldt, M., Dohlen, K., et al.: In: Ground-Based and Airborne Instrumentation for Astronomy II. Proceedings of SPIE, vol. 7014, p. 701418 (2008)
Borucki, W.J., Koch, D., Basri, G., et al.: Science **327**, 977 (2010)
Borucki, W.J., Agol, E., Fressin, F., et al.: Science **340**, 587 (2013)
Bouchy, F., Bazot, M., Santos, N.C., Vauclair, S., Sosnowska, D.: Astron. Astrophys. **440**, 609 (2005)
Bowler, B.P., Johnson, J.A., Marcy, G.W., et al.: Astrophys. J. **709**, 396 (2010)
Brewer, J.M., Fischer, D.A., Basu, S., Valenti, J.A., Piskunov, N.: Astrophys. J. **805**, 126 (2015)
Bruntt, H., Bedding, T.R., Quirion, P.-O., et al.: Mon. Not. R. Astron. Soc. **405**, 1907 (2010)
Buchhave, L.A., Latham, D.W., Johansen, A., et al.: Nature **486**, 375 (2012)
Burke, C.J., Christiansen, J.L., Mullally, F., et al.: Astrophys. J. **809**, 8 (2015)
Campante, T.L., Barclay, T., Swift, J.J., et al.: Astrophys. J. **799**, 170 (2015)
Campante, T.L., Schofield, M., Kuszlewicz, J.S., et al.: Astrophys. J. **830**, 138 (2016)
Carter, J.A., Agol, E., Chaplin, W.J., et al.: Science **337**, 556 (2012)
Chaplin, W.J., Sanchis-Ojeda, R., Campante, T.L., et al.: Astrophys. J. **766**, 101 (2013)
Chaplin, W.J., Basu, S., Huber, D., et al.: Astrophys. J. Suppl. Ser. **210**, 1 (2014)
Chatterjee, S., Ford, E.B., Matsumura, S., Rasio, F.A.: Astrophys. J. **686**, 580 (2008)
Christensen-Dalsgaard, J., Kjeldsen, H., Brown, T.M., et al.: Astrophys. J. **713**, L164 (2010)
Davies, G.R., Silva Aguirre, V., Bedding, T.R., et al.: Mon. Not. R. Astron. Soc. **456**, 2183 (2016)
Dong, S., Zhu, Z.: Astrophys. J. **778**, 53 (2013)
Dupuy, T.J., Kratter, K.M., Kraus, A.L., et al.: Astrophys. J. **817**, 80 (2016)
Fabrycky, D., Tremaine, S.: Astrophys. J. **669**, 1298 (2007)
Fabrycky, D.C., Winn, J.N.: Astrophys. J. **696**, 1230 (2009)
Fischer, D.A., Valenti, J.: Astrophys. J. **622**, 1102 (2005)
Frandsen, S., Lehmann, H., Hekker, S., et al.: Astron. Astrophys. **556**, A138 (2013)
Gaulme, P., McKeever, J., Jackiewicz, J., et al.: Astrophys. J. **832**, 121 (2016)
Gilliland, R.L., McCullough, P.R., Nelan, E.P., et al.: Astrophys. J. **726**, 2 (2011)
Gizon, L., Solanki, S.K.: Astrophys. J. **589**, 1009 (2003)
Gonzalez, G.: Mon. Not. R. Astron. Soc. **285**, 403 (1997)
Gould, A., Huber, D., Penny, M., Stello, D.: J. Korean Astron. Soc. **48**, 93 (2015)
Gratia, P., Fabrycky, D.: Mon. Not. R. Astron. Soc. **464**, 1709 (2017)
Grunblatt, S.K., Huber, D., Gaidos, E.J., et al.: Astron. J. **152**, 185 (2016)
Grundahl, F., Christensen-Dalsgaard, J., Kjeldsen, H., et al.: In: Deng, L., Chan, K.L. (eds.) IAU Symposium, vol. 252, pp. 465–466 (2008)
Guyon, O., Martinache, F., Garrel, V., et al.: In: Adaptive Optics Systems II. Proceedings of SPIE, vol. 7736, p. 773624 (2010)
Hadden, S., Lithwick, Y.: Astrophys. J. **787**, 80 (2014)
Howard, A.W., Marcy, G.W., Bryson, S.T., et al.: Astrophys. J. Suppl. Ser. **201**, 15 (2012)
Huber, D.: In: Giants of Eclipse: The Zeta Aurigae Stars and Other Binary Systems. Astrophysics and Space Science Library, vol. 408, p. 169 (2015a)
Huber, D.: (2015b). ArXiv e-prints, arXiv:1511.07441
Huber, D., Chaplin, W.J., Christensen-Dalsgaard, J., et al.: Astrophys. J. **767**, 127 (2013a)
Huber, D., Carter, J.A., Barbieri, M., et al.: Science **342**, 331 (2013b)
Huber, D., Bryson, S.T., Haas, M.R., et al.: Astrophys. J. Suppl. Ser. **224**, 2 (2016)
Jenkins, J.M., Twicken, J.D., Batalha, N.M., et al.: Astron. J. **150**, 56 (2015)
Johnson, J.A., Winn, J.N., Albrecht, S., et al.: Publ. Astron. Soc. Pac. **121**, 1104 (2009)
Kallinger, T., Mosser, B., Hekker, S., et al.: Astron. Astrophys. **522**, A1 (2010)
Kane, S.R.: Astrophys. J. **782**, 111 (2014)
Kipping, D.M.: Mon. Not. R. Astron. Soc. **407**, 301 (2010)
Lebreton, Y., Goupil, M.J.: Astron. Astrophys. **569**, A21 (2014)
Li, G., Naoz, S., Valsecchi, F., Johnson, J.A., Rasio, F.A.: Astrophys. J. **794**, 131 (2014)
Lin, D.N.C., Bodenheimer, P., Richardson, D.C.: Nature **380**, 606 (1996)

Lissauer, J.J., Marcy, G.W., Rowe, J.F., et al.: Astrophys. J. **750**, 112 (2012)
Lithwick, Y., Xie, J., Wu, Y.: Astrophys. J. **761**, 122 (2012)
Lund, M.N., Lundkvist, M., Silva Aguirre, V., et al.: Astron. Astrophys. **570**, A54 (2014)
Lundkvist, M.S., Kjeldsen, H., Albrecht, S., et al.: Nat. Commun. **7**, 11201 (2016)
Macintosh, B.A., Graham, J.R., Palmer, D.W., et al.: In: Adaptive Optics Systems. Proceedings of SPIE, vol. 7015, p. 701518 (2008)
Marois, C., Macintosh, B., Barman, T., et al.: Science **322**, 1348 (2008)
Matsakos, T., Königl, A.: Astron. J. **153**, 60 (2017)
Meléndez, J., Asplund, M., Gustafsson, B., Yong, D.: Astrophys. J. **704**, L66 (2009)
Mortier, A., Sousa, S.G., Adibekyan, V.Z., Brandão, I.M., Santos, N.C.: Astron. Astrophys. **572**, A95 (2014)
Murphy, S.J., Bedding, T.R., Shibahashi, H.: Astrophys. J. **827**, L17 (2016)
Nutzman, P., Gilliland, R.L., McCullough, P.R., et al.: Astrophys. J. **726**, 3 (2011)
Otor, O.J., Montet, B.T., Johnson, J.A., et al.: Astron. J. **152**, 165 (2016)
Petigura, E.A., Howard, A.W., Marcy, G.W.: Proc. Natl. Acad. Sci. **110**, 19175 (2013)
Pietrinferni, A., Cassisi, S., Salaris, M., Castelli, F.: Astrophys. J. **612**, 168 (2004)
Quinn, S.N., White, T.R., Latham, D.W., et al.: Astrophys. J. **803**, 49 (2015)
Quintana, E.V., Barclay, T., Raymond, S.N., et al.: Science **344**, 277 (2014)
Ramírez, I., Meléndez, J., Asplund, M.: Astron. Astrophys. **508**, L17 (2009)
Rauer, H., Catala, C., Aerts, C., et al.: Exp. Astron. (2014). arXiv:1310.0696
Ricker, G.R., Winn, J.N., Vanderspek, R., et al.: In: Space Telescopes and Instrumentation 2014: Optical, Infrared, and Millimeter Wave. Proceedings of SPIE, vol. 9143, p. 914320 (2014)
Rogers, L.A.: Astrophys. J. **801**, 41 (2015)
Sanchis-Ojeda, R., Winn, J.N., Marcy, G.W., et al.: Astrophys. J. **775**, 54 (2013)
Seager, S., Mallén-Ornelas, G.: Astrophys. J. **585**, 1038 (2003)
Silva Aguirre, V., Davies, G.R., Basu, S., et al.: Mon. Not. R. Astron. Soc. **452**, 2127 (2015)
Sliski, D.H., Kipping, D.M.: Astrophys. J. **788**, 148 (2014)
Southworth, J.: Mon. Not. R. Astron. Soc. **417**, 2166 (2011)
Southworth, J.: Mon. Not. R. Astron. Soc. **426**, 1291 (2012)
Spergel, D., Gehrels, N., Breckinridge, J., et al.: (2013). ArXiv e-prints, arXiv:1305.5425
Steffen, J.H., Fabrycky, D.C., Ford, E.B., et al.: Mon. Not. R. Astron. Soc. **421**, 2342 (2012)
Torres, G., Fischer, D.A., Sozzetti, A., et al.: Astrophys. J. **757**, 161 (2012)
Van Eylen, V., Albrecht, S.: Astrophys. J. **808**, 126 (2015)
Van Eylen, V., Lund, M.N., Silva Aguirre, V., et al.: Astrophys. J. **782**, 14 (2014)
Weiss, L.M., Marcy, G.W.: Astrophys. J. **783**, L6 (2014)
Winn, J.N., Fabrycky, D., Albrecht, S., Johnson, J.A.: Astrophys. J. **718**, L145 (2010)
Xie, J.-W., Dong, S., Zhu, Z., et al.: Proc. Natl. Acad. Sci. **113**, 11431 (2016)
Zerbi, F.M., Rodríguez, E., Garrido, R., et al.: Mon. Not. R. Astron. Soc. **303**, 275 (1999)

Tutorial: Asteroseismic Data Analysis with DIAMONDS

Enrico Corsaro

Abstract Since the advent of the space-based photometric missions such as *CoRoT* and NASA's *Kepler*, asteroseismology has acquired a central role in our understanding about stellar physics. The *Kepler* spacecraft, especially, is still releasing excellent photometric observations that contain a large amount of information not yet investigated. For exploiting the full potential of these data, sophisticated and robust analysis tools are now essential, so that further constraining of stellar structure and evolutionary models can be obtained. In addition, extracting detailed asteroseismic properties for many stars can yield new insights on their correlations to fundamental stellar properties and dynamics. After a brief introduction to the Bayesian notion of probability, I describe the code DIAMONDS for Bayesian parameter estimation and model comparison by means of the nested sampling Monte Carlo (NSMC) algorithm. NSMC constitutes an efficient and powerful method, in replacement to standard Markov chain Monte Carlo, very suitable for high-dimensional and multimodal problems that are typical of detailed asteroseismic analyses, such as the fitting and mode identification of individual oscillation modes in stars (known as peak-bagging). DIAMONDS is able to provide robust results for statistical inferences involving tens of individual oscillation modes, while at the same time preserving a considerable computational efficiency for identifying the solution. In the tutorial, I will present the fitting of the stellar background signal and the peak-bagging analysis of the oscillation modes in a red-giant star, providing an example to use Bayesian evidence for assessing the peak significance of the fitted oscillation peaks.

E. Corsaro (✉)
INAF—Osservatorio Astrofisico di Catania, via S. Sofia 78, I-95123 Catania, Italy

Laboratoire AIM Paris-Saclay, CEA/DRF—CNRS—Université Paris Diderot, IRFU/SAp, L'Orme des Merisiers, bat. 709, 91191 Gif-sur-Yvette Cedex, France

Departamento de Astrofísica, Instituto de Astrofísica de Canarias, E-38205—Universidad de La Laguna, E-38206 La Laguna, Tenerife, Spain
e-mail: enrico.corsaro@oact.inaf.it

T.L. Campante et al. (eds.), *Asteroseismology and Exoplanets: Listening to the Stars and Searching for New Worlds*, Astrophysics and Space Science Proceedings 49, DOI 10.1007/978-3-319-59315-9_7

1 Bayesian Statistics

Let us assume to consider a given physical problem, e.g., the fitting of an observational dataset through the use of a predictive model. We term the dataset D and the fitting model $\mathcal{M}_k$, the latter having a number of k free parameters that we represent with the k-dimensional parameter vector $\boldsymbol{\theta} = (\theta_1, \theta_2, \ldots, \theta_k)$. The number of free parameters sets the dimensionality of the problem, to which a k-dimensional parameter space $\Omega_{\mathcal{M}_k}$ is associated, representing the space of the solutions. Our aim is to obtain optimal estimates of each free parameter and a corresponding statistical weight of the model $\mathcal{M}_k$ that takes into account both the number of dimensions and the fit quality. This statistical inference can be properly addressed through the means of Bayesian statistics (Jeffreys 1961; Sivia and Skilling 2006; Trotta 2008; Bolstad 2013; Corsaro et al. 2013; Corsaro and De Ridder 2014). In particular, the core of the statistical representation is given by Bayes' theorem:

$$p(\boldsymbol{\theta} \mid D, \mathcal{M}_k) = \frac{\mathcal{L}(\boldsymbol{\theta} \mid D, \mathcal{M}_k)\pi(\boldsymbol{\theta} \mid \mathcal{M}_k)}{p(D \mid \mathcal{M}_k)} , \tag{1}$$

where $\mathcal{L}(\boldsymbol{\theta} \mid D, \mathcal{M}_k)$ (hereafter, $\mathcal{L}(\boldsymbol{\theta})$ for simplicity) is the likelihood function, which represents the way we sample the data, while $\pi(\boldsymbol{\theta} \mid \mathcal{M}_k)$ is the prior probability density function (PDF) that reflects our knowledge about the model parameters. The left-hand side of Eq. (1) is the posterior PDF, which has a key role in the parameter estimation problem. Through a marginalization of the posterior PDF, namely an integration over the uninteresting free parameters, we estimate the free parameters of the model. Among the different estimators for each parameter, in Bayesian statistics the median is usually preferred because it represents the most resistant estimator, namely the least sensitive to possible outliers, and because it is invariant for variable change.

The denominator on the right-hand side of Eq. (1) is instead a normalization factor, generally known as the Bayesian evidence (or marginal likelihood), which is defined as

$$\mathcal{E} \equiv p(D \mid \mathcal{M}_k) = \int_{\Omega_{\mathcal{M}_k}} \mathcal{L}(\boldsymbol{\theta} \mid D, \mathcal{M}_k)\pi(\boldsymbol{\theta} \mid \mathcal{M}_k) d\boldsymbol{\theta} . \tag{2}$$

The Bayesian evidence is used for as a statistical weight for model comparison because it encompasses the principle of the Occam's razor, meaning that models are favored if they provide a better fit to the data but are penalized if their number of free parameters is larger than that of a competitor model. For our study, the model comparison is performed by computation of the Bayes' factor $\mathcal{B}_{ij} = \mathcal{E}_i / \mathcal{E}_j$ (see also Sect. 5), in which the model corresponding to a larger Bayesian evidence is statistically more likely (Jeffreys 1961; Trotta 2008; Corsaro et al. 2013; Corsaro and De Ridder 2014).

2 Nested Sampling Monte Carlo

Since Eq. (2) is a multi-dimensional integral, with increasing number of dimensions its evaluation becomes quickly unsolvable both analytically and by numerical approximations. For overcoming this problem, a NSMC algorithm was developed (Skilling 2004). This algorithm allows for an efficient evaluation of the Bayesian evidence for any number of dimensions and provides the sampling of the posterior probability distribution (PPD) for parameter estimation as a straightforward byproduct. Detailed descriptions of the algorithm can be found in Skilling (2004), Sivia and Skilling (2006), Feroz and Hobson (2008), Feroz et al. (2009), Corsaro and De Ridder (2014).

In short, a prior mass X is defined such that

$$X(\mathscr{L}^*) = \int_{\mathscr{L}(\boldsymbol{\theta})>\mathscr{L}^*} \pi(\boldsymbol{\theta} \mid \mathscr{M}) d\boldsymbol{\theta} \,, \tag{3}$$

with $\mathscr{L}^*$ being some fixed value of the likelihood function. As a consequence, $0 \leq X \leq 1$ because $\pi(\boldsymbol{\theta} \mid \mathscr{M})$ is a PDF. Equation (3) is therefore the fraction of volume under the prior PDF that is contained within the hard constraint $\mathscr{L}(\boldsymbol{\theta}) > \mathscr{L}^*$. This means that the higher is the constraining value $\mathscr{L}^*$, the smaller is the prior mass considered. This is equivalent to considering a portion of parameter space delimited by the iso-likelihood contour $\mathscr{L}(\boldsymbol{\theta}) = \mathscr{L}^*$, in which also the maximum value $\mathscr{L}_{\mathrm{max}}$ is contained.

In the NSMC, the sampling of the posterior PDF is performed by starting with a prior mass $X = 0$ (thus considering the entire parameter space) and an initial sampling of N_{live} points that are distributed according to the prior, hence drawn from the prior PDF itself. At each new iteration, a new sampling point is drawn from the prior PDF with a corresponding likelihood value that satisfies the hard constraint $\mathscr{L} > \mathscr{L}^*$, with $\mathscr{L}^*$ the worst likelihood value of the previous iteration. The point associated to the worst likelihood value is then removed from the sample and a new iteration starts. At the end, the prior mass reached corresponds to $X = 1$ and the sampling terminates in a region that is located around the maximum (or the maxima) of the likelihood function.

2.1 *The* DIAMONDS *Code*

The high-DImensional And multi-MOdal NesteD Sampling (DIAMONDS) code[1] is a `C++11` software for Bayesian parameter estimation and model comparison that uses a version of the NSMC algorithm. A major difficulty in implementing the

[1]DIAMONDS is publicly available at https://fys.kuleuven.be/ster/Software/Diamonds/ or through its public GitHub repository at https://github.com/EnricoCorsaro/DIAMONDS.

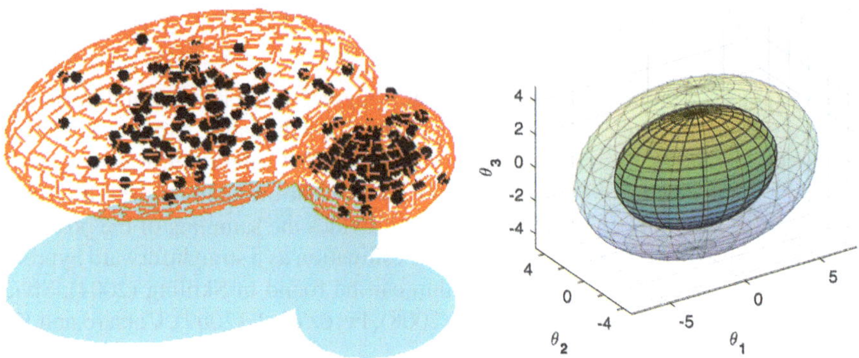

Fig. 1 *Left panel:* Three-dimensional ellipsoids containing two different clusters of sampling points in the parameter space. *Right panel:* The enlargement of an ellipsoid used to optimize the sampling efficiency throughout the nesting process

NSMC algorithm is the drawing from the prior PDF that satisfies the hard constraint in the likelihood value of the drawn point. Following on the developments made for other existing codes that implement NSMC (see, e.g., Shaw et al. 2007; Feroz and Hobson 2008; Feroz et al. 2009), DIAMONDS overcomes this problem by adopting a simultaneous ellipsoidal sampling algorithm (Corsaro and De Ridder 2014). This means that the posterior PDF is actually sampled by means of multi-dimensional ellipsoids, which decompose the parameter space $\Omega_{\mathcal{M}_k}$ into small hyper-volumes, as shown in the left panel of Fig. 1. Each ellipsoid can thus be used to easily draw new points from, and it is reduced in its volume as the nested iteration proceeds toward a termination condition. In particular, one crucial parameter to control the behavior of the ellipsoids is the initial enlargement fraction, f_0, which is used to enlarge their axes along each direction for as many dimensions as imposed by the number of free parameters. This parameter, whose effect is depicted in the right panel of Fig. 1, tunes the efficiency of the sampling throughout the nested iterations and therefore requires a careful calibration, which I show in Fig. 2 as a function of the number of dimensions, k. A calibrated relation, already implemented in DIAMONDS, reads

$$f_0 = (0.267 \pm 0.014)\, k^{0.643 \pm 0.017} \tag{4}$$

and allows for using DIAMONDS for a wide range of applications without the need to adjust the parameter f_0 every time a new model or a different number of parameters is involved in the analysis.

DIAMONDS includes a library of likelihood functions and prior PDFs that can be used for a wide range of applications. As for any inference problem, the code requires an input dataset, a model to be fit to the observations, and the adoption of a given likelihood function and of prior PDFs for each free parameter of the model. The termination condition that allows the code to finalize its computations

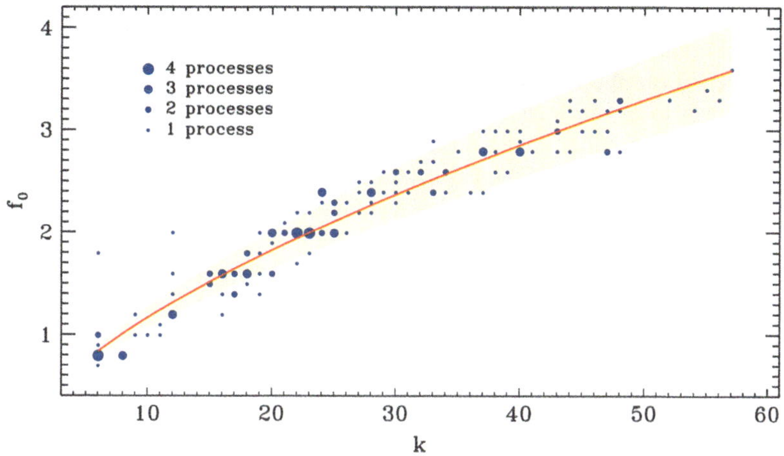

Fig. 2 The initial enlargement fraction f_0 as a function of the number of dimensions k involved in the inference problem. The 152 independent computations provided by Corsaro et al. (2015) used 4 clusters each to sample the parameter space. The size of the circles is proportional to the number of processes for which the same f_0 was used. The *colored band* shows the 68.3% confidence region for the power law fit (*thick red line*)

is based on the remaining Bayesian evidence, as described by Keeton (2011) (see also Corsaro and De Ridder 2014 for additional details). Instructions on how to configure the code and a description of its different parts can be found in the online user guide.[2] In the following examples, DIAMONDS is set up in different ways depending on the specific inference problem that is considered.

3 Fitting the Background Signal

The first step in the asteroseismic analysis process is to estimate the background signal in the power spectrum of a star.[3] This is an important phase of the analysis because if not properly performed it can introduce significant systematics in the asteroseismic parameters that characterize individual oscillation modes (Corsaro and De Ridder 2014). The first part of the tutorial is therefore focused on the estimation of the background signal in the red giant KIC 12008916, observed by NASA's *Kepler* mission (Borucki et al. 2010; Koch et al. 2010) for more than 4

---

[2]A comprehensive user guide to DIAMONDS can be found at https://fys.kuleuven.be/ster/Software/Diamonds/DIAMONDS_UserGuide.

[3]The power spectrum is usually converted into a power spectral density, PSD, to allow for direct comparisons independently of the observing length of the data. Its units are expressed in $\text{ppm}^2\,\mu\text{Hz}^{-1}$.

years. The dataset has been prepared following García et al. (2011, 2014), thus optimized for asteroseismic analysis.

In order to run the tutorial, one needs to have the DIAMONDS code already installed in a local machine. This procedure can be accomplished by following the instructions provided in the installation guide section of the code website.[4] Subsequently it is required to download the code extension for background fitting,[5] containing the specific fitting model, priors, and dataset to be used in the tutorial. The extension contains a library of `Python` routines that can be used to plot the results obtained with DIAMONDS. We note that throughout this tutorial we will adopt an exponential likelihood function, as appropriate for datasets deriving from a Fourier transform of a time series (Duvall and Harvey 1986; Corsaro and De Ridder 2014).

The background model, considered as a function of the cyclic frequency in the PSD of the star, reads

$$P_{\mathrm{bkg}}(\nu) = W + R(\nu)\left[B(\nu) + G(\nu)\right] , \tag{5}$$

where W is a flat noise level and $R(\nu)$ the response function that considers the sampling rate of the observations for *Kepler* data,

$$R(\nu) = \mathrm{sinc}^2\left(\frac{\pi\nu}{2\nu_{\mathrm{Nyq}}}\right) , \tag{6}$$

with $\nu_{\mathrm{Nyq}} = 283.212\,\mu\mathrm{Hz}$ the Nyquist frequency in the case of long-cadence data (Jenkins et al. 2010). We fit three Harvey-like profiles (Harvey 1985) given by

$$B(\nu) = \sum_{i=1}^{3} \frac{\zeta a_i^2/b_i}{1 + (\nu/b_i)^4} , \tag{7}$$

with a_i the amplitude in ppm, b_i the characteristic frequency in μHz, and $\zeta = 2\sqrt{2}/\pi$ the normalization constant (Kallinger et al. 2014). The power excess containing the oscillations is described as

$$G(\nu) = H_{\mathrm{osc}} \exp\left[-\frac{(\nu - \nu_{\mathrm{max}})^2}{2\sigma_{\mathrm{env}}^2}\right] \tag{8}$$

[4]The installation guide of DIAMONDS can be found at https://fys.kuleuven.be/ster/Software/Diamonds/installation-guide.

[5]The Background extension of DIAMONDS can be downloaded from https://fys.kuleuven.be/ster/Software/Diamonds/package/AzoresSC16_background_extension.tar.gz. Further information on how to run the tutorial can be found at http://www.iastro.pt/research/conferences/faial2016/files/presentations/TA1.pdf.

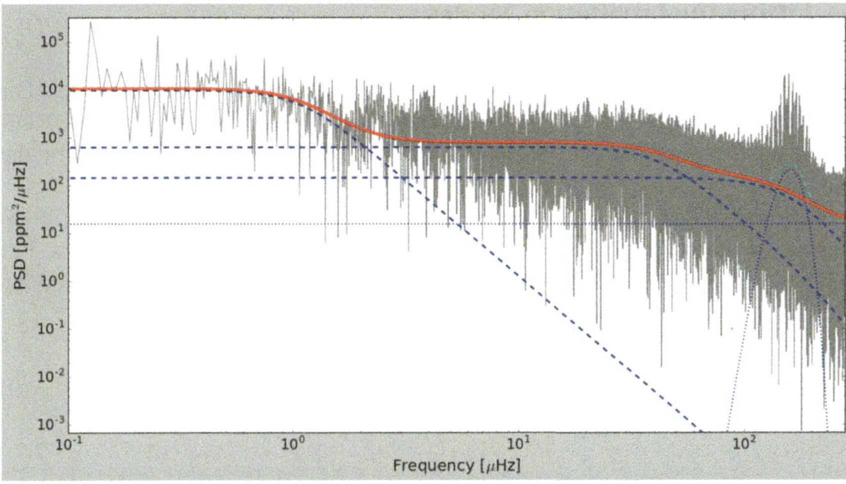

Fig. 3 Background fit of the star KIC12008916 by means of DIAMONDS. The original PSD is shown in *gray*. The *red thick line* represents the background model without the Gaussian envelope. The *cyan dotted line* accounts for the additional Gaussian component. The individual components of the background model as given by Eq. (5) are shown by *blue dot-dashed lines*

and is only considered when fitting the background model to the overall PSD of the star. The global model given by Eq. (5) therefore accounts for ten free parameters. The resulting fit obtained with DIAMONDS is shown in Fig. 3.

Questions & Problems:

- For any of the estimated free parameters, which Bayesian parameter estimator should be preferred among the mode, the median and the mean? And why?
- What is the value of ν_{max} for this star?
- Could you guess what the evolutionary stage of this red-giant star is from its ν_{max} value?
- Using your fitted ν_{max}, and assuming $\Delta\nu = 12.9\,\mu$Hz as the large frequency separation (Ulrich 1986), $T_{\mathrm{eff}} = 5100$ K, and solar reference values $\nu_{\mathrm{max},\odot} = 3100\,\mu$Hz, $\Delta\nu_{\odot} = 134.9\,\mu$Hz, and $T_{\mathrm{eff},\odot} = 5777$ K, estimate the mass and radius of the star through scaling relations.

4 Fitting the Oscillation Modes

The second part of the tutorial is related to the fitting of the oscillation modes. For this purpose it is necessary to download and install the extension of DIAMONDS related to the peak-bagging analysis,[6] similarly to what has been done for the background.

The model that is taken into account is the one presented by Corsaro et al. (2015) and includes a mixture of resolved and unresolved oscillation mode profiles. For resolved modes, i.e., modes with lifetimes much shorter than the total observing time, the typical profile is a Lorentzian expressed as

$$\mathscr{P}_{\mathrm{res},0}(\nu) = \frac{A_0^2/(\pi\Gamma_0)}{1+4\left(\frac{\nu-\nu_0}{\Gamma_0}\right)^2}, \tag{9}$$

where A_0, Γ_0, and ν_0 are the amplitude in ppm, the linewidth in μHz, and the centroid frequency in μHz, respectively, and represent the three free parameters to be estimated during the fitting process. For the unresolved modes, i.e., modes with a lifetime comparable or even longer than the total observing time, we consider the profile

$$\mathscr{P}_{\mathrm{unres},0}(\nu) = H_0 \,\mathrm{sinc}^2\left[\frac{\pi(\nu-\nu_0)}{\delta\nu_{\mathrm{bin}}}\right], \tag{10}$$

where H_0 and ν_0 are the height in PSD units and the centroid frequency in μHz of the oscillation peak, respectively, and must be estimated during the fitting process, while $\delta\nu_{\mathrm{bin}}$ is fixed as the frequency resolution of the dataset, here corresponding to 0.008 μHz.

Following Corsaro and De Ridder (2014), Corsaro et al. (2015), we fix the background parameters corresponding to the white noise, $W = \overline{W}$, and the Harvey-like profiles, $B(\nu) = \overline{B}(\nu)$, to the median values estimated in the tutorial in Sect. 3. Then, the final peak-bagging model can be represented as

$$P(\nu) = \overline{W} + R(\nu)\left[\overline{B}(\nu) + P_{\mathrm{osc}}(\nu)\right], \tag{11}$$

where

$$P_{\mathrm{osc}}(\nu) = \sum_{i=1}^{N_{\mathrm{res}}} \mathscr{P}_{\mathrm{res},i}(\nu) + \sum_{j=1}^{N_{\mathrm{unres}}} \mathscr{P}_{\mathrm{unres},j}(\nu), \tag{12}$$

[6] The PeakBagging extension of DIAMONDS can be downloaded from https://fys.kuleuven.be/ster/Software/Diamonds/package/AzoresSC16_peakbagging_extension.tar.gz. The extension contains a library of `Python` routines that can be used to plot the results obtained with DIAMONDS. Further informations on how to run the tutorial can be found at http://www.iastro.pt/research/conferences/faial2016/files/presentations/TA1.pdf.

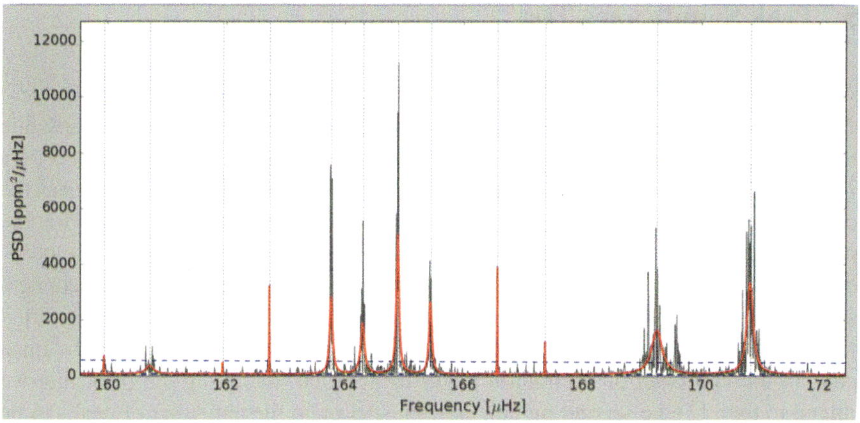

Fig. 4 Peak-bagging fit of the star KIC 12008916 by means of DIAMONDS. The original PSD is shown in *gray*. The *red thick line* represents the estimated peak-bagging model [cf. Eq. (11)], while the blue dashed lines mark the background signal and a scaled (by a factor of eight) version of it

with $N_{\rm res}$ and $N_{\rm unres}$ the number of resolved and unresolved peaks to be fitted, respectively. Clearly, any inference problem that takes into account this peak-bagging model will involve a total number of $3N_{\rm res} + 2N_{\rm unres}$ free parameters. The result of the fit for KIC 12008916 done with DIAMONDS is shown in Fig. 4.

Questions & Problems:

- In Fig. 4 spot the positions of the radial ($\ell = 0$), quadrupole ($\ell = 2$) and octupole ($\ell = 3$) modes, as follows from the asymptotic relation of the acoustic modes (Tassoul 1980).
- Which oscillation modes are the most p-dominated mixed modes?
- Compute the spacing (expressed in seconds) between the frequency $\nu_{\ell=1,m=0} = 165.178\,\mu$Hz and another frequency that has to be computed as the average between the two frequency centroids of the unresolved profiles having the largest frequency (in the range 166–168 μHz). The frequency centroids of the unresolved profiles must be those from the fitting results obtained with DIAMONDS.
- Compare the derived period spacing in the ΔP–$\Delta\nu$ diagram shown in Fig. 8 of Corsaro et al. (2012) and determine the evolutionary stage of the star assuming $\Delta\nu = 12.9\,\mu$Hz.

5 Peak Significance Test

As shown by Corsaro and De Ridder (2014) and later on applied by Corsaro et al. (2015) on red-giant stars, by means of the Bayesian evidence it is possible to perform a direct model comparison aimed at assessing the significance of a given oscillation peak. The final part of the tutorial with DIAMONDS foresees the computation of the peak significance test for one oscillation mode fitted during the peak-bagging analysis. In order to achieve this result, it is required that the peak-bagging presented in Sect. 4 is performed with two different models. By selecting a specific oscillation peak that we want to test, then the competing models to be fitted to the PSD of the star have to be defined as follows: (1) the first model, $\mathscr{M}_1$, must contain the entire set of oscillation peaks to be fitted, including the peak that we intend to test; (2) the second model, $\mathscr{M}_2$, must contain the entire set of peaks to be fitted, except the peak that we intend to test. This implies that the parameters that configure the prior PDFs of the models $\mathscr{M}_1$ and $\mathscr{M}_2$ should be identical, except for the peak to test. Using the set up of the PeakBagging extension of DIAMONDS, this can easily be achieved by removing the prior parameters of the corresponding peak when we have to fit model $\mathscr{M}_2$. Among the outputs of DIAMONDS, there will be the Bayesian evidence.[7] The best model, or statistically more likely, can be identified by computing the Bayes' factor (see Sect. 1) as $\ln \mathscr{B}_{1,2} = \ln \mathscr{E}_1 - \ln \mathscr{E}_2$. If, for example, $\ln \mathscr{B}_{1,2} > 5$, according to Jeffreys' scale of strength for the evidence (Jeffreys 1961; Trotta 2008) we then conclude that the peak is significant and that it should be considered as a real oscillation mode.

Questions & Problems:

- Why are two different models needed to test the significance of an individual peak?
- How many models are required to test the significance of two peaks?
- Perform the peak significance test for the $\ell = 3$ mode shown in Fig. 4 by means of DIAMONDS.
- Provide the value of the natural logarithm of the Bayes' factor for the aforementioned oscillation mode and assess the strength of the evidence according to Jeffreys' scale.

Acknowledgements This work has been funded by the European Community's Seventh Framework Programme (FP7/2007–2013) under grant agreement no. 312844 (SPACEINN).

[7] More details can be found at http://www.iastro.pt/research/conferences/faial2016/files/presentations/TA1.pdf.

References

Bolstad, W.: Introduction to Bayesian Statistics. Wiley, New York (2013)
Borucki, W.J., Koch, D., Basri, G., et al.: Science **327**, 977 (2010)
Corsaro, E., De Ridder, J.: Astron. Astrophys. **571**, A71 (2014)
Corsaro, E., Stello, D., Huber, D., Bedding, T.R., Bonanno, A., Brogaard, K., Kallinger, T., Benomar, O., White, T.R., Mosser, B., Basu, S., Chaplin, W.J., Christensen-Dalsgaard, J., Elsworth, Y.P., García, R.A., Hekker, S., Kjeldsen, H., Mathur, S., Meibom, S., Hall, J.R., Ibrahim, K.A., Klaus, T.C.: Astrophys. J. **757**, 190 (2012). doi:10.1088/0004-637X/757/2/190
Corsaro, E., Fröhlich, H.-E., Bonanno, A., et al.: Mon. Not. R. Astron. Soc. **430**, 2313 (2013)
Corsaro, E., De Ridder, J., García, R.A.: Astron. Astrophys. **579**, A83 (2015)
Duvall Jr., T.L., Harvey, J.W.: In: Gough, D.O. (ed.) NATO ASIC Proceedings 169: Seismology of the Sun and the Distant Stars, pp. 105–116 (1986)
Feroz, F., Hobson, M.P.: Mon. Not. R. Astron. Soc. **384**, 449 (FH08) (2008)
Feroz, F., Hobson, M.P., Bridges, M.: Mon. Not. R. Astron. Soc. **398**, 1601 (F09) (2009)
García, R.A., Hekker, S., Stello, D., et al.: Mon. Not. R. Astron. Soc. **414**, L6 (2011)
García, R.A., Mathur, S., Pires, S. et al.: Astron. Astrophys. **568**, A10 (2014)
Harvey, J.: In: Rolfe, E., Battrick, B. (eds.) Future Missions in Solar, Heliospheric & Space Plasma Physics. ESA Special Publication, vol. 235, pp. 199–208. Eruopean Space Agency, Paris (1985)
Jeffreys, H.: Theory of Probability, 3rd edn. Oxford University Press, Oxford (1961)
Jenkins, J.M., Caldwell, D.A., Chandrasekaran, H., et al.: Astrophys. J. **713**, L120 (2010)
Kallinger, T., De Ridder, J., Hekker, S., et al.: Astron. Astrophys. **570**, A41 (2014)
Keeton, C.R.: Mon. Not. R. Astron. Soc. **414**, 1418 (K11) (2011)
Koch, D.G., Borucki, W.J., Basri, G., et al.: Astrophys. J. **713**, L79 (2010)
Shaw, J.R., Bridges, M., Hobson, M.P.: Mon. Not. R. Astron. Soc. **378**, 1365 (S07) (2007)
Sivia, D., Skilling, J.: Data Analysis: A Bayesian Tutorial. Oxford Science Publications. Oxford University Press, Oxford (2006)
Skilling, J.: AIP Conf. Proc. **735**, 395 (SK04) (2004)
Tassoul, M.: Astrophys. J. Suppl. Ser. **43**, 469 (1980)
Trotta, R.: Contemp. Phys. **49**, 71 (2008)
Ulrich, R.K.: Astrophys. J. **306**, L37 (1986)

Tutorial: Asteroseismic Stellar Modelling with AIMS

Mikkel N. Lund and Daniel R. Reese

Abstract The goal of AIMS (Asteroseismic Inference on a Massive Scale) is to estimate stellar parameters and credible intervals/error bars in a Bayesian manner from a set of asteroseismic frequency data and so-called classical constraints. To achieve reliable parameter estimates and computational efficiency, it searches through a grid of pre-computed models using an MCMC algorithm—interpolation within the grid of models is performed by first tessellating the grid using a Delaunay triangulation and then doing a linear barycentric interpolation on matching simplexes. Inputs for the modelling consist of individual frequencies from peak-bagging, which can be complemented with classical spectroscopic constraints. AIMS is mostly written in `Python` with a modular structure to facilitate contributions from the community. Only a few computationally intensive parts have been rewritten in `Fortran` in order to speed up calculations.

1 Introduction

The AIMS (Asteroseismic Inference on a Massive Scale) software was developed by D. R. Reese as one of the deliverables for the SPACEINN network, a European project specialised in helio- and asteroseismology. The goal of this software is to estimate stellar parameters and reliable error bars for a given set of asteroseismic and classical constraints. The present tutorial explains how to use this software through various simple examples. Specifically, it explains how to find stellar parameters and error bars for a given set of constraints, generate a binary grid file usable by AIMS, and test the accuracy of the interpolation for a given grid. The necessary files and

M.N. Lund (✉)
School of Physics and Astronomy, University of Birmingham, Birmingham B15 2TT, UK
e-mail: mikkelnl@phys.au.dk

D.R. Reese
LESIA, Observatoire de Paris, PSL Ressearch University, CNRS, Sorbonne Universités, UPMC University Paris 06, Univ. Paris Diderot, 92195 Sorbonne Paris Citée, Meudon, France
e-mail: daniel.reese@obspm.fr

T.L. Campante et al. (eds.), *Asteroseismology and Exoplanets: Listening to the Stars and Searching for New Worlds*, Astrophysics and Space Science Proceedings 49, DOI 10.1007/978-3-319-59315-9_8

data to run the examples in this tutorial are available at the following website: http://bison.ph.bham.ac.uk/spaceinn/aims/tutorial/.

2 Getting Started

2.1 Prerequisites and Downloads for the Tutorial

The following will be needed to run the examples in this tutorial:

- **Python modules.** The following `Python` modules and utilities are needed by AIMS: `emcee`, `corner`, `dill`, `Scipy`, `Numpy`, `f2py`, `Matplotlib`. The last four are included in most distributions.
- **A grid of models.** AIMS works by comparing observational data to a grid of pre-computed models. The tutorial website provides two binary grids (`data_mesa` and `data_cestam`), which will be used for finding model fits to a set of observed stars, as well as a folder containing a non-binary subset of one of the grids, which we shall use when trying to generate a binary grid. When "unpacked" by AIMS, some of the grids take up a lot of live memory[1] (RAM). Accordingly, a "light" version of the CESTAM grid, `data_cestam_reduced`, has been provided.
- **File(s) with observational data.** The tutorial website provides files with observational data for three stars (Stars 1–3). The mode frequencies were obtained from peak-bagging of *Kepler* data for the so-called LEGACY project (Lund et al. 2017). Spectroscopic data were obtained from the Stellar Parameters Classification tool (SPC; Buchhave et al. 2012).

2.2 Downloading and Installing AIMS

The latest version of the AIMS package, currently version 1.2, can be downloaded from the following site: http://bison.ph.bham.ac.uk/spaceinn/aims/version1.2/index.html. This file is unpacked as follows:

```
tar -zxvf AIMS.tgz
```

The AIMS program itself is contained within the `AIMS` folder.

As mentioned earlier, the latest version of AIMS contains some `Fortran` subroutines which need to be compiled before running AIMS. This is done via the `f2py` program. A `Makefile` has been provided for convenience. Please edit

[1] In some cases, this problem can further be compounded by the use of parallelisation, which is activated by setting `parallel=True` in `AIMS_configure.py`.

the `Makefile` by inserting the appropriate `Fortran` compiler and compilation options. Then run the following command:

```
make
```

This will produce a file called `aims_fortran.so` which can be used by AIMS.

3 Running AIMS: Model Fit

We shall first consider running AIMS with the goal of optimising the fit to a given set of observational constraints, such as mode frequencies, ratios, and spectroscopic parameters. In Sect. 5 we will look into testing the interpolation scheme in AIMS. Figure 1.1 of the overview document[2] provides a simple schematic flowchart with the basic working components of AIMS.

3.1 Setting Up the Configuration File

Before running the AIMS program, a binary file with the grid must first be created—we will come to this in Sect. 4. Assuming in the mean time that has been done, the most important concern is to set up the configuration file: `AIMS_configure.py`.

Most parameters in the configuration file are well documented and should be self-explanatory—for instance, you can choose which asteroseismic parameters to fit[3] (individual frequencies, ratios, or average asteroseismic parameters), the name of the binary grid to use, which parameters should be output from the optimisation, control parameters for the MCMC, which grid parameters to use in the optimisation and which priors to set on these etc.

Two parameters are of special relevance:

- `write_data` should be set to `False` (see Sect. 4 for when this should be `True`).
- `test_interpolation` should be set to `False` (see Sect. 5 for when this should be `True`).

It is also very important to put the correct values for the `grid_params` and `user_params` parameters. These values will depend on the binary grid being used. They are provided on the tutorial website and can also be obtained with the `analyse_grid.py` utility,[4] which should be run in the `AIMS` folder.

[2] http://bison.ph.bham.ac.uk/spaceinn/aims/version1.2/_downloads/Overview.pdf.

[3] The user should be careful not to choose a set of asteroseismic parameters which are redundant, as this would lead to a singular covariance matrix and poor numerical results.

[4] http://bison.ph.bham.ac.uk/spaceinn/aims/tutorial/download/analyse_grid.py.

3.2 Observational Constraints

The file with observational constraints follows a very similar format to the one used with the Asteroseismic Modeling Portal[5] (AMP), apart from a few minor differences.[6] This format is described as follows:

- Characters following a "#" are ignored, and the ordering of the lines is unimportant.
- A set of lines, one per individual mode, describe the asteroseismic observables. If the keyword `read_n` is set to `True`, then four columns should be given, namely, the degree of the mode (l), the radial order (n), the frequency in cyclic μHz (f), and its associated error bar (δf). If `read_n` is `False`, then only l, f, and δf should be given. Note that in AIMS, even if you choose to work with frequency combinations such as ratios, the asteroseismic inputs are still individual frequencies—AIMS will use these to calculate the frequency combinations, associated error bars, and correlations.
- Optionally, non-asteroseismic constraints may be included in addition to the asteroseismic observables. The first column must consist of a character or a keyword, e.g., "`T` or `Teff`" ($T_{\rm eff}$), "`L` or `Luminosity`" ($L/L_{\odot}$), "`R` or `Radius`" ($R/R_{\odot}$), "`M`, `Fe_H` or `M_H`" ([M/H]), "`g` or `log_g`" ($\log g$), or "`Rho`" (ρ). This is followed by a central value and an error.
- If a non-asteroseismic constraint is included, e.g., "`Teff 5777 50`", then it is assumed that this parameter follows a Gaussian distribution. Alternatively, you may explicitly define the distribution to be used, e.g., "`Teff Uniform 5727 5837`" meaning a uniform distribution from 5727 to 5837 K. You can adopt either a "`Uniform`", "`Gaussian`" (the default), or "`Truncated_gaussian`" probability distribution.
- The average asteroseismic parameters $\nu_{\rm max}$ ("`numax`") and $\Delta\nu$ ("`Dnu`") can also be supplied here as a constraint.[7]

The following gives an example of how the constraints file may look like:

```
0 1847.63576435 0.76075810437
1 1904.67149676 0.827874359443
2 1954.85689326 0.751669075869
0 1968.18957034 0.212209047899
                ...
T 6120 80
Fe_H -0.06 0.12
numax 2763 100
```

[5]See https://amp.phys.au.dk/guide/fileformat.

[6]See http://bison.ph.bham.ac.uk/spaceinn/aims/version1.2/formats.html#format-of-a-file-with-observational-constraints.

[7]We note that this is not the preferred way of supplying $\Delta\nu$, as it does not correctly take into account correlations with other asteroseismic constraints. A better approach is to introduce the large separation via the `seismic_constraints` variable in `AIMS_configure.py`.

3.3 *Running AIMS*

Once the grid and configuration file are set up, one simply runs the program as follows:

```
./AIMS.py observational_constraints_file
```

AIMS will then import the Python configuration file, `AIMS_configure.py`, so make sure you do not modify the name of this file. Output generated from AIMS will then be saved to a run folder with the same name as your constraints file, so give your constraints file a sensible name so that you can distinguish the results from several runs for the same star. The variable `output_dir` in `AIMS_configure.py` specifies the path to the root folder which contains all of the run folders.

3.4 *Understanding the Results*

The results from AIMS are obtained from the posterior distributions of the MCMC run on the model grid, as illustrated in the bottom panel of Fig. 1. Inside the file `AIMS_configure.py`, you can define the set of stellar parameters for which you want an output, as well as which plots to create. For computational reasons, such parameters are only calculated for a subset of the MCMC samples, except for those parameters actively used in the MCMC optimisation. Accordingly, AIMS produces two files with the samples: `samples.txt` with all of the MCMC samples but only the parameters involved in the optimisation, and `samples_big.txt` with a subset of the samples but all of the stellar parameters. The corresponding files `results.txt` and `results_big.txt` provide summary statistics for the above samples, namely, the distribution averages and standard deviations, along with correlations between different model parameters. The samples files may, of course, be used to extract different summary statistics (e.g., median or mode) for the parameters. The file `best_MCMC_model.txt` gives the stellar parameters and computed mode frequencies for the best model from the MCMC, whereas `best_grid_model.txt` gives the best model from the initial grid search used to initialise the MCMC (i.e., prior to interpolation within the grid). The top panel of Fig. 1 shows the échelle diagram produced from the observations and from the best MCMC model.

4 Creating Your Own Binary Grid Files with AIMS

To compute a binary grid file, you first need to calculate a grid of models with your favourite stellar evolution code, as well as oscillation modes for each model. Information for each model should then be entered into a "model list" file. The first line of this file should contain a prefix which is typically the root folder of the

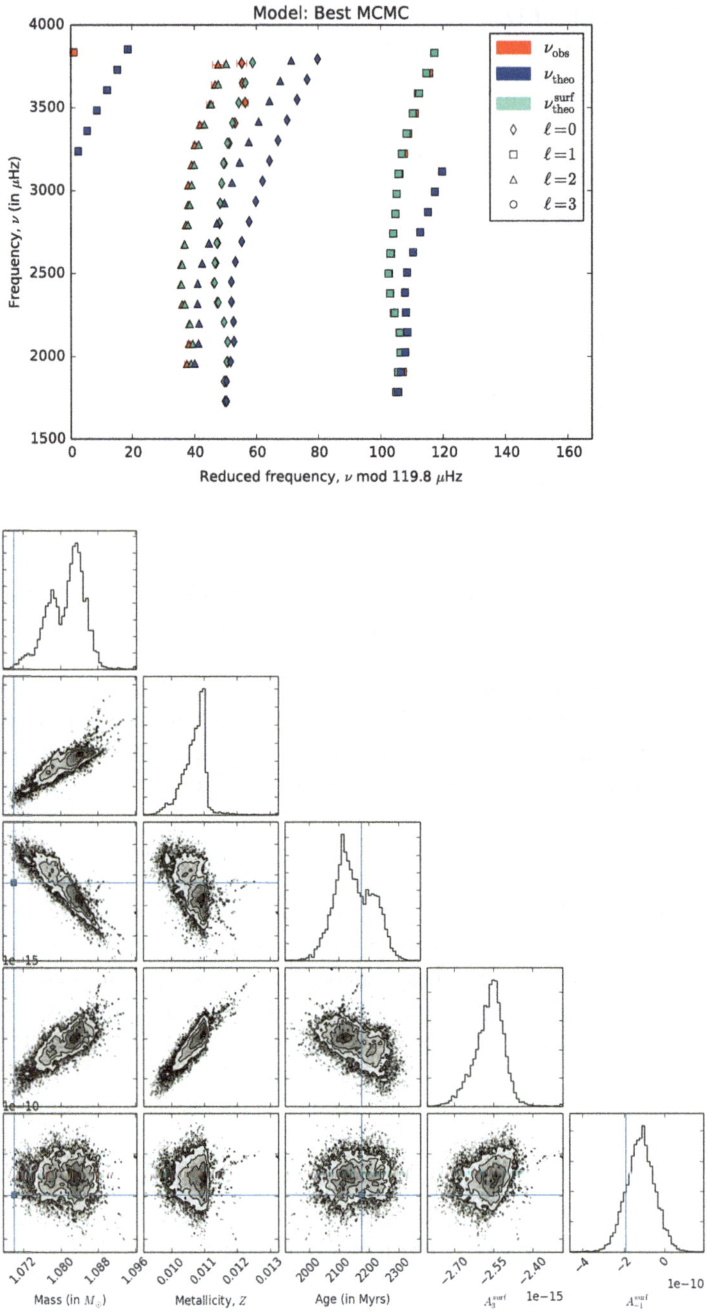

Fig. 1 Example of output plots from an AIMS run. *Top:* Échelle diagram for the best MCMC model showing the observed frequencies, theoretical frequencies, and surface-corrected theoretical frequencies. *Bottom:* Triangle plot with correlation maps between different stellar parameters for the MCMC samples. The *blue lines* indicate the results obtained from the initial full-grid search

Table 1 Columns in "model list" file

Column #	1	2	3	4	5	6	7	8	9, 10 ...
Parameter	Model name	Mass	Rad.	Lum.	Z_0	X_0	Age	$T_{\rm eff}$	`user_params`
Unit	–	g	cm	$\mathrm{g\,cm^2\,s^{-3}}$	–	–	Myr	K	–

grid of models and, optionally, a postfix giving the end part of the filenames with the mode frequencies, the default value being "`.freq`". The following lines then contain multiple columns with the information for each model in the grid. Below we show an example of a model list file (see column description in Table 1):

```
/home/dreese/models_inversions/Grid_6819/m1.6.ovh0.0.ovhe0.00.z0.01756.y0.26/
m1.6.z0.01756.y0.26_n2026.FGONG  3.18272E+33  9.411631E+11  2.689504E+35  0.01756  0.72244  2.402848E+03  4543.38696
m1.6.z0.01756.y0.26_n2093.FGONG  3.18272E+33  9.645173E+11  2.811105E+35  0.01756  0.72244  2.402920E+03  4537.93601
m1.6.z0.01756.y0.26_n1986.FGONG  3.18272E+33  9.010663E+11  2.513558E+35  0.01756  0.72244  2.402825E+03  4565.49605
m1.6.z0.01756.y0.26_n1575.FGONG  3.18272E+33  1.166064E+12  3.745062E+35  0.01756  0.72244  2.402670E+03  4434.01896
```

The prefix plus each model name in the first column, plus the postfix gives the name of a file that contains the oscillation parameters of the model. These files can come in one of two formats, as specified by the `mode_format` variable in `AIMS_configure.py`. One of the formats is a `Fortran` binary format known as the "grand summary" file from the ADIPLS[8] code (Christensen-Dalsgaard 2008) and is described on pages 32–33 of the ADIPLS documentation.[9] The other format is a text format (described below) and which is what is used in this tutorial.

The text version of the oscillation parameter files begins with a one-line header followed by five columns which correspond to l, n, frequency, `dfreq_var`, and mode inertia. Note that the `dfreq_var` column is currently discarded, as are frequencies above the estimated cut-off frequency times the value of the `cutoff` variable in `AIMS_configure.py`.

To generate a binary grid you should specify to following relevant parameters in `AIMS_configure.py`:

- `write_data` should be set to `True`.
- `list_grid` gives the filename of the model list file (see above).
- `binary_grid` gives the filename of the binary file that will be generated. If `write_data` is set to `False`, this is the binary grid that will be loaded.
- `grid_params` specifies the parameters relevant to the grid you want to generate (excluding age, which will be dealt with separately). *It is extremely important that each set of values for these parameters corresponds to a unique*

[8]http://astro.phys.au.dk/~jcd/adipack.n/.

[9]http://astro.phys.au.dk/~jcd/adipack.n/notes/adiab_prog.ps.gz.

evolutionary track, since AIMS *reconstructs these tracks based on these values.* For instance, if mass and Z are the parameters which describe your grid, then pairs of values such as (mass $= 1\,M_{\odot}, Z = 0.02$) should correspond to a unique track.
- `user_params` specifies supplementary parameters which describe your grid. This variable should contain a pair of strings for each supplementary parameter. The first string should give the parameter's name.[10] The second string should be a nice LaTeXversion of the name to be used in plot titles.
- `npositive` should be set to `True` if you only want to save $n \geq 0$ modes (i.e., acoustic modes) in the binary file.
- `cutoff`: frequencies above this value times the estimated cut-off frequency (as based on a scaling law) will be discarded. For example, if `cutoff=1.1` then only frequencies below $1.1\,\nu_c$ are kept.
- `agsm_cutoff`: this only applies to binary "grand summary" files from ADIPLS. If set to `True`, then only frequencies for which `icase=10010` (i.e., which are below the cut-off frequency when using an isothermal boundary condition) are kept.

The binary file is then simply generated by running:

```
./AIMS.py
```

5 Testing Interpolation Accuracy

5.1 Calculating Interpolation Errors

Since AIMS works by interpolating in a pre-computed grid of stellar models, it also includes a way of testing the accuracy of the interpolation. To test the interpolation you should specify to following options in `AIMS_configure.py`:

- `write_data` should be set to `False`.
- `test_interpolation` should be set to `True`.
- `interpolation_file` gives the name of the binary file that will contain the output of the test. The results saved in this file can then be plotted using `plot_interpolation_test.py`.

The interpolation test is then simply run as:

```
./AIMS.py
```

[10]The names "`Xs`" and "`Zs`" should be used for the surface hydrogen and metallicity content, as some of the functions in AIMS specifically look for these variables.

5.2 Various Interpolation Errors

There are two basic components to model interpolation in AIMS, each of which contributes to interpolation error:

- **Age interpolation:** this is interpolation along a given evolutionary track.
- **Track interpolation:** this is interpolation as a function of the other model parameters, such as mass, metallicity, mixing length parameter, or whatever parameters are relevant to your grid.

The first type of interpolation is dealt with through a simple linear interpolation between two adjacent models on the evolutionary track. The second uses Delaunay tessellation before calculating linear barycentric weights. For a more detailed description of interpolation in AIMS, we refer the reader to Chap. 4 of the overview document (see footnote 2).

The interpolation tests carried out in AIMS allow the user to estimate the error from both types of interpolation. For the age interpolation, we number the models on a given evolutionary track, starting at $n = 0$. As schematically illustrated in Fig. 2, the age interpolation tests involve combining models $n - n_{\text{incr}}$ and $n + n_{\text{incr}}$, and seeing how well the interpolated frequencies reproduce the frequencies of model n. This test is carried out throughout the entire track except for the n_{incr} models at either end. Figure 2 schematically illustrates these interpolation tests for $n_{\text{incr}} = 2$. AIMS carries out tests for $n_{\text{incr}} = 1$ and 2 in order to assess the impact of the time step on the age interpolation.

Testing track interpolation (see Figs. 3 and 4) is more complicated because it is based on a Delaunay tessellation. The approach used in AIMS involves randomly selecting half of the evolutionary tracks, creating a new tessellation from these, and using this to interpolate to the remaining tracks. Figure 4 illustrates such a partitioning of the evolutionary tracks.

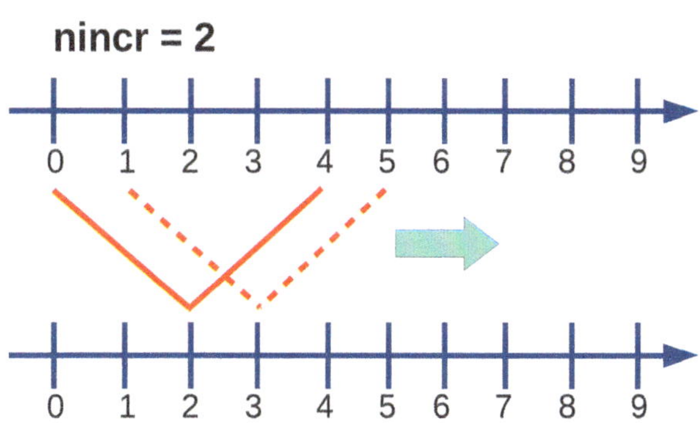

Fig. 2 Schematic plot showing how the age interpolation tests are calculated for `nincr=2`. Successive models along the track are not necessarily equally spaced in age

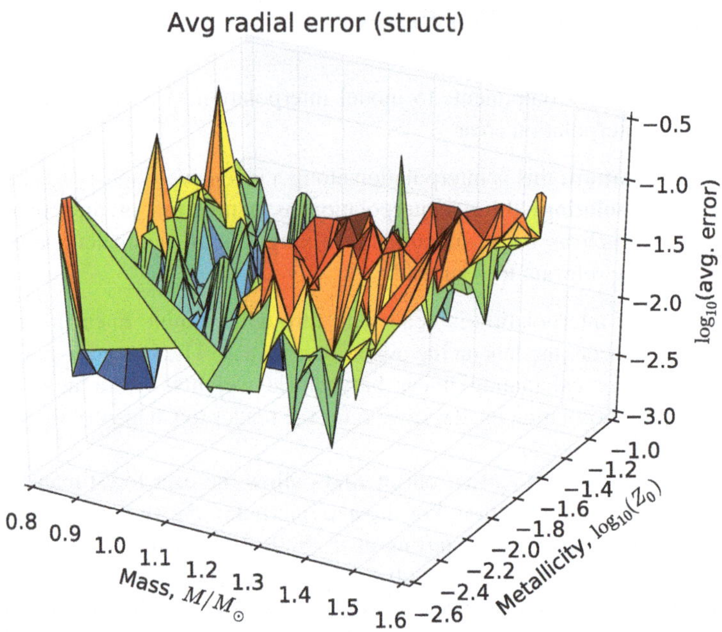

Fig. 3 Average track interpolation errors for radial modes as a function of stellar parameters

When comparing frequencies from an interpolated model with those from the original model, AIMS calculates different types of error bars. First of all, separate error bars are obtained for radial ($l = 0$) and for non-radial modes. These are further subdivided into the following categories:

- the maximum error;
- a root-mean square (RMS) error;
- an RMS error only based on the modes between $0.8\nu_{\max}$ and $1.2\nu_{\max}$, where $\nu_{\max}$ is the frequency at maximum power (this is obtained from the models via a scaling relation).

5.3 *Analysing the Results*

The results in the generated `interpolation_file` can be visualised with `plot_interpolation_test.py` by running:

```
plot_interpolation_test.py interpolation_file
```

Currently, this program only works for 3D model grids (including the age dimension).

Running this program will generate a series of plots that can be used to assess the errors introduced by the interpolation. The first 9 plots are 3D plots which show various errors as a function of the grid parameters, excluding age. These plots come

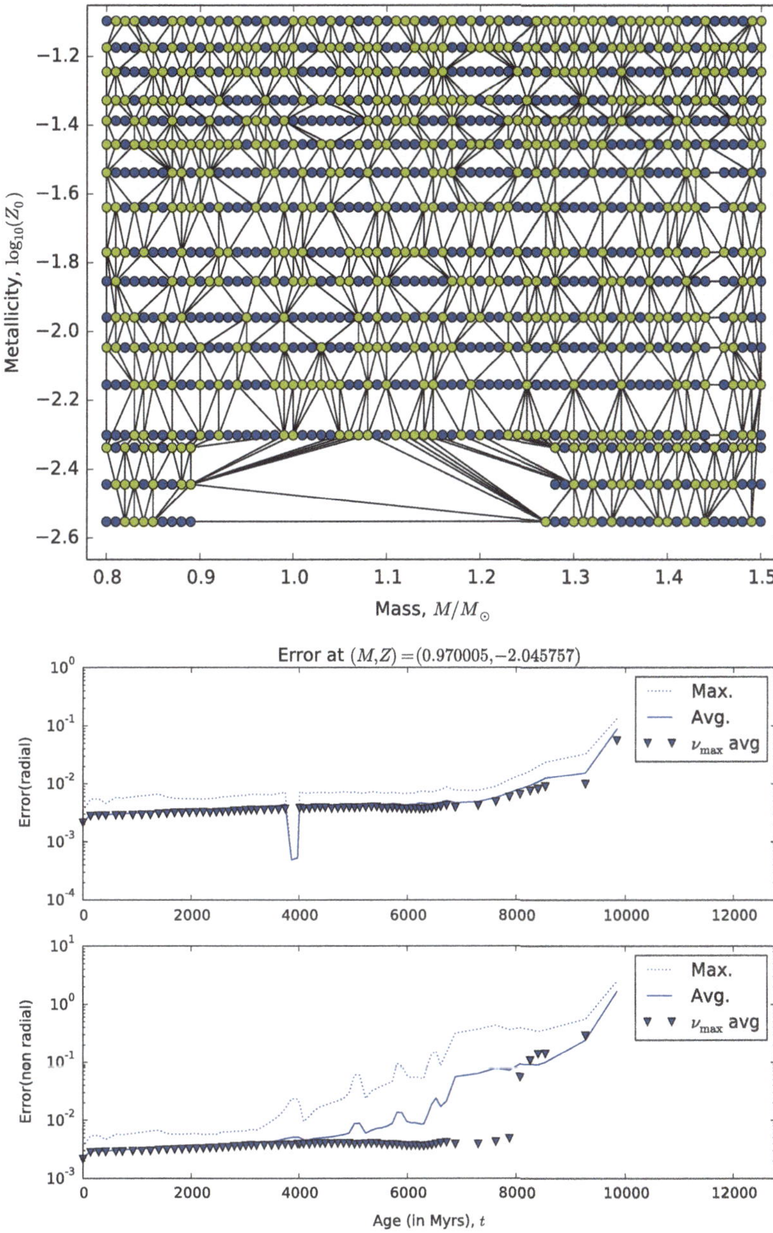

Fig. 4 Testing interpolation accuracy. *Top:* Interactive plot with the positions of the evolutionary tracks in parameter space. The *yellow dots* correspond to the tracks with which a new tessellation is created, as represented by the *connecting lines*. The *blue dots* represent tracks where the interpolation is tested. *Bottom:* Clicking on a *blue dot* produces plots with the track interpolation errors as a function of stellar age

in groups of three: the first two plots in a group show age interpolation errors for $n_{\rm incr} = 1$ and 2, and the third shows track interpolation errors. The groups of plots correspond to the following:

- **Plots 1–3:** Maximum interpolation errors for radial modes as a function of stellar parameters. For each model along a track, the maximum error is obtained. Then the maximum along the entire track is calculated.
- **Plots 4–6:** Average interpolation errors for radial modes as a function of stellar parameters. The RMS average is calculated along the entire track. Figure 3 shows such a plot for track interpolation errors.
- **Plots 7–9:** Average interpolation errors for radial modes restricted to the interval $0.8\nu_{\rm max}$ to $1.2\nu_{\rm max}$ as a function of stellar parameters. The RMS average is calculated along the entire track.

These plots are displayed in individual windows. Thanks to `Python`'s interactive capabilities, it is possible to rotate the plots and to zoom in or out.

Two additional 2D interactive plots display the positions of the evolutionary tracks in the stellar parameter space. The first of these shows all of the evolutionary tracks as blue dots. Clicking on a blue dot opens up a new window with two new plots which show how the age interpolation errors, for both radial and non-radial modes, vary as a function of stellar age. The second plot shows a partitioning of the evolutionary tracks used in the track interpolation tests as described above. An example of such a plot is shown in the top panel of Fig. 4. Clicking on a blue dot on this plot opens up a new window with two new plots with track interpolation errors as a function of age, like the ones shown in the bottom panel of Fig. 4.

More information on the individual plotting functions can be found in the comments within the `plot_interpolation_test.py` file as well as at the following website: http://bison.ph.bham.ac.uk/spaceinn/aims/version1.2/plot_interpolation_test.html.

6 Recommended Reading

For more information on the use of Bayesian inference in model optimisation, we recommend Bazot et al. (2012) and Gruberbauer et al. (2013). For details on the affine-invariant MCMC optimisation scheme used (`emcee`), we refer the reader to Goodman and Weare (2010) and Foreman-Mackey et al. (2013). For details on asteroseismic grid-based analysis in general, we refer to Gai et al. (e.g., 2011).

Acknowledgements AIMS is a software for fitting stellar pulsation data, developed in the context of the SPACEINN network, funded by the European Commission's Seventh Framework Programme. DRR wishes to thank all those who helped him in the development of AIMS, including D. Bossini, T.L. Campante, W.J. Chaplin, H.R. Coelho, G.R. Davies, B.D.C.P. Herbert, J.S. Kuszlewicz, M.W. Long, M.N. Lund, and A. Miglio.

References

Bazot, M., Bourguignon, S., Christensen-Dalsgaard, J.: A Bayesian approach to the modelling of α Cen A. Mon. Not. R. Astron. Soc. **427**, 1847 (2012)

Buchhave, L.A., Latham, D.W., Johansen, A., et al.: An abundance of small exoplanets around stars with a wide range of metallicities. Nature **486**, 375 (2012)

Christensen-Dalsgaard, J.: ADIPLS—the Aarhus adiabatic oscillation package. Astrophys. Space Sci. **316**, 113 (2008)

Foreman-Mackey, D., Hogg, D.W., Lang, D., Goodman, J.: emcee: The MCMC hammer. Publ. Astron. Soc. Pac. **125**, 306 (2013)

Gai, N., Basu, S., Chaplin, W.J., Elsworth, Y.: An in-depth study of grid-based asteroseismic analysis. Astrophys. J. **730**, 63 (2011)

Goodman, J., Weare, J.: Ensemble samplers with affine invariance. Commun. Appl. Math. Comp. Sci. **5**, 65 (2010)

Gruberbauer, M., Guenther, D.B., MacLeod, K., Kallinger, T.: Bayesian asteroseismology of 23 solar-like Kepler targets. Mon. Not. R. Astron. Soc. **435**, 242 (2013)

Lund, M.N., Silva Aguirre, V., Davies, G.R., et al.: Standing on the shoulders of dwarfs: the Kepler asteroseismic LEGACY sample. I. Oscillation mode parameters. Astrophys. J. **835**, 172 (2017)

Part II
Exoplanetary Science

Exoplanetary Science: An Overview

Nuno C. Santos and João P. Faria

Abstract Two decades ago, the first exoplanet was found orbiting a solar-type star. Since then, many surprising discoveries have been made. We now know that the architecture and properties of the Solar System's planets are more an exception than a rule. In this paper we review the main planet detection methods and present some of the most interesting discoveries as well as their connection to the study of planet-host stars.

1 Introduction

The number of known extrasolar planets is rising on an almost daily basis, with more than 3500 currently listed at The Extrasolar Planets Encyclopaedia[1] (Schneider et al. 2011). The impact of these discoveries is considerable, both scientifically and socially. They represent the first firm steps of humankind towards the detection and characterization of other planets similar to our Earth. This domain is opening new bridges between different fields in Astrophysics (e.g., stellar astrophysics, solar system research) and other areas of knowledge such as geophysics (e.g., Valencia et al. 2006) and biology (Kaltenegger and Sasselov 2011). Together these bring new hopes of finding an Earth-like planet where life may have evolved.

[1]http://exoplanet.eu.

N.C. Santos (✉)
Instituto de Astrofísica e Ciências do Espaço, Universidade do Porto, CAUP, Rua das Estrelas, 4150-762 Porto, Portugal

Departamento de Física e Astronomia, Faculdade de Ciências, Universidade do Porto, Rua do Campo Alegre, 4169-007 Porto, Portugal
e-mail: nuno.santos@astro.up.pt

J.P. Faria
Instituto de Astrofísica e Ciências do Espaço, Universidade do Porto, CAUP, Rua das Estrelas, 4150-762 Porto, Portugal
e-mail: joao.faria@astro.up.pt

T.L. Campante et al. (eds.), *Asteroseismology and Exoplanets: Listening to the Stars and Searching for New Worlds*, Astrophysics and Space Science Proceedings 49, DOI 10.1007/978-3-319-59315-9_9

The diversity of discovered planets is raising new questions and opening new pathways. The community presently agrees that planets, in particular rocky planets like our Earth, are very common around solar-type stars (FGK and M dwarfs; e.g., Udry et al. 2007; Howard et al. 2012; Mayor et al. 2014). This conclusion is fully supported by state-of-the-art planet formation models based on the core-accretion paradigm, that predict low-mass/small-radius planets to largely surmount the number of their Jovian or Neptune-like counterparts (Ida and Lin 2004; Mordasini et al. 2012).

While the number and variety of discovered planets are still important assets for exoplanet research (e.g., with an impact for the models), the focus of extrasolar planet researchers is now moving towards three main lines: (1) the detection of progressively lower-mass planets, with the goal of finding an Earth sibling, (2) the detailed characterization of known exoplanets, including their interior structures and atmospheres, and (3) the understanding of the planet formation processes, by comparing the properties of the detected worlds with expectations from theoretical models.

All these lines of research have already seen their own success. Radial-velocity (RV) and transit surveys have found an increasing number of low-mass/small-radius planets orbiting other suns (Borucki et al. 2011; Dumusque et al. 2012; Mayor et al. 2014). Some of these may even be in the habitable zone (Pepe et al. 2011; Quintana et al. 2014; Anglada-Escudé et al. 2016). The precision of the transit measurements, in combination with mass determinations from RV measurements and planet interior models, also allowed to determine the bulk composition of several planets (Léger et al. 2009; Howard 2013; Pepe et al. 2013). For the most favorable cases, exquisite measurements further allowed to detect both the emitted (infrared) and reflected (optical) light of exoplanets, as well as the presence of specific atmospheric lines (Brogi et al. 2012; Rodler et al. 2012; Martins et al. 2015). These measurements are providing a first insight into the physics of exoplanet atmospheres.

New hopes are now coming from new instruments, including ground-based optical spectrographs such as ESPRESSO (Pepe et al. 2014), capable of achieving down to sub-m/s precision in RV. A whole new generation of infrared (IR) spectrographs is also on its way, as well as a new optical and near-infrared (NIR) high-resolution spectrograph for ESO's E-ELT. To these we should add a whole new generation of ground- and space-based projects that will search and characterize transiting low-mass/small-radius planets. Many of these projects are meant to start operating in no more than 1 or 2 years, opening for the first time the possibility to address in a comprehensive way the detection and characterization of Earth-like planets orbiting nearby stars.

In this chapter we will briefly review some of the most relevant points inherent to exoplanet research. We will start by reviewing the main detection methods, their potential and limitations. The challenges imposed by astrophysical noise will then be presented, including the problems produced by stellar activity. We then briefly review the main results from exoplanet research. We will concentrate on the Doppler and transit methods, with a stronger accent on the former. The information provided by the analysis of the planet-host stars will further be discussed. We will conclude with the prospects from future missions and projects in the field. More details can be found in the cited papers.

2 Planet Detection Methods

At a moment when thousands of planets are known, we may wonder that exoplanet detection has always been an easy task. However, we must be reminded that a long way has been covered from the early frustrating efforts using the astrometric methods (e.g., van de Kamp 1983) until the detection of the first exoplanet orbiting a solar-type star (51 Peg b; Mayor and Queloz 1995). The enthusiasm that followed this first detection precipitated the discovery of other worlds. The numbers slowly started rising, but in the end of the twentieth century only a dozen or so planets were known to orbit other solar-type stars.

The detection of the first transiting planet was only due 5 years after 51 Peg b (Charbonneau et al. 2000). This detection was not only the first allowing to derive the radius of an exoplanet, but also a confirmation that the "strange" hot Jupiters (short-period giant planets, previously not expected to exist based on planet formation models) were indeed planetary in nature (see below). The years that followed were of great success: planets with lower and lower masses were being found using the RV method (Butler et al. 2004; McArthur et al. 2004; Santos et al. 2004). Several transit surveys also started to provide the first results. But only in the last few years we could finally start probing in more detail the smaller mass/radius end of the planet population. This was largely due to the launch of the *Kepler* mission (Borucki et al. 2011) and to the improvement of the methods used to extract information from the radial-velocity data (e.g., Fischer et al. 2016).

Indeed, although the present day success of the exoplanet detection and characterization efforts is based on the development of a number of different techniques and methods, the radial-velocity and transit methods are certainly the most prolific so far. In this review we will thus concentrate on a basic description of the most used or successful methods, or those that will, from our perspective, be more fruitful in the years to come. In particular, we will focus on the radial-velocity, transit, and astrometric methods, as well as on a review of their stronger aspects and challenges. For more details and a description of other methods, we point the reader to some recent reviews (Seager et al. 2010; Perryman 2014) and to some of the chapters in this book.

2.1 Radial Velocities

The method behind the detection of the first exoplanet orbiting a solar-type star is the radial-velocity technique. This method is based on the detection of the stellar motion (or wobble) of the star around the center of mass of the star-planet system.

The velocity semi-amplitude expected for a star of mass M_1 orbited by a planet of mass M_2 can be shown to be:

$$K_1 = 212.9 \left(\frac{M_1}{P}\right)^{1/3} \frac{q}{(1+q)^{2/3}} \frac{\sin i}{\sqrt{1-e^2}} \quad \left[\mathrm{km\,s^{-1}}\right] , \tag{1}$$

where $q = M_2/M_1$, and i is the inclination of the orbital axis with respect to the line of sight. In this equation, the masses (M_1 and M_2) are expressed in solar masses, and the orbital period (P) in days (for details, see, e.g., Hilditch 2001). This equation shows that the higher the planet mass and the shorter the orbital period, the stronger will be the RV signal. RVs are thus more sensitive to short-period, massive planets. It is thus with no surprise that the first detected planets orbiting Sun-like stars were short-period giants, the so called hot Jupiters[2] (Mayor et al. 2014).

The RV of the star can be measured from the Doppler shift with high-resolution spectroscopic measurements, using the Doppler equation $\frac{\Delta\lambda}{\lambda} = \frac{v}{c}$, where c is the speed of light, λ is the reference wavelength (at zero velocity; typically the reference wavelength of an absorption spectral line), $\Delta\lambda$ is the wavelength shift observed, and v is the radial velocity. The biggest challenge of this technique is that one needs to measure the stellar velocity with a very high precision. In optical wavelengths these small amplitudes translate to values of $\Delta\lambda \sim 10^{-4}$Å. For comparison, a typical high-resolution spectrograph (with a resolution $R = \lambda/\Delta\lambda = 100{,}000$) is able to resolve two adjacent wavelengths separated by ~0.1Å.

From Eq. (1) we can derive that the semi-amplitude K_1 of a star induced by the presence of a Jupiter-like planet (with a mass of 318 $M_\oplus$ and an orbital period of ~12 years) is only ~13 m/s, while for an Earth-like planet this value decreases to a mere ~8 cm/s.

To circumvent the difficulties involved in the detection of such low-amplitude signals, two main aspects must be taken into account. First, the typical spectrum of a solar-type star has thousands of well-defined absorption lines. Using this information in a statistical way we will be able to achieve the necessary precision. But this is not enough if the spectrograph itself is not stable, or if we cannot control the instrument drifts as a function of time. An accurate way to measure and control the wavelength-to-pixel calibration is thus needed. This is usually achieved using the spectrum of a calibration lamp that is obtained simultaneously with the target spectrum (e.g., Baranne et al. 1996), or using a gas cell whose spectrum is superposed on the spectrum of our star (e.g., Campbell et al. 1988).

For reference, currently the most accurate RV instruments for planet searches are able to measure long-term RVs with a precision better than 1 m/s (for a review, see Fischer et al. 2016). Future instruments such as ESPRESSO, for the ESO-VLT, will achieve the 10 cm/s level, allowing to detect Earth-like planets in the habitable zones of solar-type stars.

[2]Even if they are not the most prevalent kind of planets.

One immediate limitation of the RV technique is that we are only able to measure the projected radial velocity, i.e., the component of the radial-velocity in the direction of the line of sight. This implies that we can only estimate the "projected mass" of the companion responsible for the observed stellar wobble, i.e., its minimum mass ($M_2 \sin i$). Fortunately, it can be shown that for orbits randomly oriented in space it is much more likely to have $\sin i$ close to unity. This means that the minimum masses obtained are statistically very close to the real masses. The unambiguous determination of the true mass is, however, only possible if a value for the orbital inclination is obtained (e.g., through an astrometric detection or a transit measurement).

2.2 *Photometric Transits*

When a planet crosses the stellar disk as seen from us, it will occult part of it. This phenomenon, called a transit, can be observed if the orbital axis of the planet is closely perpendicular to our line of sight. For a given system, we can compute that the geometric probability (p) that a full transit will occur can be expressed by (to a good approximation):

$$p = \frac{R_{\rm star}}{a}, \tag{2}$$

where $R_{\rm star}$ and a are the stellar and orbital radius, respectively. This formula is valid for the case of a circular orbit. From this equation we can see that the transit technique is more sensitive to short-period planets. While for a 3-day orbit hot Jupiter p is close to 10%, for a planet at 1 AU from its parent star (orbital period close to 1 year) p goes down to only 0.5%.

If a transit event is observed, the expected luminosity variation can be derived to be of the order of:

$$\frac{\Delta L}{L} = \left(\frac{R_{\rm planet}}{R_{\rm star}}\right)^2. \tag{3}$$

For a Jupiter-like planet, $R_{\rm planet} \sim 0.1\,R_{\rm star}$, inducing thus a photometric variation of the order of 1%. A value of the order of 100 parts per million (ppm) is expected for an Earth-radius object. Large planets around small stars are thus easier to detect using the transit method.

Transits have shown to be an excellent way to detect planets orbiting other stars. At first used to complement the detections of radial-velocity planets (e.g., Charbonneau et al. 2000), large ground- and space-based surveys like WASP, HAT, *CoRoT* and *Kepler* revealed the presence of thousands of candidates, some with radii smaller than that of Earth. However, one point that is clear from above is the fact that the transit method only provides information about the radius of the planet (provided that the stellar radius is known). Except in a few cases where multi-planet systems are detected and, e.g., planet-planet interactions are strong (e.g., Ford et al. 2012), there is no information on the planet mass.

In other words, given the large diversity of planet densities already discovered, if we want to characterize the planet in more detail, we need to use a method (e.g., radial velocities) that enables us to derive the planet mass. Having both radius and mass allows one to estimate the planet density, and thus have a first idea about the planetary composition. Further to this, it is well known that different "non-planetary" phenomena can produce signals that almost perfectly mimic a photometric transit. These are responsible for the large false positive rates that affect especially the higher-mass planets (Santerne et al. 2016). Complementary observations are thus usually needed once a transit signature is detected, even if a statistical confirmation is done (Torres et al. 2011; Díaz et al. 2014).

2.3 Astrometry

Astronomers have long tried to use the dynamical effect that a planet has on the stellar motion to measure the small astrometric periodic shift of a star as it moves about the center of mass of the star-planet system. The astrometric detection of an extrasolar planet can be described, in a very basic approach, by simple physics. The semi-major axis of the orbital motion of a star around the center of mass of a two-body system can be described by:

$$M_1\, a_1 = M_2\, a_2\,, \tag{4}$$

where M_1 and M_2 are the masses of the two bodies, and a_1 and a_2 the semi-major axes of their orbits. The distance $a = a_1 + a_2$ (the semi-major axis of the relative orbit) is also related to the orbital period P through Kepler's third law,

$$P^2 = \frac{a^3}{M_1 + M_2}\,. \tag{5}$$

In principle, if we measure a_1, a_2, and P, we can solve the system above and derive the mass of the two bodies. This is the case for some visual binary stars. In practice, the measurement of the astrometric motion of the star in a star-planet system is far more complex. For instance, we can only hope to measure a_1 and the period P, since we are not able to directly observe the planet. To solve the above system we need, for example, to estimate the mass of the star (M_1) using stellar evolution models.

Given the small expected astrometric motions[3] (of the order of 1 microarcsecond for the best cases), current technology still did not allow to unambiguously detect from scratch a planet orbiting another star using the astrometric method. The only existing detections are of planets or brown dwarfs firstly detected using the radial-velocity technique (e.g., Benedict et al. 2006). More hopes come from the analysis of data from the ESA *Gaia* mission, whose exquisite astrometric precision is expected to allow the detection of thousands of giant planets (Sozzetti et al. 2001).

As we can see from Eqs. (4) and (5), the semi-major axis of the astrometric motion of the star around the center of mass of the star-planet system is proportional both to the mass of the companion and to its orbital period. This means that the astrometric technique is most sensitive to long-period companions. This makes this method complementary to the radial-velocity technique. Similarly to the latter, however, it is mostly sensitive to the detection of planets around lower-mass stars.

3 Challenges

Adding to the technical challenges mentioned above, the detection and characterization of other planets also has to deal with the "noise" induced by different astrophysical sources. Phenomena related to stellar activity, stellar granulation, and oscillations are particularly nasty for exoplanet detection and characterization efforts using the radial-velocity method. They can prevent us from finding planets, if the perturbation is larger than the orbital RV variation, or even give us false candidates, if they produce a periodic and stable signal over a few rotational periods (e.g., Figueira et al. 2010). Furthermore, these physical phenomena produce signals with different timescales: from several years (related with the long-term magnetic cycles; Santos et al. 2010) down to a few minutes (the oscillation modes of stars, that allow one to apply asteroseismic methods to probe the stellar interiors; Dumusque et al. 2011).

The acoustic modes of solar-type stars as well as the atmospheric granulation motions can induce RV amplitudes of the order of a few m/s. To circumvent this effect, long exposures (longer than the timescale of the oscillation modes) are usually taken to average out the solar-type acoustic modes. These modes have typical periods of the order of 5 min in a solar-type star. The granulation noise has longer timescales (from hours to a few days), that are more difficult to handle. Specific observational strategies are often used to minimize the problem (e.g., Dumusque et al. 2012).

The phenomena related to stellar activity, induced by strong magnetic fields in the stellar surface (e.g., spots, faculae, convective changes), also strongly affect both the photometric and RV signals, in timescales typical of the rotational period of the

[3]For example, a Jupiter-like planet in a 10-year orbit around a solar-mass star located 10 pc away from us, induces an astrometric motion of only 440 microarcseconds.

star. In RVs, measurements of different activity indicators, such as line asymmetries, through bisector analysis (e.g., Queloz et al. 2001) are often used to diagnose these effects. However, recent examples show that the effects of stellar activity features are not fully understood, leading to signals that almost perfectly mimic the signature of planets (even for non-active stars; Santos et al. 2014). Recent developments of data analysis techniques promise to solve part of these problems (e.g., Faria et al. 2016).

Stellar activity is also particularly relevant when dealing with transit searches. Not only it induces strong photometric modulations (that need to be filtered), but also they induce in-transit fluctuations that prevent us from having precise values for the transit depth, and hence the planet radius (e.g., Oshagh et al. 2013). More details about these and other effects are discussed in the chapter by M. Oshagh in this book.

In brief, different sources of noise are a strong challenge in planet detection and characterization efforts. The understanding of the different physical phenomena as well as ways to model or subtract them are now one of the most important avenues to guarantee the success of future ground- and space-based exoplanet projects.

4 Population Statistics

The explosion of exoplanet discoveries in the 1990s was soon followed by studies of the overall population of exoplanets. These studies rely on dedicated surveys that allow for the detection efficiencies to be quantified as a function of planet parameters such as orbital period, mass and eccentricity. In this section we summarize the main results of the ensemble study of the planet population, focusing on radial-velocity surveys.

The minimum masses and orbital periods of the known exoplanets detected with radial velocities are shown in Fig. 1. We can identify three main groups: massive giant planets at short and long periods, and lower-mass planets at short periods.

4.1 Gas-Giant Planets

51 Peg b (Mayor and Queloz 1995) and other short-period planets discovered soon after (e.g., Butler et al. 1997) were the first examples of a new family of planets with circular orbits, Jupiter-like masses, and orbital periods less than 10 days (upper left quadrant of Fig. 1). These so-called *hot Jupiters* are now known to be relatively rare, with an occurrence rate close to 1% (Marcy et al. 2005; Mayor et al. 2011; Wright et al. 2012), but the fact that they exist as well as their orbital properties provide important clues to their formation process (e.g., Batygin et al. 2016; and references therein).

Radial-velocity surveys also revealed a population of gas giants at orbital distances between 1 and 5 AU (upper right quadrant of Fig. 1) with an overall

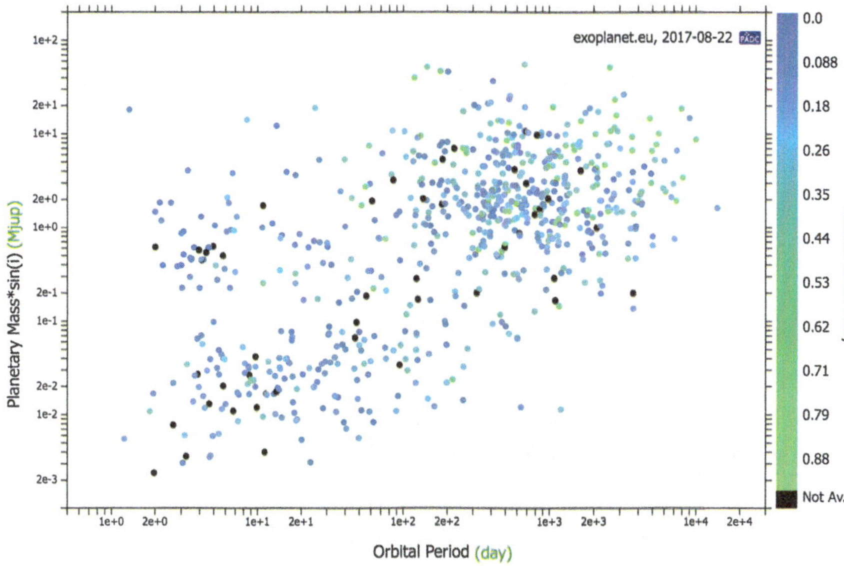

Fig. 1 Distribution of periods and minimum masses for the extrasolar planets detected with the radial-velocity method. The color code reflects the eccentricity of each planet. Obtained from http://exoplanet.eu

occurrence rate of about 15% for minimum masses larger than 50 $M_\oplus$ (0.15 M_{Jup}) and orbital periods less than 10 years (e.g., Udry and Santos 2007).

The mass distribution of the giant planets is shown in Fig. 2. The distribution peaks around 1–2 M_{Jup} and presents a long tail toward masses larger than 10 M_{Jup}. Within the brown-dwarf regime (masses between ~15 M_{Jup} and ~60 M_{Jup}) the number of detections is very small, in what has been called the "brown-dwarf desert". The number of objects with larger masses (stars) then rises again (not shown in Fig. 2). The bimodality of this mass distribution is taken as evidence of different formation mechanisms for stellar binaries and planetary systems (Sahlmann et al. 2010).

The population of giant planets shows a wide distribution of orbital eccentricities (noticeable in Fig. 1), unlike the low eccentricities seen in the Solar System. The eccentricity distribution is closer to that of binary stars (Udry and Santos 2007; their Fig. 6). This can be a signature of planet-planet scattering or interactions with bound or passing stellar companions. Many giant planets are also found in systems with various dynamical configurations (e.g., Correia et al. 2009; Wright et al. 2009). In some multi-planet systems, the gravitational interactions between planets are strong enough to be detectable in radial-velocity measurements, allowing for the orbital inclination angles and the true masses of the planets to be measured (e.g., Correia et al. 2010).

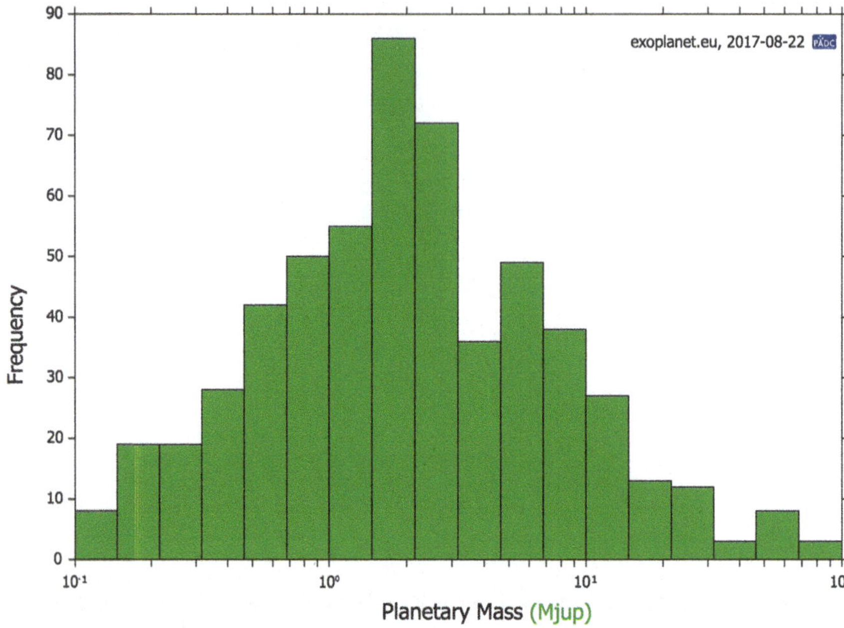

Fig. 2 Mass distribution of the population of giant planets detected with the RV method. Obtained from http://exoplanet.eu

4.2 Neptunes and Super-Earths

Some radial-velocity instruments, such as HARPS (Mayor et al. 2003; Lovis et al. 2006) or APF (Vogt et al. 2014), were built to achieve consistent precisions of 1 m/s. This opens the possibility for RV surveys to search for lower-mass planets. But since the detection of low-mass planets requires a large observational effort, only a few surveys gathered enough detections to allow for statistical studies. Results from the HARPS survey of FGK stars (Mayor et al. 2011), the HARPS survey of M dwarfs (Bonfils et al. 2013), and the HIRES survey of FGK stars (Howard et al. 2010), suggest a large population of Neptunes and super-Earths in short-period orbits. The occurrence rate of these planets, with $M \sin i < 30 M_\oplus$ and $P < 50$ days, is about 30% around FGK stars and 40% around M dwarfs.

The left panel of Fig. 3 shows the mass distribution of the planets detected in the HARPS survey (Mayor et al. 2011). After correcting the distribution for the detection biases (red histogram), we clearly see the importance of the population of low-mass planets, with a strong decrease of the number of planets between a few Earth masses and about 40 $M_\oplus$. Planet population synthesis models (e.g., Mordasini et al. 2009) predicted these features of the mass distribution to be detectable when the RV measurement precision reached 1 m/s (see Fig. 3, right panel).

The high occurrence rates tell us that systems of multiple planets with masses between 1 $M_\oplus$ and 20 $M_\oplus$, orbiting within 0.5–1 AU, are the most common type of

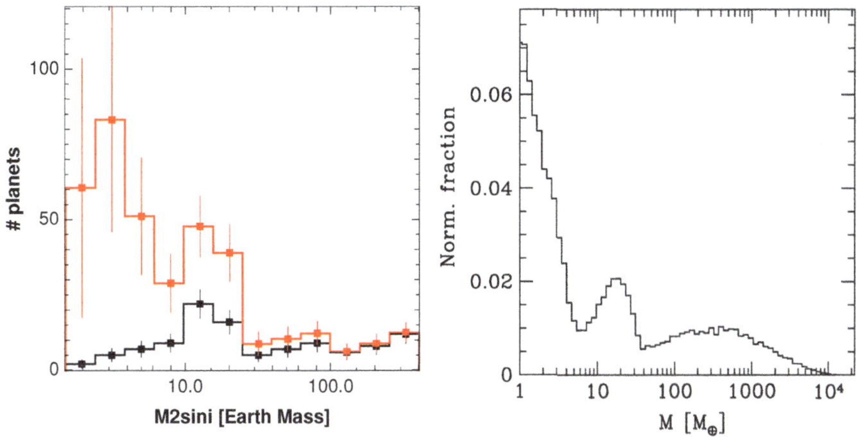

Fig. 3 *Left:* Planetary mass distributions from the HARPS planet survey, before (*black*) and after (*red*) correction for the detection bias (from Mayor et al. 2011). *Right:* Planetary initial mass function from a population synthesis model (from Mordasini et al. 2009)

planetary systems in the Galaxy. This remarkable result has been confirmed with transit surveys (see, e.g., Lissauer et al. 2014) and is in agreement with planet population synthesis models (Ida and Lin 2004; Mordasini et al. 2012, 2015).

4.3 Benchmark Planetary Systems

From the pool of exoplanet discoveries, a few systems have intrinsic interest as historical landmarks or as examples of the diversity of planetary properties. We briefly describe three such discoveries, which also showcase the interplay between the transit and radial-velocity methods.

Kepler-78 Kepler-78b was found transiting its G-type host star, with an unusually short orbital period of just 8.5 h (Sanchis-Ojeda et al. 2013). Radial-velocity follow-up observations provided, for the first time, a mass measurement for an Earth-sized planet (Pepe et al. 2013). The planet has a radius of 1.16 $R_\oplus$ and a mass of 1.86 $M_\oplus$, resulting in a density 5.57 g cm^{-3}, which is similar to that of the Earth (5.51 g cm^{-3}). This suggests that Kepler-78b is also made primarily of rock and iron. This is one of many examples where combined RV and photometric observations provide strong constraints on the internal constitution of the planets.

Even though Kepler-78 is an active star, the RV detection was possible because of the short orbital period. The separation between activity-induced and planetary signals is much easier when the orbital period of the planet is much smaller than the rotation period of the star (e.g., Hatzes 2014).

HD 10180 Lovis et al. (2011) reported on the discovery of one of the most populated exoplanet systems known to date. Up to seven planets were found orbiting the solar-type star HD 10180 (but see Tuomi 2012; who find evidence for nine planets). This system is interesting because of its complex orbital configuration, showing significant secular interactions but no mean-motion resonances.

By the time this system was discovered, it became clear that several low-mass planetary systems exhibit a "packed" orbital architecture with little or no (dynamical) space left for additional planets. These architectures can be interpreted as the signature of planet formation scenarios in which type-I migration plays a major role.

GJ 667C Located at about 22 light years away in the constellation of Scorpius, GJ 667C is the smallest member of its triple star system. This star is an M1.5V red dwarf with an estimated mass of 0.33 times that of the Sun. The discovery history of the planetary system orbiting GJ 667C is complicated (Anglada-Escudé et al. 2012, 2013; Bonfils et al. 2013; Delfosse et al. 2013), mainly because of the presence of a few candidate planets orbiting well inside the habitable zone. Some of these planets have since been put into question (Feroz and Hobson 2014; Robertson and Mahadevan 2014) as being due to the magnetic activity of the star. Finding exoplanets in the habitable zones of M dwarfs is easier than in solar-type stars, but the rotation periods are also closer to the orbital periods of such planets. Additional data for this star and more refined analysis techniques are required to definitively explain the observed variations in the radial velocity of GJ 667C.

5 Know the Stars, Know the Planets

The study of planet-host stars is paramount to the understanding of the properties and formation mechanisms of exoplanets. For example, when a planet is found transiting, the measurement precision on the planetary radius depends directly on a precise knowledge of the stellar radius (cf. Sect. 2.2; see also Torres et al. 2008; Mortier et al. 2013) In addition, the chemical compositions of the planet (both interior and atmospheric), the protostellar disk and the stellar atmosphere are linked. Therefore, precise stellar chemical abundances can provide important clues in understanding the planets and their observed properties.

The metallicity-giant planet correlation is one example of this connection: it is one of the most striking and best established results for the population of giant planets that their host stars have a higher average metallicity when compared with field stars (e.g., Gonzalez 1997; Santos et al. 2001; Fischer and Valenti 2005). The clear correlation between the presence of giant planets and metallicity is visible in the left panel of Fig. 4. This correlation is key in constraining the planet formation process and its existence provides strong evidence for core-accretion being the main process of formation of giant planets. In protoplanetary disks with higher metallicity, rocky or icy cores are able to form in time for runaway accretion before

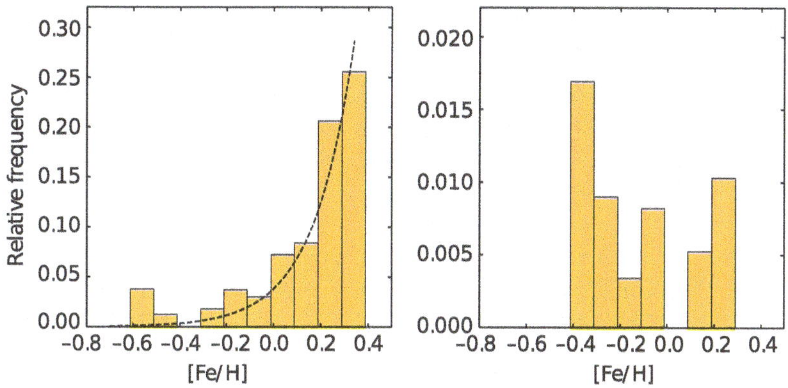

Fig. 4 The metallicity distribution of stars hosting giant planets (*left*) and Neptunes or super-Earths (*right*). There is a clear correlation between the presence of giant planets and the metallicity of the star, but this trend is not present for stars hosting lower-mass planets (Sousa et al. 2011). Adapted from Mayor et al. (2014)

disk dissipation occurs, while in lower-metallicity disks the cores do not grow fast enough to accrete gas in large quantities before disk dissipation, which results in a lower fraction of giant planets (Mordasini et al. 2009).

For the lower-mass planets (Fig. 4, right panel), there is no strong metallicity dependence of the occurrence rate (Sousa et al. 2011), indicating that these planets can form around stars with a wide range of metallicities (Buchhave et al. 2012). However, more recent results suggest that Neptunes (between 10 and 40 $M_{\oplus}$) may indeed show a (weak) metallicity correlation, while super-Earths ($< 10\,M_{\oplus}$) do not (Adibekyan et al. 2012; Courcol et al. 2016; Mulders et al. 2016).

The study of specific elemental abundances gives further insight into the planet-formation process. The abundance of α elements, for example, plays an important role in the formation of planetary systems, especially in metal-poor environments (Adibekyan et al. 2012). Abundances of other chemical elements, such as lithium (Reddy et al. 2002; Israelian et al. 2009; Baumann et al. 2010) and refractory elements, are also possible signatures of planet engulfment or terrestrial planet formation (González Hernández et al. 2010; Ramírez et al. 2010).

Stellar metallicity also plays an important role in the architectures of planetary systems: planets around metal-poor stars show longer periods and lower eccentricities than those with metal-rich hosts (Adibekyan et al. 2013; Dawson and Murray-Clay 2013). These trends point to the importance of planet-disk interaction and orbital migration, and provide constraints for the models and numerical simulations of planet formation and evolution.

6 A Bright Future Ahead

The diversity of exoplanet worlds is overwhelming. What else can the future bring? In the next few years we will have access to an array of instruments that will allow the study of entire planetary systems around nearby bright stars. This will mean precise measurements of both stellar and planetary parameters, from a combination of different observational techniques.

The legacy of *Kepler* in the search for transits of small planets will be in the hands of the Next Generation Transit Search (NGTS; Wheatley et al. 2013), the MEarth project (Irwin et al. 2009), and the *TESS* (Ricker et al. 2016), *CHEOPS* (Broeg et al. 2013) and *PLATO* (Rauer et al. 2014) space missions.

Radial-velocity surveys will continue to explore planetary systems in the solar neighborhood, with high-precision spectrographs like HARPS, HARPS-N, HIRES and APF. The cm/s level is now on sight with ESPRESSO (Pepe et al. 2014), and the NIR domain is open for exploration with CARMENES/Calar Alto (Quirrenbach et al. 2010), SPiROU/CFHT (Artigau et al. 2014), HPF/HET (Mahadevan et al. 2010) and GIANO/TNG (Oliva et al. 2004). Follow-up of transit detections with these instruments will provide precise densities and internal compositions for a large number of planets.

The *Gaia* mission already started delivering high-accuracy fundamental stellar parameters for all of the planet-host stars (Lindegren et al. 2016) and will also detect giant planets at intermediate semi-major axes. The *James Webb Space Telescope* (Gardner et al. 2006) and the future ground-based extremely large telescopes will study the atmospheric composition of the planets with both transmission and emission spectroscopy.

In summary, the instrumentation of the next few years will answer many of the most important questions about exoplanets. Many other new surprises will come as new discoveries arise. No one can tell with certainty if one of these instruments or missions will not discover the first Earth orbiting another Sun.

References

Adibekyan, V.Z., Delgado Mena, E., Sousa, S.G., et al.: Astron. Astrophys. **547**, A36 (2012)
Adibekyan, V.Z., Figueira, P., Santos, N.C., et al.: Astron. Astrophys. **560**, A51 (2013)
Anglada-Escudé, G., Arriagada, P., Vogt, S.S., et al.: Astrophys. J. **751**, L16 (2012)
Anglada-Escudé, G., Tuomi, M., Gerlach, E., et al.: Astron. Astrophys. **556**, A126 (2013)
Anglada-Escudé, G., Amado, P.J., Barnes, J., et al.: Nature **536**, 437 (2016)
Artigau, É., Kouach, D., Donati, J.-F., et al.: Ground-based and airborne instrumentation for astronomy V. In: Proceedings of SPIE, vol. 9147, 914715 (2014)
Baranne, A., Queloz, D., Mayor, M., et al.: Astron. Astrophys. Suppl. Ser. **119**, 373 (1996)
Batygin, K., Bodenheimer, P.H., Laughlin, G.P.: Astrophys. J. **829**, 114 (2016)
Baumann, P., Ramírez, I., Meléndez, J., Asplund, M., Lind, K.: Astron. Astrophys. **519**, A87 (2010)
Benedict, G.F., McArthur, B.E., Gatewood, G., et al.: Astron. J. **132**, 2206 (2006)
Bonfils, X., Delfosse, X., Udry, S., et al.: Astron. Astrophys. **549**, A109 (2013)

Borucki, W.J., Koch, D.G., Basri, G., et al.: Astrophys. J. **736**, 19 (2011)
Broeg, C., Fortier, A., Ehrenreich, D., et al.: EPJ Web Conf. **47**, 03005 (2013)
Brogi, M., Snellen, I.A.G., de Kok, R.J., et al.: Nature **486**, 502 (2012)
Buchhave, L.A., Latham, D.W., Johansen, A., et al.: Nature (2012). doi:10.1038/nature11121
Butler, R.P., Marcy, G.W., Williams, E., Hauser, H., Shirts, P.: Astrophys. J. **474**, L115 (1997)
Butler, R.P., Vogt, S.S., Marcy, G.W., et al.: Astrophys. J. **617**, 580 (2004)
Campbell, B., Walker, G.A.H., Yang, S.: Astrophys. J. **331**, 902 (1988)
Charbonneau, D., Brown, T.M., Latham, D.W., Mayor, M.: Astrophys. J. **529**, L45 (2000)
Correia, A.C.M., Udry, S., Mayor, M., et al.: Astron. Astrophys. **496**, 521 (2009)
Correia, A.C.M., Couetdic, J., Laskar, J., et al.: Astron. Astrophys. **511**, A21 (2010)
Courcol, B., Bouchy, F., Deleuil, M.: Mon. Not. R. Astron. Soc. **461**, 1841 (2016)
Dawson, R.I., Murray-Clay, R.A.: Astrophys. J. **767**, L24 (2013)
Delfosse, X., Bonfils, X., Forveille, T., et al.: Astron. Astrophys. **553**, A8 (2013)
Díaz, R.F., Almenara, J.M., Santerne, A., et al.: Mon. Not. R. Astron. Soc. **441**, 983 (2014)
Dumusque, X., Udry, S., Lovis, C., Santos, N.C., Monteiro, M.J.P.F.G.: Astron. Astrophys. **525**, A140 (2011)
Dumusque, X., Pepe, F., Lovis, C., et al.: Nature, **491**, 207 (2012)
Faria, J.P., Haywood, R.D., Brewer, B.J., et al.: Astron. Astrophys. **588**, A31 (2016)
Feroz, F., Hobson, M.P.: Mon. Not. R. Astron. Soc. **437**, 3540 (2014)
Figueira, P., Marmier, M., Bonfils, X., et al.: Astron. Astrophys. **513**, L8 (2010)
Fischer, D.A., Valenti, J.: Astrophys. J. **622**, 1102 (2005)
Fischer, D.A., Anglada-Escude, G., Arriagada, P., et al.: Publ. Astron. Soc. Pac. **128**, 066001 (2016)
Ford, E.B., Fabrycky, D.C., Steffen, J.H., et al.: Astrophys. J. **750**, 113 (2012)
Gardner, J.P., Mather, J.C., Clampin, M., et al.: Space Sci. Rev. **123**, 485 (2006)
Gonzalez, G.: Mon. Not. R. Astron. Soc. **285**, 403 (1997)
González Hernández, J.I., Israelian, G., Santos, N.C., et al.: Astrophys. J. **720**, 1592 (2010)
Hatzes, A.P.: Astron. Astrophys. **568**, A84 (2014)
Hilditch, R.W.: An Introduction to Close Binary Stars. Cambridge University Press, Cambridge (2001)
Howard, A.W.: Science **340**, 572 (2013)
Howard, A.W., Marcy, G.W., Johnson, J.A., et al.: Science **330**, 653 (2010)
Howard, A.W., Marcy, G.W., Bryson, S.T., et al.: Astrophys. J. Suppl. Ser. **201**, 15 (2012)
Ida, S., Lin, D.N.C.: Astrophys. J. **616**, 567 (2004)
Irwin, J., Charbonneau, D., Nutzman, P., Falco, E.: 15th cambridge workshop on cool stars, stellar systems, and the sun. In: Stempels, E. (ed.) American Institute of Physics Conference Series, vol. 1094, pp. 445–448 (2009)
Israelian, G., Delgado Mena, E., Santos, N.C., et al.: Nature **462**, (2009). doi:10.1038/nature08483
Kaltenegger, L., Sasselov, D.: Astrophys. J. **736**, L25 (2011)
Léger, A., Rouan, D., Schneider, J., et al.: Astron. Astrophys. **506**, 287 (2009)
Lindegren, L., Lammers, U., Bastian, U., et al.: Astron. Astrophys. **595**, A4 (2016)
Lissauer, J.J., Dawson, R.I., Tremaine, S.: Nature **513**, 336 (2014)
Lovis, C., Pepe, F., Bouchy, F., et al.: Society of Photo-Optical Instrumentation Engineers (SPIE) Conference Series, 6269 (2006). doi:10.1117/12.669991
Lovis, C., Ségransan, D., Mayor, M., et al.: Astron. Astrophys. **528**, A112 (2011)
Mahadevan, S., Ramsey, L., Wright, J., et al.: Ground-based and airborne instrumentation for astronomy III. In: Proceedings of SPIE, vol. 7735, 77356X (2010)
Marcy, G., Butler, R.P., Fischer, D., et al.: Prog. Theor. Phys. Suppl. **158**, 24 (2005)
Martins, J.H.C., Santos, N.C., Figueira, P., et al.: Astron. Astrophys. **576**, A134 (2015)
Mayor, M., Queloz, D.: Nature **378**, 355 (1995)
Mayor, M., Pepe, F., Queloz, D., et al.: The Messenger **114**, 20 (2003)
Mayor, M., Marmier, M., Lovis, C., et al.: arXiv:1109.2497 (2011)
Mayor, M., Lovis, C., Santos, N.C.: Nature **513**, 328 (2014)
McArthur, B.E., Endl, M., Cochran, W.D., et al.: Astrophys. J. **614**, L81 (2004)
Mordasini, C., Alibert, Y., Benz, W., Naef, D.: Astron. Astrophys. **501**, 1161 (2009)

Mordasini, C., Alibert, Y., Benz, W., Klahr, H., Henning, T.: Astron. Astrophys. **541**, A97 (2012)
Mordasini, C., Mollière, P., Dittkrist, K.-M., Jin, S., Alibert, Y.: Int. J. Astrobiol. **14**, 201 (2015)
Mortier, A., Santos, N.C., Sousa, S.G., et al.: Astron. Astrophys. **558**, A106 (2013)
Mulders, G.D., Pascucci, I., Apai, D., Frasca, A., Molenda-Żakowicz, J.: Astron. J. **152**, 187 (2016)
Oliva, E., Origlia, L., Maiolino, R., et al.: Ground-based instrumentation for astronomy. In: Moorwood, A.F.M., Iye, M. (eds.) Proceedings of SPIE, vol. 5492, pp. 1274–1279 (2004)
Oshagh, M., Boisse, I., Boué, G., et al.: Astron. Astrophys. **549**, A35 (2013)
Pepe, F., Lovis, C., Ségransan, D., et al.: Astron. Astrophys. **534**, A58 (2011)
Pepe, F., Cameron, A.C., Latham, D.W., et al.: Nature **503**, 377 (2013)
Pepe, F., Molaro, P., Cristiani, S., et al.: Astron. Nachr. **335**, 8 (2014)
Perryman, M.A.C.: The Exoplanet Handbook, 1st edn. Cambridge University Press, Cambridge (2014)
Queloz, D., Henry, G.W., Sivan, J.P., et al.: Astron. Astrophys. **379**, 279 (2001)
Quintana, E.V., Barclay, T., Raymond, S.N., et al.: Science **344**, 277 (2014)
Quirrenbach, A., Amado, P.J., Mandel, H., et al.: Ground-based and airborne instrumentation for astronomy III. In: Proceedings of SPIE, vol. 7735, 773513 (2010)
Ramírez, I., Asplund, M., Baumann, P., Meléndez, J., Bensby, T.: Astron. Astrophys. **521**, A33 (2010)
Rauer, H., Catala, C., Aerts, C., et al.: Exp. Astron. **38**, 249 (2014)
Reddy, B.E., Lambert, D.L., Laws, C., Gonzalez, G., Covey, K.: Mon. Not. R. Astron. Soc. **335**, 1005 (2002)
Ricker, G.R., Vanderspek, R., Winn, J., et al.: Society of photo-optical instrumentation engineers (SPIE) conference series. In: Proceedings of SPIE, vol. 9904, 99042B (2016)
Robertson, P., Mahadevan, S.: Astrophys. J. **793**, L24 (2014)
Rodler, F., Lopez-Morales, M., Ribas, I.: Astrophys. J. **753**, L25 (2012)
Sahlmann, J., Ségransan, D., Queloz, D., Udry, S.: Proc. Int. Astron. Union **6**, 117 (2010)
Sanchis-Ojeda, R., Rappaport, S., Winn, J.N., et al.: Astrophys. J. **774**, 54 (2013)
Santerne, A., Moutou, C., Tsantaki, M., et al.: Astron. Astrophys. **587**, A64 (2016)
Santos, N.C., Israelian, G., Mayor, M.: Astron. Astrophys. **373**, 1019 (2001)
Santos, N.C., Bouchy, F., Mayor, M., et al.: Astron. Astrophys. **426**, L19 (2004)
Santos, N.C., Mayor, M., Benz, W., et al.: Astron. Astrophys. **512**, A47 (2010)
Santos, N.C., Mortier, A., Faria, J.P., et al.: Astron. Astrophys. **566**, A35 (2014)
Schneider, J., Dedieu, C., Le Sidaner, P., Savalle, R., Zolotukhin, I.: Astron. Astrophys. **532**, A79 (2011)
Seager, S., Dotson, R., Lunar and Planetary Institute (eds.): Exoplanets, The University of Arizona space science series. University of Arizona Press: Tucson; In collaboration with Lunar and Planetary Institute: Houston (2010). oCLC: ocn617461672
Sousa, S.G., Santos, N.C., Israelian, G., Mayor, M., Udry, S.: Astron. Astrophys. **533**, A141 (2011)
Sozzetti, A., Casertano, S., Lattanzi, M.G., Spagna, A.: Astron. Astrophys. **373**, L21 (2001)
Torres, G., Winn, J.N., Holman, M.J.: Astrophys. J. **677**, 1324 (2008)
Torres, G., Fressin, F., Batalha, N.M., et al.: Astrophys. J. **727**, 24 (2011)
Tuomi, M.: Astron. Astrophys. **543**, A52 (2012)
Udry, S., Santos, N.C.: Annu. Rev. Astron. Astrophys. **45**, 397 (2007)
Udry, S., Bonfils, X., Delfosse, X., et al.: Astron. Astrophys. **469**, L43 (2007)
Valencia, D., O'Connell, R.J., Sasselov, D.: Icarus, **181**, 545 (2006)
van de Kamp, P.: Astron. Nachr. **304**, 97 (1983)
Vogt, S.S., Radovan, M., Kibrick, R., et al.: Publ. Astron. Soc. Pac. **126**, 359 (2014)
Wheatley, P.J., Pollacco, D.L., Queloz, D., et al.: EPJ Web Conf. **47**, 13002 (2013)
Wright, J.T., Upadhyay, S., Marcy, G.W., et al.: Astrophys. J. **693**, 1084 (2009)
Wright, J.T., Marcy, G.W., Howard, A.W., et al.: Astrophys. J. **753**, 160 (2012)

Deriving High-Precision Radial Velocities

Pedro Figueira

Abstract This chapter describes briefly the key aspects behind the derivation of precise radial velocities. I start by defining radial velocity precision in the context of astrophysics in general and exoplanet searches in particular. Next I discuss the different basic elements that constitute a spectrograph, and how these elements and overall technical choices impact on the derived radial velocity precision. Then I go on to discuss the different wavelength calibration and radial velocity calculation techniques, and how these are intimately related to the spectrograph's properties. I conclude by presenting some interesting examples of planets detected through radial velocity, and some of the new-generation instruments that will push the precision limit further.

1 Precise Radial Velocities

Radial velocities are, by definition, the velocities measured along a given line of sight, and often refer to a velocity calculated through the measurement of the Doppler shift of a given spectral line. In its non-relativistic form, the well-known Doppler shift formula,

$$\frac{\Delta\lambda}{\lambda} = \frac{v}{c}, \tag{1}$$

relates the displacement $\Delta\lambda$ of a line of wavelength λ to a radial velocity v, with c being the speed of light in vacuum. Measuring a radial velocity (henceforth RV) is fundamentally different from measuring directly a velocity on the plane of the sky, in the sense that the error on the RV does not depend geometrically on the distance to the source as it does for physical velocities measured on the plane of the sky.

P. Figueira (✉)
Instituto de Astrofísica e Ciências do Espaço, Universidade do Porto, CAUP, Rua das Estrelas, 4150-762 Porto, Portugal
e-mail: pedro.figueira@astro.up.pt

T.L. Campante et al. (eds.), *Asteroseismology and Exoplanets: Listening to the Stars and Searching for New Worlds*, Astrophysics and Space Science Proceedings 49, DOI 10.1007/978-3-319-59315-9_10

Instead, the error on the RV depends only on the noise present in the spectrum, and how this noise translates to an uncertainty on the line shift value.

At this point, it is extremely important to clarify what we mean by *precision*, and to characterize the associated type of error. The precision of a measurement system, also called reproducibility or repeatability, is the degree to which repeated measurements performed under unchanged conditions lead to the same results. This repeatability error can be associated to a value measured by the system, becoming the precision of the value. This is conceptually different from the accuracy of a measurement system or value, which is how close a measurement of a quantity is to its real—true—value. In several astronomical studies, one is concerned about accurate RVs; here we are only concerned with precise RVs.

Precise RVs have been used to calculate the velocities of stars and study, for instance, Galactic kinematics, stellar binarity, and determine stellar masses. For these scientific objectives, an overall precision of the order of the km/s was enough. More recently, the presence and characterization of exoplanets was possible when the precision threshold crossed the level of 50–100 m/s. On the other hand, to measure stellar oscillations, differential line shifts, and line profile variations, the asteroseismic studies routinely require a precision on the order of 1–100 m/s, often measured in individual lines as opposed to the whole spectrum, as has been done for exoplanetary searches or binary characterization. These examples serve to illustrate how different scientific objectives require different precision level. But to understand how to derive precise RVs, we will have to understand how a spectrograph works.

2 Breaking Down a Spectrograph

A spectrograph is a scientific instrument that receives the light collected by a telescope, disperses it, forming a spectrum, and records this spectrum on a detector. But what is inside it? In a conceptual way, a spectrograph is composed of four types of components:

- The light interface/feeding with the telescope—a slit or a fiber;
- The dispersive elements (main and secondary)—like prisms, grisms;
- The detector (usually a CCD or CMOS) and its camera;
- The optics.

The light interface or light feed of the spectrograph has a double function. The first one is to select the target (or targets) of interest in the field of view of the telescope, so that only the selected target's light is fed into the spectrograph. The second one is to define spatially the spectrograph's resolution element: it is the image fed to the spectrograph that will be dispersed as a function of wavelength, and ultimately projected onto the detector. The first element of the spectrograph defines this first (crucial) image. The light interface can be a *slit* or a *fiber*.

A slit is a mechanical aperture with two parallel jaws, allowing one to select a rectangular image from the field of view of the telescope to feed into the spectrograph. This rectangular image is often longer than wider, and its dimensions can be adjusted mechanically; for instance, often one of the slit jaws is fixed and the position of the second one is adjustable, allowing one to change the slit width on the fly. The rectangular image created by the slit will be dispersed along the direction of the slit width—named *dispersion direction*. Importantly, this setup preserves one direction of the image, the direction perpendicular to the dispersion—named *spatial direction*. This opens an interesting set of possibilities, allowing one to position the slit in creative ways to feed several different objects and thus record simultaneously different spectra with the spectrograph. However, it is important to remember that the same target (or ensemble of targets), when positioned in different ways on the slit, will lead to different light distributions on the rectangular image of the slit. These different images will then by dispersed by the spectrograph, leading to different spectra as recorded on the detector. As such, it is important to bear in mind that different illumination and light collection patterns of the same targets on the slit will lead to different recorded spectra.

The fiber addresses this illumination aspect directly. A fiber is simply a waveguide that works based on the total internal reflection principle (see, e.g., Avila and Singh 2008; Chazelas et al. 2010). It is usually composed of a fused silica core and a protective cladding. One end of the fiber is placed at the image created on the focal plane of the telescope. The other end of the fiber will feed the light to the spectrograph. In this setup, it is the image formed at the exit of the fiber that will be dispersed by the spectrograph. As the light rays undergo several reflections inside the fiber, the light distribution is scrambled, losing the memory of its initial distribution (for a circular fiber this scrambling is more efficient in the azimuthal direction, being rather imperfect in the radial one). This scrambling is quantified by the *scrambling gain*, one of the main properties of the fibers, the others being their spectral transmission window and efficiency, the attenuation as a function of fiber length and the focal ratio degradation. Other than the reduction of the impact of illumination variation on the recorded spectra, by using a fiber one can move the spectrograph away from the telescope focus. This allows one to develop heavier, larger, and more stable spectrographs, as done in recent years. Also, by using different fibers one can inject light from several sources onto the same spectrograph simultaneously, exploring interesting concepts like that of UVES+Flames (e.g., Pasquini et al. 2002), or simply allowing to use a calibration cell at the same time as we observe our target star. This said, the main disadvantage of using a fiber follows from its ability to scramble light efficiently: all the light sources inside the field of view of the fiber are scrambled and fed to the spectrograph simultaneously, creating a composite spectrum. One loses the ability to identify the spectra associated to each of the targets. Moreover, when adding an extra optical element, there is a fraction of light that is lost at the fiber interface due to reflection from it. This is minimized by using anti-reflection coatings, but a small loss is always present.

The second type of components that exist in all spectrographs are, very naturally, the dispersive elements. These dispersive elements can be *prisms* or *gratings*. A prism is the simplest dispersive element one can conceive. It is a refracting optical element that, through Snell's law and the fact that the refraction index $n = n(\lambda)$, disperses the light that strikes one of its faces. For a detailed description on how the angle of dispersion depends on the wavelength $\theta(\lambda)$, the reader is referred to, e.g., Schroeder (1987). Unfortunately, the angle of dispersion that can be achieved with such optical devices is very low, and as a consequence the ability to disperse light into different wavelengths is rather limited. As such, one has to resort to more complex optical devices to create spectrographs with the ability to disperse light to resolve fine-scale spectra.

The most efficient dispersive element is arguably the diffraction grating. As written in the *Newport Diffraction Grating Handbook*[1] (one of the most renowned gratings manufacturers):

> A diffraction grating is a collection of reflecting (or transmitting) elements separated by a distance comparable to the wavelength of light under study. It may be thought of as a collection of diffracting elements, such as a pattern of transparent slits (...) or reflecting grooves (...).

This device allows for much larger separation angles than prisms, and from the diffraction equation for gratings we have that

$$m\lambda = d(\sin\alpha + \sin\beta)\,, \tag{2}$$

from which it follows

$$\beta(\lambda) = \arcsin\frac{m\lambda}{d} - \sin\alpha\,, \tag{3}$$

where α and β are the incident and diffracted rays' angles as measured relative to the vertical of the grating, λ the wavelength of the light, m the order of interference, and d the separation between groves. The angular dispersion can be found by differentiating the last equation:

$$\frac{d\beta}{d\lambda} = \frac{m}{d\cos\beta} = \frac{\sin\beta + \sin\alpha}{\lambda\cos\beta}\,. \tag{4}$$

Equation (4) shows that for a given λ the angular dispersion depends only on α and on β. It is clear, however, that Eq. (2) and those derived from it have multiple solutions. One is of particular interest for us, called *Littrow*, for which $\alpha = \beta$, i.e., the light is dispersed along the same direction as that of incoming rays. However, the existence of solutions with different m leads to a superposition of different orders along a given β. One can get around this issue and select the order of interest (for

[1]Which can be found, e.g., at http://optics.hanyang.ac.kr/~shsong/Grating%20handbook.pdf.

instance, through the usage of filters, or by making the grating much more efficient for a given m than for others), but a more efficient and elegant solution can be achieved by using a second dispersive element. One can *cross-disperse* the orders, dispersing the overlapping orders in a direction perpendicular to the first dispersion (done by the grating). The first dispersion is called *main dispersion* and the second the *secondary dispersion*. The secondary dispersion has the objective of separating physically the already dispersed orders, and as such a less powerful dispersive element can be used. By applying the two dispersions and focusing the image of the dispersed orders on the spectrograph, one obtains a "ladder-like" pattern, in which the orders are (approximately) parallel to each other. The wavelengths increase along each order, but also from order to order. This pattern gave the name to one of the most used types of high-dispersion gratings in the market, the *echelle grating*.

For a grating for which the grooves are perfectly aligned on a plane, and applying some simple ray-tracing geometry to Eq. (2) and the configuration behind it, one gets that the maximum efficiency of the grating occurs for $m=0$ (i.e., reflection), decreasing fast as $|m|$ increases (i.e., the interference with high orders). Since we are often interested in working in *Littrow* or *quasi-Littrow* condition (i.e., the refracted angle is only slightly different for the incident angle, for practical purposes), this forces us to work at high m. To avoid working in a very low-efficiency regime of the grating, one can change the geometry of the grooves, by adjusting their angle relative to the grating surface. To make the *Littrow* condition angle the angle with the highest transmission, we can introduce a so-called *Blaze* angle δ between the grooves and the grating surface. The efficiency is maximized for the *Littrow* angle when $\alpha = \beta = \delta$, and this is called the *Littrow Blaze condition*.

A third and very important part of the spectrograph—and any astronomical instrument, for that matter—is the detector. The detector is very simply a device that transforms the incident light into electric charge, usually by photoelectric effect. They are often 2-dimensional, being plane or approximately so, and thus allowing one to record the 2-dimensional image focused by the convergence of the dispersed light. The detectors work by photoelectric effect, through which the arrival of a photon at a given pixel will lead to the production of an electric charge. The measurement of the electric charge as a function of position allows one to map the incident photons, and create an electronic image. The main property to take into account is then, very naturally, the photosensitive material. Different elements and mixtures of elements will have different valence gaps and valence energies, i.e., they will have minimum energies by which they are sensitive to photoelectric effect. For a long while Silicon-based architectures dominated the market of detectors. The ability to create homogeneous grids of photosensitive Silicon (often with Boron), along with the ability to store and transfer the charge generated, opened the way to the *Charge-Coupled Device* (CCD) architecture. In this architecture the charge generated in the pixels is transferred to a common amplifier and register, where it is amplified and read. The usage of a single reading port greatly homogeneizes the detector, allowing the characterization with a single gain value (ability to transform electric charge in readable digital units), readout noise (error introduced by the reading process), along with other key properties. However, the Silicon atom is only sensitive to photons with wavelengths shorter than 1.1 μm (and detectors made of it are often of very low efficiency for $\lambda > 1\,\mu$m). To observe at longer wavelengths one must then go for different photosensitive materials. The first and obvious drawback is that these detectors have to be operated at a much lower temperature, for they are more sensitive to their own thermal radiation, and low-energy photons and electrons in general. The second much less obvious feature is that there is no equivalent of Silicon for these longer wavelengths, i.e., there is no material that can be manipulated electronically to transfer the charge to a common amplifier and reader. This means that near-infrared detectors have to perform the charge amplification and reading in-pixel. This leads to much more complex electronics, and the architecture behind it is called *Composite Metal Oxide Semiconductors* (CMOS). While significantly more complex, the fact that the charge is manipulated in-pixel provides very interesting options. For instance, the charge can be read multiple times to reduce readout noise, or one can read only specific parts of the detector, two options that are unavailable when reading implies a clocked transfer of charge, as for CCDs. The price for the local conversion of charge into voltage is a lower degree of homogeneity in the gain and error, and the more complex electronics often conduce to higher readout noise and spurious currents. However, the increasing demand of CMOS led to a fast development of the technology associated to it, and in several situations CMOS, when available, are already being preferred over the CCD technology. A common example is that of acquisition or guiding cameras, in which the readout time is a critical aspect; one can easily accept a noisy detector if it can be read much faster (through customized window reading, for instance) than its less noisy counterpart.

The last important part of a spectrograph is its optics. The optics accomplish different functions:

- to transform the convergent rays of light, focused at the entrance of the spectrograph by the telescope optics, into collimated light;
- to transform the dispersed (but still collimated) light into focused light that can be recorded in the detector;
- to create a spectral format such that all the orders and wavelengths can be recorded on the detector.

These correspond, basically, to the collimation of the light after it enters the spectrograph, and the transformation of the dispersed light into a focused image on the camera. These tasks have to be performed in such a way that the recording of the light is practical and eases data analysis, while maintaining the most desirable properties of the spectrograph and spectra. And this is exactly the point we will address next.

3 Spectrograph's Properties

The spectrograph's design will define its properties, which in turn will be translated directly to the spectra it forms. The most important spectrograph properties, that can be considered as properties of the associated spectra, are:

- Wavelength range;
- Transmission (or efficiency);
- Resolution;
- Sampling;
- Instrumental Profile Characteristics (especially shape and stability).

The first two properties are rather self-explanatory. The wavelength range is the wavelength domain in which the spectrograph operates, and the domain of the formed spectra. The choice of wavelength has very important consequences not only on the detectors, as seen before, but on all optical and dispersive elements (requiring different coatings to reduce unwanted reflections and scattered light, for instance). Associated to this first property is the transmission of the spectrograph or its efficiency, which can be defined in terms of total energy or photon number. The transmission is the fraction of energy or photons that traverses the spectrograph and is recorded by the detector.[2]

The property that follows is a key one: the resolution. The resolution, defined as $R \equiv \Delta\lambda/\lambda$, relies on the *Rayleigh criterion* to establish $\Delta\lambda$ for a given λ: it is the smallest difference in wavelength between two lines of equal intensity that can be discernible using the spectrograph, i.e., it is the smallest difference in wavelength for which two lines of equal intensity are resolved. Fortunately, there is a different way of understanding the resolution of a spectrograph. A spectrograph can be thought of as a device that convolves an infinite-resolution spectrum, coming from the source, with the instrumental profile (IP) that characterizes the spectrograph. The width $\Delta\lambda$ of this instrumental profile, as measured at a given λ, defines its resolution R. This means that, in practice, to measure the resolution of a spectrograph we can use a line with a full width at half maximum *FWHM* such that $FWHM/\lambda \ll 1/R$. In other words, we are feeding into the spectrograph a line with a width much smaller than the spectrograph's IP. As a consequence the width of the line that results from the convolution, is defined by the spectrograph's IP only, and its *FWHM* defines the resolution. It is never too much to stress the impact of the resolution on the final spectra, and the reader is invited to look at several examples by him/herself.

A point which is often overlooked is that of *sampling* (also referred to as *numerical resolution* or *numerical sampling*, or even more obscure names). Sampling is the number of pixels used to record the wavelength interval covering one resolution element of the spectrograph. An application of the Nyquist theorem to this situation

[2]Importantly, spectrographs with variable slit width separate the transmission of the spectrograph into *transmission of the spectrograph* × *transmission of the slit*, and detail the transmission of the latter as a function of its (tunable) properties.

informs us that the sampling should be larger than 2 pixels in order to avoid losing a significant fraction of the information. Most modern spectrographs opt for a sampling value of 3 or similar.

The final point in our list is the most intricate one. The instrumental profile (IP) is the profile that represents the broadening introduced by the spectrograph relative to a conceptual infinite-resolution spectrum emitted by the source. By construction, all elements that constitute the spectrograph have an impact on the IP. An ideal spectrograph should introduce a broadening that depends on its resolution, but no other deformation to the spectrum; from that it follows that the IP should be as symmetric as possible, maintaining the original profile of the lines. It is often conceptualized as a positive-definite Gaussian function of unit area. The IP shape should also be independent of wavelength, a condition which is not respected to some degree due to the presence of optical aberrations. Finally, and very importantly, this IP should be as stable as possible, being virtually independent of time. Any variation on the IP will be imprinted on the observed lines, and as such will have an impact on the measurement of the characteristics of the lines, like the RV. This IP stability is arguably the most difficult condition to characterize and ensure, and has a significant impact on the RV, as we will soon see.

4 Radial-Velocity Precision

We finally arrive at the issue of RV precision. What are the instrumental factors and stellar parameters that have an impact on it? And how can we design both a spectrograph and observations for the best achievable RV? These are the two questions that we ask ourselves in this chapter.

4.1 Precision Achievable on a Given Spectrum

The first and most fundamental question to ask is what is the ultimate precision one can achieve when calculating the RV on a given spectrum. This "floor level" of precision is the value one achieves when one considers as the only source of error the noise present in the spectrum; for high-*SNR* spectra, this is the stellar photon noise. The spectrograph measuring the RV is considered as perfect, and the act of measurement introduces no noise.

Based on the work of Connes (1985), and assuming that a spectrum experiences a differential shift $\delta\lambda$ relative to its own noise-free reference copy, Bouchy et al. (2001) calculated that the optimal weight $W(i)$ to be given to a pixel i when calculating the RV is

$$W(i) = \frac{\lambda^2(i)(\delta A_0(i)/\delta\lambda)^2}{A_0(i) + \sigma_D^2}, \tag{5}$$

where the spectrum (perceived simply as a group of consecutive pixels) is represented by the function $(x(i), y(i)) = (\lambda(i), A_0(i))$, and σ_D represents the readout error.[3] It is important to notice that the denominator of Eq. (5) is the variance of the flux A_0 at a given pixel, when considering both photon noise and readout noise. The error (or scatter) on the measured RV, δv, can then be calculated as

$$\delta v = \frac{c}{\sqrt{\sum_i W(i)}}. \tag{6}$$

Arguably, the most striking aspect about these equations is that the weight of a pixel, for a given λ, depends on the absolute value of the slope of the spectrum $(\delta A/\delta\lambda)$. A larger slope value leads to a larger weight, or to put it differently, the RV information content in a spectrum is contained in the slope of its lines. Our ability to measure the position of a spectrum relative to its own copy depends on the slope of its lines. The sharper the lines, the higher our precision.

However, this is just one way of measuring the ultimate precision achievable on a given spectrum. Hatzes and Cochran (1992) took a rather different approach: assuming that the noise on the spectrum is photon noise only, and that the spectrum is characterized by an uniform density of lines, the authors concluded that the RV error $\sigma_{\rm RV}$ is given by

$$\sigma_{\rm RV} \propto \frac{1}{\sqrt{F}\,\sqrt{\Delta\lambda}\,R^{1.5}}, \tag{7}$$

where F is the average flux level, $\Delta\lambda$ is the wavelength coverage and R the resolution. The term $\sqrt{F}$ represents the photon noise error calculated from the flux, and the $\sqrt{\Delta\lambda}$ represents the increase in statistics represented by including independent measurements of lines. The error on the average RV calculated from N different lines can be thought of $\overline{\sigma} \propto \sqrt{N}$ and as such $\propto \sqrt{\Delta\lambda}$. The new insight from this formula comes from the term $R^{1.5}$: the RV precision depends more steeply on the resolution of the spectrograph than on any of the other mentioned factors. This brings us back to the concept of RV information content and how it is contained in the slope of the lines. When one increases the resolution, the slope of the spectral lines is increased due to an increase in both the line's contrast and a reduction of the line's width.

[3]This error can be represented as a function of pixel i, becoming σ_i and even characterize other sources of error, without loss of generality.

These dependencies on key parameters are very informative, but it is also interesting to understand how precision changes for a given spectral line, and the simplest assumption one can make about a spectral line is to approximate it by a Gaussian function. Assuming a Gaussian-shaped line and applying the formalism of Bouchy et al. (2001), one can calculate that the RV precision is given by

$$\sigma_{RV} = \frac{(\pi \ln 2)^{-1/4}}{2} \frac{\sqrt{FWHM}}{SNR} \frac{\sqrt{PXLSC}}{C} F(C_{\mathrm{eff}}) \left[\mathrm{m\,s^{-1}}\right], \tag{8}$$

where C is the contrast of the Gaussian line, *SNR* the signal-to-noise ratio of the spectrum at hand, and *PXLSC* the pixel scale of the spectrograph (i.e., the dimension of the pixel as measured in velocity). $F(C_{\mathrm{eff}})$ is a polynomial function of the effective contrast $C_{\mathrm{eff}} = C/(1 + \sigma_D^2/A_0)$. This equation shows us very important basic properties, namely:

- The RV precision increases linearly with the *SNR* with which we measure the spectrum;
- The RV precision is proportional to the contrast C of a line, and inversely proportional to the square root of the *FWHM* of the line.

These two aspects were already represented in Eq. (7). Yet, now the *SNR* dependence is written explicitly (and one can consider noise contributions other than photon noise), and the impact of resolution is broken down into the two characteristics of the lines that it changes: *FWHM* and C. And of course these two characteristics are exactly the key parameters that regulate the slope present in a line. We have made full circle, coming back to the first conclusion brought by Bouchy et al. (2001).

We have been looking at the impact of the spectrograph's resolution on the achievable RV. As said before, the line shape associated to observing with a given resolution can be seen as the result of the convolution of the stellar spectrum with the IP. This said, another effect has exactly the same impact on the line shape, and as such on the RV: stellar rotation. Stellar rotation, and the line broadening associated to it, is often modeled through the convolution of (non-rotating) stellar spectra with a rotational kernel, a function depending solely on the projected rotational velocity, $v \sin i$, of the star. As such, it comes as no surprise that $\sigma_{RV} \propto (v \sin i)^{1.5}$, when other line broadening mechanisms are negligible. And this is the reason why on fast rotators we always obtain very poor RV precision.

4.2 *Spectrographs for Precise Radial Velocities*

We have seen how the properties of a spectrograph define the properties of the spectra acquired with it. Of all these properties, we discussed at length the non-trivial impact of the resolution on the recorded line shape, and in turn the impact of the final line shape on the achievable RV. We saw how the impact of resolution results or can be understood as the convolution with an IP that represents a spectrograph. We also stressed a key point: *Any change in the IP will lead to a change in line shape, which can translate into a measured RV variation.* The realization of this key aspect of the spectrograph's operation led to two schools of thought on how to handle IP-induced RV:

1. Control the IP as much as possible, reducing to a minimum its variations as a function of time so that one can reduce its impact on the RV;
2. Allow the IP to vary but model its variation and remove its effect on the measured spectrum.

The two approaches lead to completely different technical choices, and we discuss each one of them in detail.

4.2.1 IP Control

When trying to control the IP variation, one has to act across the whole spectrograph. As seen in Sect. 2, the first important aspect is that of light injection. Instruments aiming at controlling their IP should use light-scrambling devices, such as fibers, to reduce the spatial effects of variable illumination on RV precision. These variable illumination effects can come from imperfect centering, guiding problems, or simply variable seeing.

On top of the special care taken with illumination and light-feeding aspects, the whole instrument is designed to ensure its IP is as stable as a function of time as possible. This translates into building instruments that operate under vacuum, and are pressure- and temperature-controlled. For reference, HARPS is stabilized in pressure and temperature at 0.01 mbar and 0.01 K, respectively.

While the whole instrument is carefully monitored to reduce the IP changes, very subtle profile variations can occur, especially over long timescales, over which the physical parameters control can exert a smaller leverage. To monitor these comparatively small IP variations, the instruments are often built so that one can record the spectrum of a calibration source simultaneously with the scientific target. This simultaneous calibration is obtained through two sets of orders, recorded interweaved on the detector. One of the two sets is fed by light collected by the telescope on a science target, and the other set comes from a calibration lamp or device located in a calibration unit. The reference spectrum coming from this second fiber allows one to define on real time a wavelength calibration on the detector, and evaluate how this calibration changes with time. If one assumes that the two sets of orders (or fibers, as one prefers) experience the same IP changes, one can evaluate the wavelength calibration changes in the reference and apply them to the scientific channel. In its simplest form, this corresponds to using the reference to measure an RV drift that can be applied to each exposure to correct the wavelength calibration relative to wavelength solution obtained at the beginning of the night, on the first fiber.

If we have the spectrum recorded in our spectrograph with a wavelength calibration adjusted to correct for IP variations, then to calculate the RV of the star we simply need to find a way of calculating the RV from all the lines in the spectrum, and of averaging them in an optimal way (in the statistical sense). While one could in principle model each line independently, extract the wavelength corresponding to the center of each line from the model and calculate its RV relative to the theoretical wavelength, since there are 3000–4000 sharp lines per spectrum, this would be a computationally-heavy procedure. Instead, one condenses the information from all stellar lines present in the spectrum in an average stellar line, which is representative of the star. This is done by calculating the *Cross-Correlation function* (CCF) between the spectra and a line list containing all lines selected for the RV calculation (Baranne et al. 1996). In practice, the cross-correlation function is the convolution between the recorded spectra and a binary mask containing the wavelengths of the lines of interest, performed in the RV space (i.e., after shifting the mask over a range of RV). The binary mask can be upgraded to a mask containing the depth of each line so that the contribution of the depth on the precision is considered in the optimal construction of the average line (Pepe et al. 2002); when doing so we are assuming that all lines have very similar *FWHM*. The resulting average line is also called *CCF* in what is an obvious abuse on nomenclature. For a slowly-rotating G- or K-type star (as are often the stars considered for high-precision RV searches), the CCF has a Gaussian shape, and the fit of a Gaussian function can efficiently deliver the center of the line in RV. At this point it is important to remember that a mismatch between the stellar line and the fitting function will not introduce an error on the value of the measured center, *as long as the spectral line shape remains the same, i.e., as long as the IP does not change with time*. Any mismatch or systematic error introduced by the line shape in the calculation of the RV will be present in exactly the same way on every RV measurement, and will not impact our study of the RV variation, and of the RV *precision*.

As one moves to later spectral types, i.e., to M dwarfs, the spectra become overpopulated with lines to the point the average distance between lines becomes smaller than the resolution of the spectrograph. The lines become *blended* and the recorded spectra show a dense forest of overlapping lines, often creating regions of strong absorption and even a pseudo-continuum (see, e.g., Figueira et al. 2016). An analysis of the spectra will still reveal that the information is in the slope of the spectra, naturally; however, cross-matching it with a mask will create an average line polluted by blends, with deep wings; more importantly, the procedure will not deliver the best precision. For M stars it is preferable to cross-correlate the recorded spectrum with a template derived either from a theoretical model or from an average spectrum calculated iteratively from the observations (e.g., Astudillo-Defru et al. 2015; Anglada-Escudé et al. 2016).

4.2.2 IP Modeling

The IP control is a conceptually straightforward approach, but with its requirements on light injection stability and environmental control, it is impossible to apply on a general-purpose spectrograph, which is seldom built with these constraints in mind. The impossibility of implementation on typical slit spectrographs motivated the development of an alternative approach: to use a spectrograph that allows the IP to vary, but to devise observations and data analysis so that one can characterize the IP variation and correct for it. For this one needs a wavelength calibration device called a *gas-cell*. A gas-cell is a container with a gas species (or group of species) with a well-characterized, high-resolution absorption spectrum, and that can be mechanically inserted before the slit of the spectrograph. This cell will then superimpose a wavelength reference on the stellar spectrum, and the product of the two will be registered by the spectrograph. The objective is to use the gas-cell spectrum to define the wavelength scale on top of the science spectrum, defining a wavelength calibration for each observation. This is possible because our gas-cell spectrum will be subject to the IP variations, and by comparing our gas-cell observations with the cell high-resolution wavelength spectrum, one can fully characterize the IP. One can then *deconvolve* the IP from the measured spectrum to recover the stellar spectrum and measure its position relative to the reference. For a detailed description of the procedure the reader is referred to Butler et al. (1996).

The equation that represents the observed spectrum $A(\lambda)$ is

$$A(\lambda) = \int [I_2(\lambda')\, S(\lambda + \delta\lambda)]\, IP(\lambda - \lambda')\, d\lambda' , \tag{9}$$

where $A(\lambda)$ is the relative intensity of the final spectrum as measured by the spectrograph, $I_2(\lambda)$ is the iodine cell spectrum, $S(\lambda)$ is the source spectrum and $IP(\lambda)$ is the instrumental profile, all as a function of wavelength λ. $\delta\lambda$ is the relative wavelength shift between the science and reference spectra. This equation represents how the observed spectrum is the product of the scientific/source spectrum by the reference spectrum, convolved with the spectrograph's instrumental profile. Our final scientific objective is to determine $\delta\lambda$. However, with the exception of $I_2(\lambda)$, all elements on the right-hand side of Eq. (9) are unknown. As such, we will have to devise a clever observational scheme to determine each one of these unknowns, or spectra. The most common used recipe is as follows:

1. Measure the $I_2(\lambda)$ with a *Fourier Transform Spectrograph* (FTS) to obtain a spectrum with a much higher resolution than that recorded with our spectrograph (usually $R = 500{,}000$ or larger).
2. Observe a line-less emission spectrum (e.g., lamp or bright hot star) with the spectrograph + cell and deconvolve the $I_2(\lambda)$ to obtain the $IP(\lambda)$.
3. Observe the science target with a very high *SNR* and *without the I_2 cell*, to deconvolve the $IP(\lambda)$ from these observations and get $S(\lambda)$.
4. Observe the science target *with the I_2 cell*, and recover $\delta\lambda$ from the evaluation of Eq. (9).

Deriving RV through this method is clearly a complex process. It is subject to the fidelity with which each of the intermediate data products is obtained or determined. In particular the *IP* reconstruction is a very delicate process; a careful parameterization should characterize the *IP* as both a function of time and λ, with particular attention on how it depends on the position of the spectrum on the detector. Labour-intensive as it is, this methodology has been widely used to transform general-purpose spectrographs into efficient planet-hunting machines.

We can summarize the previous two sections in the following way:

- The **IP control** technique requires a stable spectrograph, both in light injection and thermo-mechanical stability. One can get the best of it by using a second channel to simultaneously record the spectrum of a reference calibration. **It minimizes the presence of instrumental RV shifts**.
- The **IP modeling technique** can be used on a general-purpose slit spectrograph, and **models and subtracts the IP variations that induce an RV shift**. It requires several on-sky calibrations and as such it is observationally expensive.

When we are talking of the RV precision required to detect planets, of m/s (or no larger than 10 times that), we have to remember that these correspond to shifts of spectral lines at 1/1000 of the pixel size. This type of precision is incredible, and both techniques are undoubtedly successful by reaching this mark. Also, before comparing the two techniques it is important to note that many instruments can only use one of the methodologies due to the practical requirements they impose on the instrumentation.

It is impossible to state which technique is capable of delivering the most precise RV without resorting to observational data. One technique minimizes the RV shifts without characterizing them, while the other characterizes them but through a complex process that has its own practical limitations. Probably the only way to settle this argument is to look at the best precision achieved on the two instruments that best embody the two techniques described here: HARPS (Mayor et al. 2003) and HIRES (Vogt et al. 1994). While HARPS reached a precision of 1 m/s and better, HIRES floored at 2–3 m/s of precision. These results were the subject of hot debates for a long time, but it is now solidly established the IP control technique is the only one able to deliver sub-m/s precision.

So far we have been debating how to achieve the best precision on very general terms, but we have not mentioned that our own Earth is traveling in space, and the projection of our own RV along the line of sight of our stellar observations will shift the recorded spectra of the target. In order to correct the measured RV for this effect, one has to use the ephemerides of our own solar system to calculate the position of the Earth with great accuracy (e.g., Bretagnon and Francou 1988).

5 Current and Future Planet-Hunting Machines

The great motivation for the development of precise RV spectrographs was the detection of extrasolar planets. Today the most precise planet-hunting machines are the spectrograph HARPS (Mayor et al. 2003), installed at the 3.6-m telescope located at La Silla, and its northern twin HARPS-N (Cosentino et al. 2012), installed at the TNG telescope at La Palma. These two spectrographs were developed to minimize the IP variation at an extreme level, and today yield a precision of 50–60 cm/s. The main dispersive element is an echelle grating R4, with 31.6 gr/mm, operating at a blaze angle of 75°, and the cross dispersing is done with a grism. The spectrographs are fed by octagonal fibers, which have shown to have improved scrambling properties over the typical circular ones.

Until recently, the simultaneous reference and wavelength calibration on HARPS and HARPS-N was provided by a ThAr emission lamp. However, these are far from being ideal calibrators. The large dynamics and very different spatial density of the lines led to very different wavelength calibration stability as a function of wavelength, which translated ultimately in a different RV precision as a function of wavelength. Moreover, with the increased rarity of these lamps, there has been an active search for wavelength reference alternatives.

The characteristics of the perfect wavelength reference are very clear:

- should cover the whole spectral range of the spectrograph;
- the lines used in the calibration should have a *FWHM* smaller than the spectrograph's resolution;
- the source should provide a high density of lines, up to one per 2–3 times the resolution element, and at a constant spacing;
- the wavelength of the lines should be precisely known and stable;
- the line intensities should be homogeneous, being close to the saturation but with a high dynamic range.

It is easy to conclude that neither the ThAr lamp nor the I_2 cell get even close to these specifications, and nature cares not to provide such a level of homogeneity and fine-tuning in the form of atomic or molecular transitions. As such, the most recent advances focus on the development of back-illuminated Fabry–Perot cavities (e.g., Wildi et al. 2011; Halverson et al. 2014; Reiners et al. 2014), or of laser-frequency combs, in which a femtosecond laser is stabilized with an atomic clock to produce a series of modes that are subsequently filtered by a Fabry–Perot cavity (e.g., Lo Curto et al. 2012).

With the current instrumentation, remarkable discoveries were made, like the Earth-mass planet in the habitable zone around our neighbor Proxima Centauri (Anglada-Escudé et al. 2016). The rocky Earth-mass planets around Kepler-78 (Pepe et al. 2013) and HD 219134 (Motalebi et al. 2015) show how precise RV can be used in transiting planets to help determine a planet's bulk composition. Very importantly, the sub-m/s RV precision allowed us to start to uncover and characterize the population of Earth-mass planets, and we have already located stars that contain systems of exoplanets with up to 7 planets, like HD 10800 (e.g., Lovis et al. 2011).

When several planets are present around a star and with orbital semi-amplitudes at the level of instrumental precision, it is very difficult to characterize their orbits, due to the large number of parameters to fit. This leads to the need of a large number of RV points per system, and motivated the development of more precise instruments. The forthcoming planet-hunter ESPRESSO (Pepe et al. 2010, 2014) spearheads this quest, and will be able to reach an intrinsic RV precision of 10 cm/s; along with the improved collecting capability of the VLT and associated photon noise contribution, ESPRESSO will be able to detect an Earth-mass planet inside the habitable zone around a solar-type star. This corresponds to a significant jump relative to the current 50 cm/s precision that allows us to detect Earth-mass planets in orbits of only a couple of days, at most.

A different way of looking at the challenge of detecting Earth-mass planets inside the habitable zone of their host stars is to turn to M dwarfs. The lightest stars experience a reflex motion $\sim$3 times larger than their GK companions, and their lower energy output draws the habitable zone roughly 3 times closer. This means that in order to detect an Earth-mass planet orbiting inside the habitable zone around an M dwarf, one needs only a precision of roughly 1 m/s. However, since M dwarfs are faint in optical wavelengths, this scientific objective spurred the development of near-infrared instruments like SPIRou (Artigau et al. 2014), NIRPS (Conod et al. 2016), and CARMENES (Quirrenbach et al. 2014), just to cite a few among the many spectrographs of this category. With these spectrographs we will be able to extend our studies of planetary frequency from the first 100 M dwarfs surveyed with HARPS to a more complete sample in terms of stars and precision. Last but not least, in a more distant future, the European Extremely Large Telescope (E-ELT) will have at least an instrument—HIRES—capable of delivering precise RV at a 10 cm/s precision or better, and will enable RV studies on stars that are otherwise discarded due to photon noise limitations.

This contributions tries to cover the main aspects behind RV precision and its association to planetary studies. However, with the number of developments in the latest couple of decades, with 3 h of lectures we could only skim the surface of all the topics that are there to discuss. And after the derivation of precise RVs comes its interpretation, for which the points discussed here will be of great use. But that's a story for another time.

Acknowledgements I acknowledge support by Fundação para a Ciência e a Tecnologia (FCT) through Investigador FCT contract of reference IF/01037/2013/CP1191/CT0001, and POPH/FSE (EC) by FEDER funding through the program "Programa Operacional de Factores de Competitividade - COMPETE". I further acknowledge support from FCT in the form of an exploratory project of reference also IF/01037/2013/CP1191/CT0001. I acknowledge *Peter Sport Café* for showing me what a good gin tonic tastes like, and Mário João Monteiro, Mahmoudreza Oshagh and Vardan Adibekyan for reminding me that good science requires good statistics.

References

Anglada-Escudé, G., Amado, P.J., Barnes, J., et al.: Nature **536**, 437 (2016)
Artigau, É., Kouach, D., Donati, J.-F., et al.: In: Society of Photo-Optical Instrumentation Engineers (SPIE) Conference Series, vol. 9147, 15 (2014)
Astudillo-Defru, N., Bonfils, X., Delfosse, X., et al.: Astron. Astrophys. **575**, A119 (2015)
Avila, G., Singh, P.: Advanced optical and mechanical technologies in telescopes and instrumentation. In: Proceeding of SPIE, vol. 7018, 70184W (2008)
Baranne, A., Queloz, D., Mayor, M., et al.: Astron. Astrophys. Suppl. Ser. **119**, 373 (1996)
Bouchy, F., Pepe, F., Queloz, D.: Astron. Astrophys. **374**, 733 (2001)
Bretagnon, P., Francou, G.: Astron. Astrophys. **202**, 309 (1988)
Butler, R.P., Marcy, G.W., Williams, E., et al.: Publ. Astron. Soc. Pac. **108**, 500 (1996)
Chazelas, B., Pepe, F., Wildi, F., et al.: Modern technologies in space- and ground-based telescopes and instrumentation. In: Proceeding of SPIE, vol. 7739, 773947 (2010)
Connes, P.: Astrophys. Space Sci. **110**, 211 (1985)
Conod, U., Blind, N., Wildi, F., Pepe, F.: ArXiv e-prints, arXiv:1608.01124 (2016)
Cosentino, R., Lovis, C., Pepe, F., et al.: Ground-based and airborne instrumentation for astronomy IV. In: Proceeding of SPIE, vol. 8446, 84461V (2012)
Figueira, P., Adibekyan, V. Z., Oshagh, M., et al.: Astron. Astrophys. **586**, A101 (2016)
Halverson, S., Mahadevan, S., Ramsey, L., et al.: Publ. Astron. Soc. Pac. **126**, 445 (2014)
Hatzes, A.P., Cochran, W.D.: In: Ulrich, M.-H. (ed.) European Southern Observatory Astrophysics Symposia, vol. 40, 275+ (1992)
Lo Curto, G., Manescau, A., Avila, G., et al.: In: Society of Photo-Optical Instrumentation Engineers (SPIE) Conference Series, vol. 8446 (2012)
Lovis, C., Ségransan, D., Mayor, M., et al.: Astron. Astrophys. **528**, A112 (2011)
Mayor, M., Pepe, F., Queloz, D., et al.: The Messenger **114**, 20 (2003)
Motalebi, F., Udry, S., Gillon, M., et al.: Astron. Astrophys. **584**, A72 (2015)
Pasquini, L., Avila, G., Blecha, A., et al.: The Messenger **110**, 1 (2002)
Pepe, F., Mayor, M., Galland, F., et al.: Astron. Astrophys. **388**, 632 (2002)
Pepe, F.A., Cristiani, S., Rebolo Lopez, R., et al.: In: Society of Photo-Optical Instrumentation Engineers (SPIE) Conference Series, vol. 7735 (2010)
Pepe, F., Cameron, A.C., Latham, D.W., et al.: Nature **503**, 377 (2013)
Pepe, F., Molaro, P., Cristiani, S., et al.: Astron. Nachr. **335**, 8 (2014)
Quirrenbach, A., Amado, P.J., Caballero, J.A., et al.: In: Society of Photo-Optical Instrumentation Engineers (SPIE) Conference Series, vol. 9147, 1 (2014)
Reiners, A., Banyal, R.K., Ulbrich, R.G.: Astron. Astrophys. **569**, A77 (2014)
Schroeder, D.J.: Astronomical Optics. Academic Press, San Diego (1987)
Vogt, S.S., Allen, S.L., Bigelow, B.C., et al.: Instrumentation in astronomy VIII. In: Crawford, D.L., Craine, E. R. (eds.) Proceeding of SPIE, vol. 2198, 362 (1994)
Wildi, F., Pepe, F., Chazelas, B., Lo Curto, G., Lovis, C.: In: Society of Photo-Optical Instrumentation Engineers (SPIE) Conference Series, vol. 8151 (2011)

Modelling Light and Velocity Curves of Exoplanet Hosts

Rodrigo F. Díaz

Abstract Research in extrasolar-planet science is data-driven. With the advent of radial-velocity instruments like HARPS and HARPS-N, and transit space missions like *Kepler*, our ability to discover and characterise extrasolar planets is no longer limited by instrumental precision but by our ability to model the data accurately. This chapter presents the models that describe radial-velocity measurements and transit light curves. I begin by deriving the solution of the two-body problem and from there, the equations describing the radial velocity of a planet-host star and the distance between star and planet centres, necessary to model transit light curves. Stochastic models are then presented and I delineate how they are used to model complex physical phenomena affecting the exoplanet data sets, such as stellar activity. Finally, I give a brief overview of the processes of Bayesian inference, focussing on the construction of likelihood functions and prior probability distributions. In particular, I describe different methods to specify ignorance priors.

1 Introduction

In the scientist dialogue with Nature data play a crucial role. It is the language by which we receive information from our experiments and observations. Nature communicates with scientists exclusively through data. Without data, there is no science. To be able to interpret the messages encoded in the language of physical observations, we use mathematical models. It is probably useless to stare at a series of data points or samples from a distribution without an underlying idea, however vague, of what the mechanism producing them could be. This idea can be flexible and allow for modifications, but it is an unavoidable step in translating data to

Rodrigo F. Díaz (✉)
Département d'Astronomie, Université de Genève, Versoix, Switzerland

Universidad de Buenos Aires, Facultad de Ciencias Exactas y Naturales, Buenos Aires, Argentina

Instituto de Astronomía y Física del Espacio (IAFE), CONICET-Universidad de Buenos Aires, Buenos Aires, Argentina
e-mail: rodrigo.diaz@unige.ch

T.L. Campante et al. (eds.), *Asteroseismology and Exoplanets: Listening to the Stars and Searching for New Worlds*, Astrophysics and Space Science Proceedings 49, DOI 10.1007/978-3-319-59315-9_11

knowledge. Roughly speaking, the idea of how data are produced by Nature is a model. More precisely, when the model is expressed using mathematical concepts and terminology it becomes a mathematical model.

Mathematical models consist of one or more equations relating independent to dependent variables, for example time and the radial velocity of a star at that time. Both of these can be measured in an experiment or by an astronomical observation. These equations have constants that represent properties of the system being described, and are usually referred to as the model *parameters*. The basic task of data analysis is to obtain information on model parameters from a set of observations or measurements.

Here we are interested in studying mathematical models of extrasolar-planet data. In particular, we will focus on radial-velocity data and transit light curves. To fully describe these data sets, one needs not only to consider the physics describing the movement of the planets and stars, as well as the geometry of the occultation of a fraction of the stellar disc by the planet, but also the mechanisms producing the actual observed data and their uncertainties, plus all sources of additional radial velocity or photometric variability not of primary interest to us. Actually, at the present stage of exoplanet research, the data are of good enough quality to permit the detection and characterisation of planets of very small mass and size, similar to Earth. However, the signals produced by this type of planetary companions are usually comparable or even smaller than the effects produced by other sources, such as stellar activity, which in addition are usually much more difficult to model. As we will see, the signals we are interested in are relatively simple to model but the determination of the parameters can be hindered or biased by the more complicated effects. Our ability to detect and characterise extrasolar planets is no longer limited by the precision of our instruments but by our ability to model and analyse the data.

Generally speaking, the mathematical model of datum d_i (which we assume taken at time t_i, although this is not fundamental to our formulation) consists of two parts: a prediction of the value the datum d_i should take, m_i, and an error term, e_i, that accounts for the uncertainties in the data and quantifies the possible discrepancies between the m_i and the datum d_i. In general, we can write:

$$d_i = m_i + e_i\,. \tag{1}$$

As we will describe in detail in Sect. 3, the mathematical formulation leading to m_i can be deterministic or statistical. In the deterministic case, for a certain set of parameters on which the model depends there is no uncertainty in the predicted value m_i. The second term, e_i, will be assumed to come exclusively from a statistical model of the data. This model can be as simple as the mathematical model corresponding to a series of independent Gaussian variables, but it can be as complicated as one needs to correctly represent the data. This includes models with correlation between the measurements, mixture models etc. (see Sect. 3.2).

In this chapter, I will discuss the mathematical models describing the production of two of the most important data types in exoplanet science: radial-velocity measurements and transit light curves (although the focus will be mostly on the radial-velocity models). In Sect. 2, I describe the physical models of the orbiting planetary companions, focussing in the case of single-planet systems, that permit an exact analytical description (under certain assumptions). I discuss in Sect. 3 the stochastic models needed to describe the error term e_i and more complicated phenomena not easily amenable to an analytical description. Finally, Sect. 4 presents a brief introduction to Bayesian inference and the processes involved in obtaining information on the model parameters from a set of observations.

2 Physical Models

In this section, I describe physical models used to reproduce the radial-velocity time series and light-curve data of a star hosting a planetary companion. In both cases, to reach an analytical expression we need to make a series of assumptions and approximations. Those common to both types of data are:

1. that the movement of the bodies follows the laws of Newtonian dynamics, and
2. that the bodies do not have electrical charge or any other property besides mass that allows them to interact with each other.

These approximations are usually good in the range of velocities, masses, and distances we will be dealing with.

2.1 Radial Velocities

To model radial-velocity data we make one additional assumption: the bodies are dimensionless point particles. This means that the bodies have no internal structure and therefore do not experiment any kind of tidal force. Although this is of course wrong, as we know stars and planets have interiors and processes may occur inside them, the resulting forces are usually much weaker than the gravitational attraction between the bodies and therefore in most cases only produce observable effects over very long time scales (e.g., Zahn 1977; Hut 1981).

Under these assumptions, the movement of the planet and the star reduces to the two-body problem, a traditional problem in classical mechanics. In particular, to obtain a model for the observed radial velocities, we need to obtain the position of the bodies in time. This is known as the Kepler problem. In this section, I follow closely the presentation of the two-body problem by Murray and Dermott (2000) and Murray and Correia (2010). An approach using Lagrangian mechanics is given by Goldstein (1980).

2.1.1 Elliptical Motion

Let us consider two objects of masses m_1 (the star) and m_2 (the planet) interacting gravitationally. Their equations of motion are described in an inertial frame S with origin at O:

$$\vec{F}_1 = m_1\ddot{\vec{r}}_1 = +G\frac{m_1 m_2}{r^3}\vec{r}\,, \tag{2}$$

$$\vec{F}_2 = m_2\ddot{\vec{r}}_2 = -G\frac{m_1 m_2}{r^3}\vec{r}\,, \tag{3}$$

where G is the Universal gravitational constant and $\vec{r} = \vec{r}_2 - \vec{r}_1$ is the relative position vector, pointing from the star to the planet (see Fig. 1). Dividing Eq. (2) by m_1 and Eq. (3) by m_2, these equations can be combined to produce the equation of relative motion:

$$\ddot{\vec{r}} + G(m_1 + m_2)\frac{\vec{r}}{r^3} = 0\,. \tag{4}$$

Taking the vector product of Eq. (4) with the relative position vector $\vec{r}$, and using the fact that $\vec{r} \times \vec{r} = 0$, we find the first conserved magnitude of the two-body problem: $\vec{r} \times \ddot{\vec{r}} = 0$, which is promptly integrated to give

$$\vec{r} \times \dot{\vec{r}} = \vec{h}\,. \tag{5}$$

The vector $\vec{h}$ is constant and, because it is the vector product of $\vec{r}$ and $\dot{\vec{r}}$, it is perpendicular both to the position and velocity vectors. This means that the motion of the system happens in a plane perpendicular to the constant vector $\vec{h}$, known as the orbital plane. It is then practical to express the position and velocity vector in polar coordinates on this plane: $\vec{r} = r\hat{r}$, $\dot{\vec{r}} = \dot{r}\hat{r} + r\dot{\theta}\hat{\theta}$, where $\hat{r}$ is the unit vector pointing from the star to the planet and $\hat{\theta}$ is a unit vector perpendicular to $\hat{r}$. We then

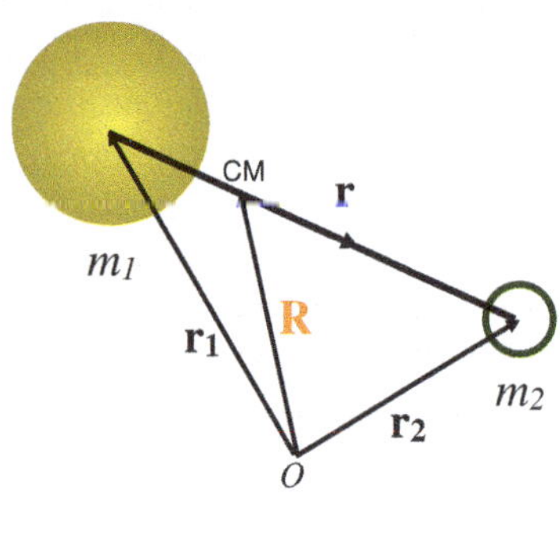

Fig. 1 Schematic view of the forces and positions in the two-body problem

find that

$$\vec{h} = r^2\dot{\theta}\hat{z}, \tag{6}$$

where $\hat{z}$ is a unit vector forming a right-handed triad with $\hat{r}$ and $\hat{\theta}$. This is a good approximation of angular momentum of the system when[1] $m_2 \ll m_1$. From the conservation of $\vec{h}$, one can deduce the second law of planetary motion discovered by Kepler (see Murray and Dermott 2000; Sect. 2.2), which states that equal areas are swept out by the position vector in equal times. Additionally, it can be shown that the area swept by unit time is

$$\frac{\mathrm{d}A}{\mathrm{d}t} = \frac{h}{2}. \tag{7}$$

With this conserved quantity, we can express the general solution of Eq. (4):

$$r = \frac{p}{1 + e\cos(\theta - \varpi)}, \tag{8}$$

where

$$p = \frac{h^2}{G(m_1 + m_2)} \tag{9}$$

and e and ϖ are two constants of integration. Equation (8) is the general equation of a conic section in polar coordinates. In particular, whenever $0 < e < 1$ and $p = a(1 - e^2)$, the equation describes an ellipse of eccentricity e and semi-major axis a:

$$r = \frac{a(1 - e^2)}{1 + e\cos(\theta - \varpi)}.$$

This is one possible solution for the movement of a planet around a star. The two extreme distances, called periapsis and apoapsis occur when the $\theta = \varpi$ and $\theta = \pi + \varpi$, respectively, and are equal to $a(1-e)$ and $a(1+e)$ (see Fig. 2). We can define a new polar angle, ν, such that $\nu = \theta - \varpi$, i.e., we measure the angles starting at the periapsis. This angle usually receives the name of true anomaly. Then,

$$r = \frac{a(1 - e^2)}{1 + e\cos\nu}. \tag{10}$$

[1]As we are describing the relative motions of the bodies, the *actual* angular momentum of the system contains a term related to the motion of the star. Whenever $m_2 \ll m_1$, this term can be neglected and h is equal to the total angular momentum of the system per unit mass of the body m_2.

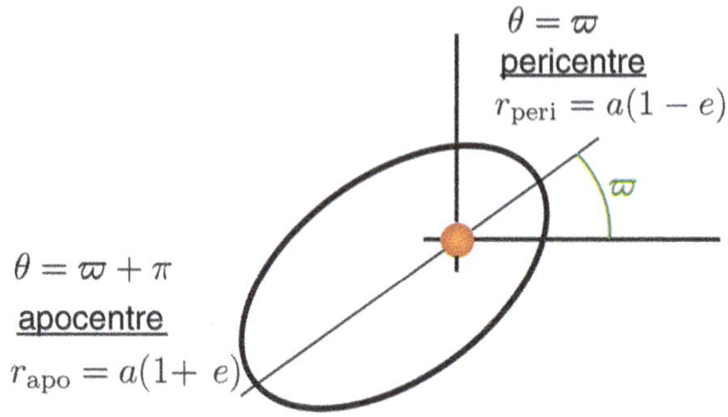

Fig. 2 Schematic view of the trajectory of an orbiting companion to the central star (at the origin of coordinates), as seen vertically from above the orbital plane. The pericentre, apocentre, and their corresponding distances are indicated

The area enclosed by the ellipse, πab, where b is the semi-minor axis of the ellipse, $b = a(1 - e^2)$, is swept by the position vector in one orbital period, P. From Kepler's second law, and the fact that $h^2 = G(m_1 + m_2)a(1 - e^2)$ (cf. Eq. 9), we arrive at Kepler's third law of planetary motion:

$$P^2 = \frac{4\pi^2}{G(m_1 + m_2)} a^3 \,. \tag{11}$$

2.1.2 Barycentric Orbits

Combining Eqs. (2) and (3), one obtains the equation of motion for the centre of mass (CM) of the system:

$$m_1 \ddot{\vec{r_1}} + m_2 \ddot{\vec{r_2}} = 0 \,, \tag{12}$$

which can be easily integrated to obtain the equation describing the position of the CM in time:

$$\vec{R}(t) = \frac{\alpha t + \beta}{m_1 + m_2} \,, \tag{13}$$

where $\vec{R} = (m_1\vec{r_1} + m_2\vec{r_2})/(m_1 + m_2)$ is the position of the CM of the system (Fig. 1). As expected from the conservation of momentum, the CM of the system moves with a constant velocity with respect to the origin O. Since the original frame of reference S was assumed inertial, this means that a reference frame with origin in the CM is also inertial. We will find it useful to describe the motions of the planet and the star in this reference frame. We can then write the positions of the masses with respect to the CM as

$$\vec{R}_1 = -\frac{m_2}{m_1 + m_2}\vec{r} \,, \tag{14}$$

$$\vec{R}_2 = +\frac{m_1}{m_1 + m_2}\vec{r} \,. \tag{15}$$

This means that $\vec{R}_1$ and $\vec{R}_2$ always have opposite directions and that the CM is in the line joining the two vectors. Besides, it implies that the trajectories of the star and the planet with respect to the barycentre of the system are simply scaled down versions of the conic section describing the relative motion of the bodies (Eq. 10). In the case of the ellipse, the semi-major axis of the star and planet orbital trajectories are scaled down by a factor $m_2/(m_1 + m_2)$ and $m_1/(m_1 + m_2)$, respectively. Note that in the typical case where $m_2 \ll m_1$, the stellar orbit with respect to the CM becomes very small while the planet orbit resembles the relative orbit.

2.1.3 Projection and Radial Velocity

The orientation of the orbit in three-dimensional space can be described by three angles. The angle of the orbital plane with respect to the plane of the sky is the orbital inclination I. The intersection between the plane of the sky and the orbital plane is called the line of nodes. The longitude of the ascending node, Ω, is the angle between a reference direction in the plane of the sky and the radius vector at the ascending node, where the planet crosses the plane of the sky from below to above. Finally, the argument of periapsis, ω, is the angle between that same radius vector and the orbital periapsis measured on the orbital plane. For $I \sim 0$, we have $\varpi = \omega + \Omega$, where ϖ is the two-body constant already found in Eq. (8).

To obtain the expected radial velocity of the star in the system, let us project the expression of the orbit to a cartesian coordinate system centred in the CM of the system and oriented so that the positive z axis points towards an observer on Earth and the xy plane corresponds to the plane of the sky. To simplify the resulting expressions, we further assume that the reference direction defined by the positive x axis coincides with the line of nodes (i.e., $\Omega = 0$). Under these conditions, the cartesian components of the barycentric movement of the stellar body are:

$$x = r_1 \cos(\omega + \nu) \,, \tag{16}$$

$$y = r_1 \sin(\omega + \nu) \cos I \,, \tag{17}$$

$$z = r_1 \sin(\omega + \nu) \sin I \,, \tag{18}$$

with

$$r_1 = a \frac{m_2}{m_1 + m_2} \frac{(1 - e^2)}{1 + e \cos \nu}$$

the distance between the star and the CM of the system. Deriving Eq. (18) with respect to time and using the fact that

$$\dot{\nu} = \dot{\theta} = h/r^2 = \frac{2\pi}{P} \frac{a^2\sqrt{1-e^2}}{r^2} ,$$

we get the expression of the stellar radial velocity V as a function of the true anomaly:

$$V = V_0 + K\left[\cos\left(\nu + \omega\right) + e\cos\omega\right] , \tag{19}$$

with

$$K = \left(\frac{2\pi G}{P}\right)^{1/3} \frac{1}{\sqrt{1-e^2}} \frac{m_2 \sin I}{(m_1 + m_2)^{2/3}} , \tag{20}$$

and V_0 and integration constant corresponding to the velocity of the CM with respect to the observer on Earth.

If we neglect m_2 with respect to m_1 in the denominator of Eq. (20), which is reasonable in the typical case $m_2 \ll m_1$, we see that the radial-velocity signal scales linearly with the mass m_2 of the orbiting companion, and inversely with $P^{1/3}$. These two facts introduce strong biases in the sample of detected planets and in the shapes of the distributions of their orbital parameters. Note that the RV amplitude depends on the combination of m_2 and $\sin I$. Using RV data alone it is therefore not possible to measure the real mass of the planets, but only the so-called "minimum mass", $m_2 \sin I$.

2.1.4 Kepler Problem

Up to this point we have obtained a description of the trajectories followed by the bodies of a star-planet system and have reached an expression for the radial velocity of the star as a function of the position of the planet in the orbit, ν. What remains to be done to link this expression (Eq. 19) with the observations is to describe the motion of the bodies in time as they traverse their orbits. In other words, we need to find the function $\nu = \nu(t)$. This is known as the Kepler problem and is much more involved than obtaining the equation of the orbit or the radial-velocity expression. We give here a summarised description of the solution to the Kepler problem and refer the interested reader to specialised literature (e.g., Goldstein 1980; Murray and Dermott 2000).

The integrals involved in finding $\nu(t)$ are more easily solved using two auxiliary variables: the eccentric anomaly, ψ, defined through the expression

$$r = a(1 - e\cos\psi) , \tag{21}$$

and the mean anomaly, μ,

$$\mu = \frac{2\pi}{P}(t - \tau),$$

where τ is the time of passage through the periapsis.[2] Note that the time t enters explicitly in the definition of μ.

It can be shown that μ and ψ are related through the Kepler equation:

$$\mu = \psi - e \sin \psi, \tag{22}$$

which is transcendental and requires iterative methods to solve it.[3] By inverting the Kepler equation, one can find ψ and the radial distance r as a function of time (Eq. 21). Furthermore, comparing the defining equation (Eq. 21) with the equation of the elliptical orbit (Eq. 10) one can find, after some algebra, an expression for ν as a function of ψ:

$$\tan\left(\frac{\nu}{2}\right) = \sqrt{\frac{1+e}{1-e}} \tan\left(\frac{\psi}{2}\right). \tag{23}$$

We have shown how to obtain the time dependence of the true anomaly ν, and therefore of the radial-velocity expression (Eq. 19), which can be promptly used to compare the model with the observations (see Sect. 4).

2.2 *Transits*

When the orbital inclination I is close to 90°, the orbiting planet, as seen from Earth, passes in front of the stellar disk, causing a slight dimming of the star known as a transit. This event is extremely rich in information about the star-planet system. In particular, it allows lifting the $m_2 \sin I$ degeneracy and measure the real planet mass.

Under certain assumptions, the model of the light curve of a transiting planet can be written analytically. We need to abandon the assumption of point masses used for the radial-velocity model, but assume both the planet and the star are perfectly spherical with radii R_p and R_s, respectively. Additionally, we assume that the planet is completely opaque and emits no radiation.

---

[2] Of course, exactly as we have redefined the polar angle so that ν is 0 at periapsis, we could also measure time starting at the moment of periastron passage, and get rid of τ in the definition of the mean anomaly. However, usually τ is unknown and including it as a model parameter allows us to measure it.

[3] A `Python` code to compute the true anomaly from the mean anomaly and eccentricity has been made available at https://github.com/exord/faial/blob/master/trueanomaly.py.

2.2.1 Uniform Source

Further assuming that the star is a uniform source, we can write the expression of the ratio between obscured to unobscured flux, $F(k,\delta) = 1-\lambda(k,\delta)$, where λ is the flux loss as a function of the radius ratio $k = R_p/R_s$ and $\delta = d/R_s$ is the projected centre-to-centre distance between the star and the planet, normalised by the radius of the star. Following Mandel and Agol (2002), and assuming $k < 1$, we find:

$$\lambda(k,\delta) = \begin{cases} 0, & \delta > 1+k\,, \\ \frac{1}{\pi}\left[k^2\kappa_0 + \kappa_1 - \sqrt{\frac{4\delta^2-(1+\delta^2-k^2)^2}{4}}\right], & 1-k < \delta \leq 1+k\,, \\ k^2, & \delta \leq 1-k\,, \end{cases} \tag{24}$$

where

$$\kappa_0 = \cos^{-1}\left[\frac{k^2+\delta^2-1}{2k\delta}\right], \qquad \kappa_1 = \cos^{-1}\left[\frac{1-k^2+\delta^2}{2\delta}\right]. \tag{25}$$

In other words, there is no flux loss as long as $d > R_p + R_s$, and the loss is constant and equal to $(R_p/R_s)^2$ when the planet disc is completely inside the stellar disc (see Fig. 3).

If the planet is on a circular orbit, Seager and Mallén-Ornelas (2003) showed there is a unique relation between the four parameters describing the model (see Fig. 3) and some combinations of the physical parameters of the system. This allows obtaining, in this simple case, information about the planetary system by measuring

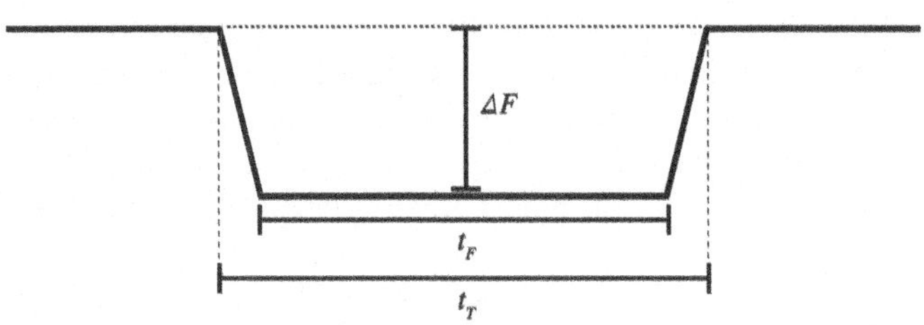

Fig. 3 Schematic view of a flat-bottomed transit. Three of the four parameters describing the model are indicated: the transit depth ΔF, the total transit duration, t_T, and the duration of the flat part, t_F. The remaining parameter is the orbital period P

the transit properties:

$$\frac{R_p}{R_s} = \sqrt{\Delta F}, \tag{26}$$

$$b = \frac{a}{R_s}\cos I = \left[\frac{\sin^2\omega_T\left(1-\sqrt{\Delta F}\right)^2 - \sin^2\omega_F\left(1+\sqrt{\Delta F}\right)^2}{\sin^2\omega_T - \sin^2\omega_F}\right]^{1/2}, \tag{27}$$

$$\frac{a}{R_s} = \left[\frac{\left(1+\sqrt{\Delta F}\right)^2 - b^2\left(1-\sin^2\omega_T\right)}{\sin^2\omega_T}\right]^{1/2}, \tag{28}$$

where $\omega_F = t_F\pi/P$ and $\omega_T = t_T\pi/P$. Note that, combining Eqs. (27) and (28), one can measure the orbital inclination I. Combining this information with RV data providing $m_2\sin I$ one can measure the true planet mass m_2. Interestingly, using Kepler's third law, we can write an expression involving the mean densities of the planet and the star:

$$\rho_s + k^3\rho_p = \frac{3\pi}{G}\frac{1}{P^2}\left(\frac{a}{R_s}\right)^3, \tag{29}$$

where the mean density is $\rho_s = M_s\left(4\pi R_s^3/3\right)^{-1}$, and an equivalent expression for ρ_p. In the usual case in which $k \ll 1$ and $M_p \ll M_s$, the mean density of the star can be approximately obtained using Eq. (28):

$$\rho_s \approx \frac{3\pi}{GP^2}\left[\frac{\left(1+\sqrt{\Delta F}\right)^2 - b^2\left(1-\sin^2\omega_T\right)}{\sin^2\omega_T}\right]^{3/2}, \tag{30}$$

valid only in the case of circular orbits.

The transit duration is usually measured in hours while the orbital period P is usually of the order of days. Therefore, we see that the parameters ω_T and ω_F are necessarily small. In this case the sine function can be approximated $\sin^2 x \approx x^2$, and Eqs. (27) and (28) can be simplified:

$$b \approx \left[\frac{t_T^2\left(1-\sqrt{\Delta F}\right)^2 - t_F^2\left(1+\sqrt{\Delta F}\right)^2}{t_T^2 - t_F^2}\right]^{1/2}, \tag{31}$$

$$\frac{a}{R_s} \approx \frac{2P}{\pi}\frac{\Delta F^{1/4}}{\sqrt{t_T^2 - t_F^2}}. \tag{32}$$

In the case of non-circular orbits, complicated algebra arises and it is no longer possible to find exact analytical expressions for the physical system parameters. Winn (2010) gives approximate expressions valid when $R_p \ll R_s \ll a$. The expression for the radius ratio (Eq. 26) and for the impact parameter (Eq. 27) remain the same, but the normalised semi-major axis is scaled by a factor $\sqrt{1-e^2}/\,(1+e\sin\omega)$:

$$\frac{a}{R_s} \approx \frac{2P}{\pi}\frac{\Delta F^{1/4}}{\sqrt{t_T^2 - t_F^2}}\left(\frac{\sqrt{1-e^2}}{1+e\sin\omega}\right). \quad (33)$$

Depending on the orientation of the orbit in the sky, the semi-major axis can be smaller or larger than the one measured in the circular case. As a consequence, the mean stellar density computed by means of Eq. (30) can be under- or overestimated if the orbit is incorrectly assumed circular. Further analysis in the case of eccentric orbits is presented by Kipping (2008).

2.2.2 Non-Uniform Source

In reality, stars are not observed as uniform disks of light. Instead, their brightness appears to decrease from the centre to the limb of the disc, an effect known as limb darkening. Mandel and Agol (2002) provide analytical equations of the flux drop, F, as a function of the normalised star-to-planet centre separation, δ, for a quadratic and non-linear limb-darkening laws, which most accurately describe the observed darkening effect (Claret 2000). Their codes are available to the community and are widely used by researchers in the field.

When limb darkening is included in the transit model, the physical parameters become strongly covariate and the model is degenerate, which complicates the process of inferring the system parameters from data (see Sect. 4).

2.2.3 Projected Distance as a Function of Time

The model of the drop in flux produced by the transit of a planetary object is given as a function of the projected centre-to-centre normalised distance, $\delta = d/R_s$, where d is the sky-projected distance between the centres of the planet and the star. To obtain a model that we can compare directly[4] to a time series of flux measurements, we need to express δ as a function of time.

This is promptly achieved by considering the projection of the relative orbit into cartesian coordinates described in Sect. 2.1.3. Recalling that we had chosen the axes

[4]Special care must be taken when the timescale of the variability is comparable to the integration time of individual points (Kipping 2010).

so that the xy plane coincided with the plane of the sky, we can write:

$$\delta = \frac{1}{R_s}\left(x^2 + y^2\right) = \frac{a}{R_s}\frac{1 - e^2}{1 + e\cos\nu}\sqrt{1 - \sin^2(\omega + \nu)\sin^2 I}\,, \tag{34}$$

where we have replaced the distance r_2 by the relative distance r (a is the semi-major axis of the relative orbit). Substituting this expression in the models for the drop of flux, and using the dependence of the true anomaly ν with time we can compute a model transit light curve to compare with the data time series.

2.3 Non-Keplerian Models

Up to here, we have assumed the modelled planetary companion moves in a Keplerian orbit around its host star. This is a good approximation when tidal forces originating in the interior of the objects can be neglected, which is often the case, as discussed above. However, the assumption of Keplerian motion is invalidated by the presence of additional objects in the system. Indeed, if the star is orbited by multiple planets, their mutual gravitational interactions will make their motion depart from the Keplerian orbit. The effect of the planet-planet interactions is usually very small on the radial-velocity data. It requires measurements with extreme precision spanning a long time for these effects to be detectable (see Correia et al. 2010). If detected, planet-planet interactions can lift the degeneracy between the RV-measured minimum masses and the orbital inclination I.

Gravitational interactions between the planets are more easily detected in transit light curves through the measurement of the timing of the transits, and their departure from perfect periodicity. This effect, known as transit timing variations (TTV) was predicted and studied for years (e.g., Agol et al. 2005; Holman and Murray 2005) before the detection of the first transit system with clear TTV (Holman et al. 2010). Besides, planet-planet interactions can change the duration of the transits, as the planets cross the stellar discs at different latitudes at different times. This gives rise to transit duration variations or TDV. Timing methods have been used to measure the mass of transiting planets for which no radial velocity measurement was possible (e.g., Jontof-Hutter et al. 2015) and to accurately measure the physical parameters of systems independently of stellar models (Almenara et al. 2015; Almenara et al. 2016). TTV of a transiting planet can also be employed to detect non-transiting planets (e.g., Ballard et al. 2011). However, unless TDV are also present (see Nesvorný et al. 2013; Barros et al. 2014), the parameters of the unseen planet are usually degenerate (Ballard et al. 2011).

Most of the analysis on TTV and TDV rely on the measurement of timings and durations of individual transits, and are limited by the precision obtained in these measurements. A way to dramatically improve the precision obtained in the parameters of systems analysed through transit timing is to perform a full photodynamical model (e.g., Carter et al. 2011; Doyle et al. 2011; Almenara et al.

2015; Almenara et al. 2016). The full photodynamical model uses a dynamical N-body simulation to compute distances between all the planets and the star at each time step. This is then input in the light curve models discussed above. In this way, the timing of each individual transit is constrained by the data from the entire light curve and not just by the points taken during that particular transit. As a consequence, analyses using photodynamical models produce superior results than those using measurements from individual transits (Almenara et al. 2016), but at the expense of the increased computational time required to perform the dynamical simulation.

3 Statistical Models

Let us go back to Eq. (1):

$$d_i = m_i + e_i \,. \tag{35}$$

Last section was dedicated to obtaining deterministic physical models for the term m_i. To do this we made a series of simplifying assumptions to render the problem tractable and arrived at deterministic expressions for the radial velocity or light curve of a star with an orbiting planetary companion. In other words, for each set of model parameters at a given time, the models from the previous sections provide a precise value for the radial velocity or flux drop.

In this section we describe a second type of models used to describe (exoplanet) data: statistical models. In a statistical (or stochastic) model there is not a single model output value for a given set of input parameters and independent variables (time), but rather a probability distribution of values. Stochastic models are strictly necessary when the system being modelled is probabilistic in nature. This is of course the case of quantum mechanics, but also of all problems involving uncertain measurements.

In Eq. (35), the error term e_i represents the discrepancy between the model m_i and the observed data. For data with non-zero uncertainty (real data) this term exhibits a stochastic behaviour. Its probabilistic nature comes from a combination of actual probabilistic processes—such as the emission of radiation by the atoms in the stellar atmosphere—with very complex, but in principle deterministic, systems—such as the detailed behaviour of the telescope and instruments used to acquire the data or weather conditions at the time of observation. These complex systems are better described in probabilistic terms instead of using complicated deterministic models with a large number of parameters.

Besides the probabilistic nature of the error term, e_i, a second source of randomness comes from uncertainties in the physical model term m_i. We have already mentioned that m_i is not necessarily deterministic. Indeed, m_i may contain a probabilistic part related either to uncertainty in the model itself or in the independent variable used to compute it. The latter is discussed in some detail in

Gregory (2005; Sect. 4.8.2). Here, we will deal exclusively with the former, and will break down the model term as

$$m_i = m_i^{\mathrm{d}} + m_i^{\mathrm{s}} , \tag{36}$$

where the *d* and *s* indices stand for *deterministic* and *stochastic*, respectively.

In general, the term m_i^{s} can include complicated processes affecting the data and on which we are not primarily interested. A prime example of this is stellar activity. As we will describe in this section, it is possible to produce a relatively simple model of the effects of stellar activity using statistical models. As we are not primarily interested in studying stellar activity, the parameters of this part of the model can be marginalised out to fully account for their effect on the uncertainty of the remaining model parameters. A deterministic alternative would require producing a model that provides the precise effect of stellar activity on the data for a given time. This usually requires a large number of parameters to describe, for example, the position, size, temperature and spectrum of active regions in time. Although this type of models exist and have been successfully put to practice (Boisse et al. 2012; Kipping 2012; Dumusque et al. 2014), they are plagued with degeneracies, because the available data are usually not sufficient to determine all model parameters independently. Besides, these models tend to require a substantial amount of computing time, which renders them impractical for iterative algorithms such as Markov chain Monte Carlo (MCMC). Another typical case where statistical models provide a simpler description than their deterministic counterparts is instrument systematics. In summary, the term m_i^{s} includes all "complex physics", defined here as all processes or systems too complicated to be described by a deterministic model with a reasonable number of parameters.

3.1 Uncorrelated Errors

The simplest possible model for the error term e_i arises when the observed data points are assumed uncorrelated and their errors are normally distributed with known variance σ_i^2, not necessarily the same for all measurements. The expression for the distribution function of the error term, $f(e_i)$ is then simply:

$$f_{\sigma_i}(e_i) = N(0, \sigma_i) = \frac{1}{\sqrt{2\pi\sigma_i^2}} \exp\left[-\frac{e_i^2}{2\sigma_i^2}\right] , \tag{37}$$

where $N(\mu, \sigma)$ is a Gaussian distribution with mean value μ and variance σ^2.

A slightly more sophisticated model consists in assuming the errors are over- or underestimated, and therefore introducing a multiplicative correction factor α:

$$f_{\sigma_i,\alpha}(e_i) = N(0, \alpha\sigma_i) = \frac{1}{\sqrt{2\pi}\alpha\sigma_i} \exp\left[-\frac{e_i^2}{2\alpha^2\sigma_i^2}\right] . \tag{38}$$

Note that this model already has an additional nuisance parameter α, which in principle is not of primary interest. The α term can be assumed to be the same for all measurements (for all i), or depend on subindex i, with the subsequent multiplication of nuisance parameters.

3.2 Gaussian Processes

A stochastic process can be roughly described as a generalisation of a probability distribution to functions (Rasmussen and Williams 2005). In other words, it is the probability distribution of an infinite-dimensional random variable, one that describes the values of the function in all possible points in input space. A Gaussian process (GP) can be seen as a collection of such random variables, any finite number of which have a joint multivariate normal distribution (Rasmussen and Williams 2005). Because of their great flexibility, attractive mathematical properties and computational tractability, GPs have been used as models by the machine learning community. Only recently GP regression has made its way to the exoplanet community. A thorough treatment of GPs is outside the scope of this chapter and we therefore give only a brief introduction to the subject in the context of exoplanetary science.

In general, GP regression is used in exoplanetary science to model complicated signals that are difficult to describe analytically. These include, but are not limited to, systematic effects originating in the instruments and signals produced by stellar activity. Sometimes GPs are said to model the covariate noise of a given dataset. This is equivalent, but means the GP is included in the error term e_i instead of in the stochastic model term m_i^s.

A GP is completely specified by its mean function and covariance function, which we denote $m(\vec{x})$ and $k(\vec{x}, \vec{x'})$, respectively, where $\vec{x}$ and $\vec{x'}$ are input vectors of the process. An input vector can have any number of dimensions, as the process can have any number of input variables. For example, when modelling stellar activity signals, one would naturally include time as an input variable, but may also find it useful to include activity proxies, such as the Ca II H & K lines index, $\log R'_{\mathrm{HK}}$. For any real process $f(\vec{x})$, the mean and covariance functions are defined as

$$m(\vec{x}) = \mathrm{E}[f(\vec{x})] \quad \text{and} \quad k(\vec{x}, \vec{x'}) = \mathrm{E}[(f(\vec{x}) - m(\vec{x}))(f(\vec{x'} - m(\vec{x'})] .$$

The mean function is usually the physical model included in addition to the GP model (i.e., m_i^{d}). For a pure GP model, the mean function can be considered as zero. We need, then, to specify only the covariance function $k(\vec{x}, \vec{x}')$, which needs to produce a valid covariance (i.e., a positive semi-definite symmetric) matrix.

One very popular example of covariance function is the squared-exponential:

$$k(t_i, t_j) = \sigma_f^2 \exp\left[-\frac{(t_i - t_j)^2}{2\tau^2}\right]. \tag{39}$$

For simplicity we assumed time is the only input variable and changed the notation accordingly. The function has two free parameters, σ_f and τ, called hyperparameters. We can sample from the GP defined by the covariance function in Eq. (39), and without including yet any data point. We call this the *prior* GP. The obtained sample is shown in Fig. 4 for three different values of the length scale parameter τ. In fact, to draw the functions in Fig. 4 we have evaluated the covariance at a finite time vector $\vec{t}_*$ of length n_*, known as the input vector. The corresponding matrix of covariance values is denoted $K(t_*, t_*)$, where for simplicity we left out the vector notation for the input vector.

The functions drawn in Fig. 4 are samples from the prior GP. Usually, we are interested in knowing how these functions change when we condition the GP to some data. For example, let us assume we have obtained a number of observations at times t_i, for $i = 1, 2, \ldots, n$. At these points we know the values of the function, f_i, with some noise, which we assume normally distributed with variance σ_i^2, for $i = 1, 2, \ldots, n$, i.e., the observation at time t_i is $y_i = f_i + \epsilon_i$, where ϵ_i is distributed as $N(0, \sigma_i)$. This data set is called the training set. We want to incorporate the new information and see how our process changes. Rasmussen and Williams (2005) explain that one way to do this is to produce a large number of samples from

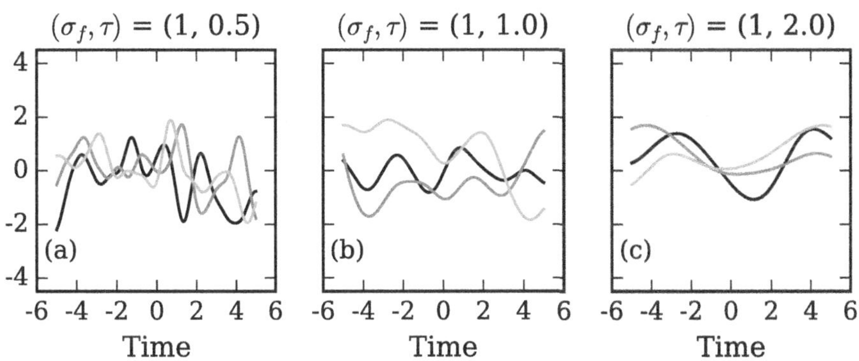

Fig. 4 Functions drawn from a GP with a squared-exponential covariance. The length scale parameter changes from 0.5 (**a**) to 1.0 (**b**) to 2.0 (**c**). As the length scale increases, the typical functions generated by the GP vary more slowly. In other words, the distance between two points in time that differ significantly increases with length scale

the prior GP and reject those that do not agree with the data. Of course, this is very inefficient computationally. Fortunately, the properties of GPs permit giving analytical formulae for the mean and covariance function of the posterior GP, f_*, conditioned to the data (see Eqs. 2.22–2.24 from Rasmussen and Williams 2005):

$$E[f_*|\vec{t},\vec{y}] = K(X_*,X)\left[K(X,X) + \sigma_n^2 I\right]^{-1}\vec{y}, \tag{40}$$

$$cov(f_*) = K(X_*,X_*) - K(X_*,X)\left[K(X,X) + \sigma_n^2 I\right]^{-1} K(X,X_*), \tag{41}$$

where the symbol $K(X,X_*)$ represents the $n \times x_*$ matrix of covariances evaluated at all pairs of training and input points.

Figure 5 shows three draws from the posterior GP conditioned to two observations, separated by 3.3 "time units". When the decay time is much smaller than this separation (panel (a)), the GP has a large variance between the observations, while its variance is much smaller when the separation of the training points is comparable with the decay time (panel (c)). In other words, if the covariance function falls off rapidly, having observed at one point in time gives little information about the rest of the points. Note that independently of the chosen hyperparameter values, the posterior GPs always manage to produce functions that accurately reproduce the training points. This attests the flexibility of the GPs, and warns us about their ability to potentially absorb coherent signals such as the signature of planets.

Another function widely used to model the effect of stellar activity is the quasi-periodic covariance,

$$k(t_i,t_j) = \sigma_f^2 \exp\left[-\frac{(t_i-t_j)^2}{2\tau^2} - \frac{2\sin^2\left(\pi(t_i-t_j)/\mathscr{P}\right)}{\eta^2}\right], \tag{42}$$

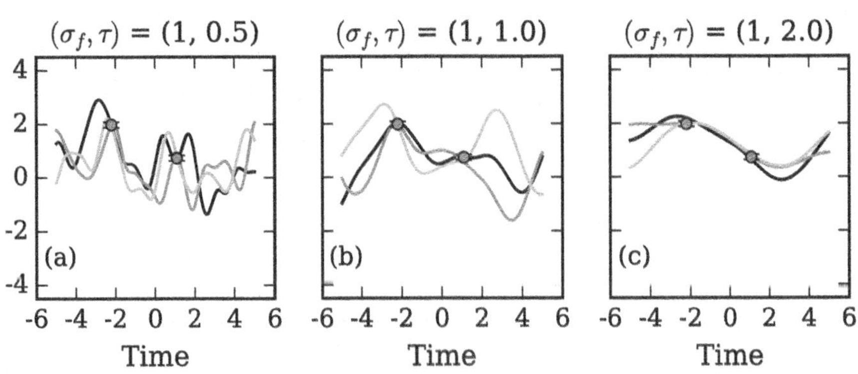

Fig. 5 Three draws from a GP obtained by conditioning the processes from Fig. 4 with two training points, with time coordinates, $t_1 = -2.2$, $t_2 = 1.1$, and observed values $y_1 = 1.98$, $y_2 = 0.74$. The variance of the noise term ϵ is 0.01 for both points. As the distance between the point becomes comparable with the decay time scale, the variance of the GP becomes smaller between the points

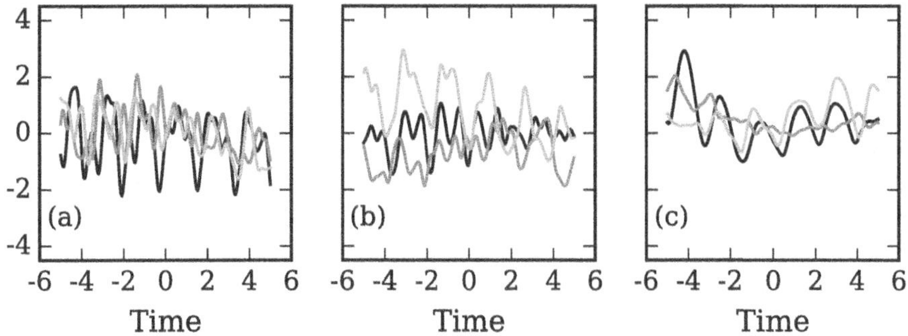

Fig. 6 Functions drawn from a GP with a pseudoperiodic covariance function, with amplitude $\sigma_f^2 = 1$, decay time scale $\tau = 3$, pseudo period $\mathscr{P} = 1.75$. The structure parameter is $\eta = 0.75$ (**a**), $\eta = 1.0$ (**b**), and $\eta = 2.0$ (**c**). We can see that as the structure (hyper)parameter increases, the functions resemble a sinusoidal function more closely, and can more accurately describe a light curve dominated by a single long-lived spot. Functions with smaller structure parameter have a larger number of contributions and resemble more closely stars with more than one active region. For example, in panel (**b**) functions clearly have two periodic components that evolve independently

with four hyperparameters: the covariance amplitude σ_f^2, the decay timescale τ, the pseudo-period $\mathscr{P}$ and the structure parameter η. We present samples from this process in Fig. 6.

The quasi-periodic covariance produces functions which resemble remarkably the light curve of spotted stars, and has been used to model both light curves and RV variations of active stars (e.g., Haywood et al. 2014; Rajpaul et al. 2015). The pseudo-period, or recurrence timescale, $\mathscr{P}$, can be identified with the stellar rotation period, which is often known from the light curve and in some situations from the radial-velocity time series. The decay time is associated with the time of the evolution of the active regions, which is harder to measure, but has been calibrated for solar-type stars using *Kepler* photometry. Additionally, as shown in Fig. 6, the structure parameter can be tuned to describe stars with a varying number of active regions. This covariance function allows therefore to describe a large variety of activity signals, making it very useful.

4 Bayesian Inference

This chapter on data modelling would not be complete without a description, however brief and summarised, of the techniques and methods used to obtain information on the model parameters from the data. The models presented so far are relatively simple and can in most cases be expressed by closed analytical formulae. However, the non-linear dependence on the model parameters and strong covariances between them makes the task of statistical inference challenging. In this section, we present a brief overview of the concepts involved in Bayesian statistical inference and provide a necessarily short introduction to a family of widely used algorithms: Markov chain Monte Carlo.

Statistical inference is the process by which we recover quantities of interest from available data, which are noisy or otherwise uncertain (see, e.g., Gregory 2005 and Trotta 2017 for an introduction addressed at astrophysicists and astronomers). The notion of probability is central in statistical inference. Statistics, sometimes called "inverse probability", is focussed on learning about the underlying probabilistic processes that give rise to the observed data. This is done by learning about the numerical values of the model parameters. Unlike the classical (frequentist) concept of probability, which is related to the number of times an event occurs, and is therefore defined only for random variables, the Bayesian theory allows us to define the probability of any proposition. Indeed, the Bayesian concept of probability is related to the degree of belief in a given proposition, with the extremes being the concepts of True or False used in classical Aristotelian logic. Actually, Bayesian probability theory can be seen as an extension of logic allowing us to reason in the face of uncertainty. A wonderful description of the concepts underlying Bayesian probability theory and the process leading to considering it an extension of logic is given by Jaynes (2003). Many other advantages of the Bayesian approach have been identified and are usually quoted (see, e.g., Trotta 2017).

In Bayesian inference the prior knowledge of any proposition—such as "the orbital period of this planet is between 3.5 and 3.51 days", or "this star is orbited by six planetary companions"—is encoded in the prior probability distribution, $p(\theta|I)$, where θ represents a proposition, which can represent a given hypothesis or be related to the value of the model parameters for a given hypothesis. In any case, I represents the knowledge available before taking into consideration the data. In what follows, we assume for simplicity that θ represents the parameter vector of a given model.

Once the data are incorporated in the analysis, the prior distribution transforms into the posterior distribution, $p(\theta|D, I)$. All the information provided by the data is encoded in this distribution. In general, the knowledge on θ changes when (new) data are incorporated. The way in which this change takes place is described by Bayes' theorem:

$$p(\theta|D, I) = \frac{p(D|\theta, I)p(\theta|I)}{p(D|I)} , \tag{43}$$

where D represents the data. The function $p(D|\theta, I) \equiv \mathscr{L}(\theta)$ is called the likelihood function and describes how probable the current data are given a certain value of the model parameters, θ. In the denominator, we find the marginal likelihood, $p(D|I)$, which is a constant term with respect to θ and ensures the correct normalisation of the posterior distribution. For the purpose of parameter inference, this can be safely ignored in most cases. On the other hand, the marginal likelihood has a central role in model comparison problems, which we do not cover here. The posterior distribution is therefore proportional to the likelihood times the prior:

$$p(\theta|D,I) \propto p(D|\theta,I)\,p(\theta|I)\,.$$

4.1 Likelihood Function

The likelihood function has therefore an important role in the inference process. A detailed description on how to construct likelihood functions is presented by Gregory (2005; Sect. 4.8). In particular, readers are referred to this book for a description on how to construct likelihoods with stochastic models.

Here we describe the simple case where the model is deterministic and the data errors are uncorrelated. Under these assumptions, one can deduce from Eq. (1) that the probability distribution of a given datum, d_i, is equal to that of the error term[5] e_i. Proposition D is simply the logical conjunction of propositions concerning individual data points: $D = d_1, d_2, \cdots, d_N$. Therefore, for independent errors, we have:

$$\begin{aligned}\mathcal{L}(\theta) = p(D|\theta,I) &= p(d_1, d_2, \cdots, d_N|\theta, I)\\ &= p(e_1, e_2, \cdots, e_N|\theta, I)\\ &= \prod_{i=1}^{N} p(e_i|\theta, I)\,. \end{aligned} \tag{44}$$

In the case of normally distributed errors (Eq. 37), the likelihood function has the form:

$$\mathcal{L}(\theta) = \prod_{i=1}^{N} \frac{1}{\sqrt{2\pi}\sigma_i} \exp - \frac{(d_i - m(t_i, \theta))^2}{2\sigma_i^2}\,, \tag{45}$$

where $m(t_i, \theta)$ is the model function, where we have explicitly indicated its dependence on the independent variable (time), t_i, and the model parameter vector, θ. Because of the large dynamical range of this likelihood function, it is usual to use its natural logarithm instead:

$$\log \mathcal{L}(\theta) = -\frac{N}{2}\log(2\pi) - \frac{1}{2}\sum_{i=1}^{N} \log \sigma_i^2 - \frac{1}{2}\sum_{i=1}^{N} \frac{(d_i - m(t_i, \theta))^2}{\sigma_i^2}\,. \tag{46}$$

[5]The derivation is not as trivial as one may think from considering Eq. (1).

The first term on the right-hand side of Eq. (47) is a constant and is usually irrelevant for inference problems. The last term is proportional to the usual χ^2 statistics. Note that if no model parameters appear in the data uncertainties, σ_i, maximising the likelihood function is equivalent to minimising the χ^2 statistics. However, if data errors risk being underestimated, one might wish to include an additional (nuisance) parameter in the model to account for this (see Eq. 38). In that case the second term on the right-hand side acts as a counterbalance impeding the likelihood from becoming artificially large by increasing the size of the error bars.

When the uncertainties are correlated, or when a GP is included as part of the model, the log-likelihood function is:

$$\log \mathscr{L}(\theta) = -\frac{N}{2}\log(2\pi) - \frac{1}{2}\log|K| - \frac{1}{2}\vec{r}^T \cdot K \cdot \vec{r}, \tag{47}$$

where K is the covariance matrix, $|K|$ is its determinant and $\vec{r}$ is the vector of residuals obtained by subtracting the model prediction from the observed data points.

4.2 Priors

The prior distribution of θ is the second element needed to compute the posterior distribution. Prior distributions (or simply "priors") are a controversial feature of Bayesian statistics. While some see in them a way to naturally include relevant prior information in the analysis others deem prior distributions as an inherently subjective element of the theory. In any case, defining the priors of a given problem is usually difficult and tricky. Priors need to include all relevant information on a problem, which can range from the results of a previous analysis to the educated guess of a renown scientific figure. As priors need to have the form of probability distributions, a non-trivial problem of codification arises. While there is no universally accepted method to assign priors that convey information, there is a vast body of literature on ways to define ignorance priors.

Ignorance priors are intended to convey the least possible information about θ. One could argue that absolute ignorance is an abstract concept and that some degree of information on a given parameter value is always available. To this Jaynes (2003) replies that ignorance priors play the role of the "zero" in Bayesian statistics, i.e., no information, and as such have an important theoretical role. In more practical terms, ignorance priors are usually handy when one needs to specify the priors for parameters for which little can be said before considering the data.

The literature contains a large number of digressions on how ignorance priors should be defined. We give a brief description of the most popular of them here:

1. *Principle of indifference.* Laplace (1812) introduced a principle for assigning probabilities to a set of n discrete hypotheses $H_1, H_2, \ldots, H_n$ when nothing allows to prefer one over the others. If the set is complete (i.e., they cover the

entire set of possible outcomes), we have $p(H_i|I) = 1/n$, for all i. Although this method seems intuitive, it is not easily extended to continuous variables, besides lacking invariance to transformation.

2. *Invariance rules.* For a certain type of parameters, "total ignorance" can be represented as invariance under certain transformations (Gregory 2005). For example, when the parameter represents a location, as for example the position X of the largest tree along a river, total ignorance requires that the problem be invariant under a translation: $X' = X + c$, for any constant c. This leads to the uniform prior:

$$p(X|I) = \text{constant}\,. \tag{48}$$

For scale parameters, such as the Poisson rate of an unknown source or the lifetime of new bacteria, ignorance is reflected by a lack of information on the order of magnitude of the parameter (the bacteria can live from a few minutes to a few days or weeks). This can be translated to invariance under rescaling, $X' = \alpha X$, which leads to a log-flat prior:

$$p(X|I) = \frac{\text{constant}}{X}\,. \tag{49}$$

Although these rules are useful, not all parameters can be classified in any of these two categories. Besides, the invariance under transformations gives us the functional form of the prior distribution, but not its boundaries (normalisation). One needs still to define the extrema of the distributions, if proper priors are needed.

3. *Jeffreys-rule prior.* The rule advocated by Jeffreys (1946, 1961) uses the expected Fisher information for a given problem:

$$I(\theta) = E_D\left[-\frac{\partial^2}{\partial\theta_i\partial\theta_j}\log\mathscr{L}(\theta)\right]\,, \tag{50}$$

where the expectation is over the data.[6] Jeffreys's rule defines the prior for θ as

$$p(\theta|I) = \sqrt{I(\theta)}\,. \tag{51}$$

Jeffreys's rule produces the optimal prior for one-dimensional models. Besides, it reproduces the results from the invariance rules. For example, the prior for the mean value of a normal model with known variance is a flat prior, and that of the scale of a normal with known mean value is a log-flat prior. However, Jeffreys-rule priors run into serious issues when applied to multi-dimensional parameters (Berger et al. 2015), reducing drastically their applicability.

[6]Real data D are not available at the moment of specifying priors for the model parameters.

Other methods involve the principle of maximum entropy (Jaynes 1957, 1968). More recent developments include reference priors (Berger et al. 2009, 2015), where the Kullback–Leibler divergence between prior and posterior is maximised seeking a prior that maximises the expected difference between posterior and prior distributions. In one-dimensional problems, reference priors reduce to Jeffreys-rule priors, but their properties in multi-dimensional problems are better.

In any case, sensitivity analyses are warranted and should be performed systematically. However, as expressed by Trotta (2017; Sect. 3.3):

> A sensitivity analysis should always be performed, i.e., change the prior in a reasonable way and assess how robust the ensuing posterior is. Unfortunately, this is seldom done in the astrophysics and cosmology literature.

4.3 Sampling the Posterior: Markov Chain Monte Carlo

Once the prior distribution is specified and the likelihood function is constructed, the posterior probability distribution (or simply the "posterior") can be obtained. In most practical cases it is not possible to perform inference on the posterior distribution analytically. For this reason, a series of algorithms to explore the posterior exist. Chief among them, Markov chain Monte Carlo (MCMC) is a family of algorithms that can be employed to produce a sample of arbitrary size from the posterior. With a sample at hand, one can make inferences on the model parameters, define credible intervals, explore the marginalised distributions, etc. An introduction to MCMC can be found in Gregory (2005). Another useful reading is the appendix of Tegmark et al. (2004). A vast body of literature and online resources exist on the subject and are promptly available.

The basic idea behind MCMC is to construct a Markov chain starting at a given point in parameter space. New links are added to the chain following some stochastic algorithm. Under quite general conditions it can be shown that after some steps, the links of the chain are samples from the posterior distribution (e.g., Gregory 2005). MCMC algorithms differ in the way the new chain links are produced. Many algorithms use a random walk process to generate a candidate link which is later accepted as part of the chain or not following some criterion.

The most popular Random Walk MCMC algorithm in exoplanetary science is the Metropolis–Hastings (MH) algorithm (Metropolis et al. 1953; Hastings 1970). In this algorithm, new steps for the chain are proposed by means of a random walk, usually the addition of a multivariate normal random variable to the current parameter vector, θ, to produce a candidate link, θ'. The proposed new state, θ', is then accepted with probability

$$r = \min\left(1, \frac{p(\theta'|D,I)}{p(\theta|D,I)} \frac{q(\theta|\theta')}{q(\theta'|\theta)}\right) ,$$

where $q(y|x)$ is the probability of proposing state y from state x. In the case of the normal proposal distribution, the ratio $q(\theta, \theta')/q(\theta', \theta)$ cancels out. In other words, if at the proposed step the posterior density increases with respect to θ, it is accepted automatically. Otherwise, the algorithm accepts θ' with probability $p(\theta'|D, I)/p(\theta|D, I) < 1$.

The efficiency of the MH algorithm is often given by the choice of the proposal distribution $q(\cdot|\cdot)$. A naïve choice as the one described above usually makes the algorithm very inefficient for sampling correlated parameter space. More sophisticated adaptive algorithms are described in the literature (e.g., Haario et al. 2001; Goodman and Weare 2010; Díaz et al. 2014). Another well-known issue of this MCMC algorithm is its inefficiency for sampling from multi-modal distributions. Readers are advised to consult the vast literature on this subject before developing their own MCMC code or using those available online.

Acknowledgements The author thanks the organisers of the IVth Azores International Advanced School in Space Sciences and acknowledges the participants—both lecturers and students—for the quality of their work. The preparation of this lecture was carried out within the frame of the Swiss National Centre for Competence in Research "PlanetS" funded by the Swiss National Science Foundation (SNSF). The author acknowledges support by the Argentinian National Council for Research and Technology (CONICET).

References

Agol, E., Steffen, J., Sari, R., Clarkson, W.: Mon. Not. R. Astron. Soc. **359**, 567 (2005)
Almenara, J.M., Díaz, R.F., Bonfils, X., Udry, S.: Astron. Astrophys. **595**, L5 (2016)
Almenara, J.M., Díaz, R.F., Mardling, R., et al.: Mon. Not. R. Astron. Soc. **453**, 2644 (2015)
Ballard, S., Fabrycky, D., Fressin, F., et al.: Astrophys. J. **743**, 200 (2011)
Barros, S.C.C., Díaz, R.F., Santerne, A., et al.: Astron. Astrophys. **561**, L1 (2014)
Berger, J.O., Bernardo, J.M., Sun, D.: Ann. Stat. **37**, 905–938 (2009)
Berger, J.O., Bernardo, J.M., Sun, D., et al.: Bayesian Anal. **10**, 189 (2015)
Boisse, I., Bonfils, X., Santos, N.C.: Astron. Astrophys. **545**, A109 (2012)
Carter, J.A., Fabrycky, D.C., Ragozzine, D., et al.: Science **331**, 562 (2011)
Claret, A.: Astron. Astrophys. **363**, 1081 (2000)
Correia, A.C.M., Couetdic, J., Laskar, J., et al.: Astron. Astrophys. **511**, A21 (2010)
Díaz, R.F., Almenara, J.M., Santerne, A., et al.: Mon. Not. R. Astron. Soc. **441**, 983 (2014)
Doyle, L.R., Carter, J.A., Fabrycky, D.C., et al.: Science **333**, 1602 (2011)
Dumusque, X., Boisse, I., Santos, N.C.: Astrophys. J. **796**, 132 (2014)
Goldstein, H.: Classical Mechanics, 2nd edn. Addison-Wesley, Reading (1980)
Goodman, J., Weare, J.: Commun. Appl. Math. Comput. Sci. **5**, 65 (2010)
Gregory, P.C.: Bayesian Logical Data Analysis for the Physical Sciences: A Comparative Approach with 'Mathematica' Support. Cambridge University Press, Cambridge (2005)
Haario, H., Saksman, E., Tamminen, J.: Bernoulli **7**, 223–242 (2001)
Hastings, W.: Biometrika **57**, 97 (1970)
Haywood, R.D., Collier Cameron, A., Queloz, D., et al.: Mon. Not. R. Astron. Soc. **443**, 2517 (2014)
Holman, M.J., Murray, N.W.: Science **307**, 1288 (2005)
Holman, M.J., Fabrycky, D.C., Ragozzine, D., et al.: Science **330**, 51 (2010)
Hut, P.: Astron. Astrophys. **99**, 126 (1981)

Jaynes, E.T.: Phys. Rev. **106**, 620 (1957)
Jaynes, E.T.: IEEE Trans. Syst. Sci. Cybern. **4**, 227 (1968)
Jaynes, E.T.: Probability Theory: The Logic of Science. Cambridge University Press, Cambridge (2003)
Jeffreys, H.: In: Proceedings of the Royal Society of London A: Mathematical, Physical and Engineering Sciences, vol. 186, pp. 453–461. The Royal Society (1946)
Jeffreys, H.: Theory of Probability, 3rd edn. Clarendon Press, Oxford (1961)
Jontof-Hutter, D., Rowe, J.F., Lissauer, J.J., Fabrycky, D.C., Ford, E.B.: Nature **522**, 321 (2015)
Kipping, D.M.: Mon. Not. R. Astron. Soc. **389**, 1383 (2008)
Kipping, D.M.: Mon. Not. R. Astron. Soc. **408**, 1758 (2010)
Kipping, D.M.: Mon. Not. R. Astron. Soc. **427**, 2487 (2012)
Laplace, P.: Theorie Analytique des Probabilités. Courcier, Paris (1812)
Mandel, K., Agol, E.: Astrophys. J. **580**, L171 (2002)
Metropolis, N., Rosenbluth, A.W., Rosenbluth, M.N., Teller, A.H., Teller, E.: J. Chem. Phys. **21**, 1087 (1953)
Murray, C.D., Correia, A.C.M.: In: Seager, S. (ed.) Exoplanets, pp. 15–23. University of Arizona Press, Tucson (2010)
Murray, C.D., Dermott, S.F.: Solar System Dynamics. Cambridge University Press, Cambridge (2000)
Nesvorný, D., Kipping, D., Terrell, D., et al.: Astrophys. J. **777**, 3 (2013)
Rajpaul, V., Aigrain, S., Osborne, M.A., Reece, S., Roberts, S.: Mon. Not. R. Astron. Soc. **452**, 2269 (2015)
Rasmussen, C.E., Williams, C.K.I.: Gaussian Processes for Machine Learning (Adaptive Computation and Machine Learning). The MIT Press, Cambridge (2005)
Seager, S., Mallén-Ornelas, G.: Astrophys. J. **585**, 1038 (2003)
Tegmark, M., Strauss, M.A., Blanton, M.R., et al.: Phys. Rev. D **69**, 103501 (2004)
Trotta, R.: ArXiv e-prints, arXiv:1701.01467 (2017)
Winn, J.N.: In: Seager, S. (ed.) Exoplanets, pp. 55–77. University of Arizona Press, Tucson (2010)
Zahn, J.-P.: Astron. Astrophys. **57**, 383 (1977)

Characterization of Exoplanet-Host Stars

Vardan Adibekyan, Sérgio G. Sousa, and Nuno C. Santos

Abstract Precise and, if possible, accurate characterization of exoplanets cannot be dissociated from the characterization of their host stars. In this chapter we discuss different methods and techniques used to derive fundamental properties and atmospheric parameters of exoplanet-host stars. The main limitations, advantages and disadvantages, as well as corresponding typical measurement uncertainties of each method are presented.

1 Introduction

The discovery of the first[1] extrasolar planet orbiting a main-sequence star, 51 Peg b (Mayor and Queloz 1995), marks the start of observational exoplanetology. Exoplanet research experienced huge progress during the last two decades and has surely become a solid research field in contemporary astrophysics. Thanks to the fast progress in the development of instrumentation and observational techniques during the past decades, as of today (January 2017) there are more than 3500 planets detected, while several thousand candidates still await validation (Coughlin et al. 2016).

[1]The detection of two terrestrial-mass companions around the pulsar PSR1257+12 had already been announced in 1992 (Wolszczan and Frail 1992).

V. Adibekyan (✉) • S.G. Sousa
Instituto de Astrofísica e Ciências do Espaço, Universidade do Porto, CAUP, Rua das Estrelas, 4150-762 Porto, Portugal
e-mail: vadibekyan@astro.up.pt; Sergio.Sousa@astro.up.pt

N.C. Santos
Instituto de Astrofísica e Ciências do Espaço, Universidade do Porto, CAUP, Rua das Estrelas, 4150-762 Porto, Portugal

Departamento de Física e Astronomia, Faculdade de Ciências, Universidade do Porto, Rua das Estrelas, 4150-762 Porto, Portugal
e-mail: nuno.santos@astro.up.pt

T.L. Campante et al. (eds.), *Asteroseismology and Exoplanets: Listening to the Stars and Searching for New Worlds*, Astrophysics and Space Science Proceedings 49, DOI 10.1007/978-3-319-59315-9_12

Today, the main efforts in exoplanet research are moving towards the precise characterization of detected planets, including their statistical properties, as well as the detection of planets with progressively lower masses. Despite the aforementioned progress, the study of extrasolar planets' properties via direct observations is still a very difficult task, and their precise study and characterization cannot be dissociated from the study of the host stars. For example, we should be aware that transit measurements only provide us with the planet-to-star radius ratio, and the mass provided by radial-velocity measurements is dependent on the stellar mass. Thus, the characterization of planets (e.g., mass, radius, density, and age) requires characterization of their hosts, and the accuracy of the planets' properties fundamentally depends on the achieved accuracy of the hosts' properties.

It is very clear for the exoplanet scientific community that poor characterization of planet hosts and planets themselves is an important limitation, which cannot be always compensated even with large number statistics. A good example is the *Kepler* mission, which provided thousands of stars with exoplanet candidates and an extremely large sample of stars with no detected planets that can be used for comparison analyses. However, the vast majority of these stars are poorly characterized, which obviously decelerates the—though still revolutionary—fast advance in the field. The example of *Kepler* and that of other ongoing (e.g., *Gaia*, K2; Perryman et al. 2001; Howell et al. 2014) and upcoming (e.g., *TESS*, *PLATO*; Ricker 2014; Rauer et al. 2014) space missions motivated the community to start coordinating efforts to characterize the planet hosts. The importance of coordinating the exoplanet follow-up efforts has been addressed and intensively discussed in several recent meetings,[2] and during the past few years several dedicated communities[3] and web interfaces[4] have been created with the goal of optimizing and coordinating the resources in exoplanet follow-up studies and characterization of their host stars.

Regarding the accuracy of the characterization of exoplanet hosts, many groups all over the world are intensively working on pushing down the precision limits and on developing methods that are less model-dependent and are most time-efficient. Unfortunately, direct measurements of physical properties of stars—including exoplanet hosts—are very rare and are possible for only specific targets. The physical properties of the host stars are usually derived by using theoretical stellar evolutionary models and/or models of atmospheres. The uncertainties in the stellar model parameters can highly influence the final accuracy with which properties of the stars and their planets are measured (e.g., Soderblom 2010; Basu et al. 2012). Asteroseismology is the tool that comes to help on improving our knowledge of fundamental properties of stars. It can provide properties for bright exoplanet-host stars (solar-type and red-giant stars, but not the cool dwarfs) with very high accuracy. *PLATO* will take full advantage of asteroseismic analyses to characterize all the

[2]For example, during ExoPAG 11 (https://exoplanets.nasa.gov/exep/events/14/) and the K2 meeting that took place in Porto in 2016 (http://www.iastro.pt/research/conferences/k2meeting/).

[3]SAG-14 (https://exoplanets.nasa.gov/exep/exopag/sag/).

[4]ExoFOP (https://exofop.ipac.caltech.edu/).

planet hosts brighter than 11th magnitude (Rauer et al. 2014). Few dozen exoplanet hosts detected by *TESS* will also benefit from asteroseismology (Campante et al. 2016). One should also not underestimate the importance of the *Gaia* mission in the characterization of exoplanet hosts. Combined with ground-based, high-resolution spectroscopy, *Gaia* will provide precise fundamental properties (radii, luminosities, distances, and surface gravities) of exoplanet hosts.

We should also note that, in exceptional cases, planetary properties can be derived without using stellar models. For example, the surface gravity of transiting exoplanets can be directly determined from the spectroscopic orbit of the parent star and the parameters measured directly from the transit light curve (Southworth et al. 2007). Absolute masses and radii of planets can be also determined with very high precision (down to 1–2%) for multi-planet systems—with detectable gravitational interactions between planets—when precise light curves of transits and radial-velocity (RV) data are available (Almenara et al. 2015). Interestingly, it was proposed that the masses of transiting planets can be estimated based solely on the transmission spectrum (de Wit and Seager 2013).

It is very interesting to realize that not only knowledge about the host star helps to better understand the planet, but also sometimes observations of exoplanets help characterizing the stars. For example, the stellar density can be directly derived from the transit light curve alone (Seager and Mallén-Ornelas 2003, but see Kipping 2014), and spatially-resolved stellar photospheres can be studied in detail when transiting planets are observed spectroscopically (e.g., Collier Cameron et al. 2010; Cegla et al. 2016).

In this manuscript we present the main methods and techniques that are widely used to characterize exoplanet-host stars. Together with the description of different methods and techniques we will also discuss the main limitations and achievable precision.

2 Fundamental Properties of Stars: Mass, Radius and Age

The mass of RV-detected planets scales as $M^{2/3}$, where M is the mass of the stellar host, while the radius of transiting planets is derived from the depth of the transit event and the radius of the parent star. Since planet formation is a relatively fast process compared to the lifetime or age of most of the planet hosts, stellar age can be used as a proxy for the age of planetary systems. Thus, basic characterization of exoplanets implies basic characterization of their hosts.

Stellar Masses Very precise dynamical masses can be derived for double-lined and single-lined (if the RV is derived for each component) eclipsing binaries, and for non-eclipsing double-lined spectroscopic binaries if astrometric orbits of the stars are known (usually through long-baseline interferometry). These techniques are quite well known (for a recent comprehensive review, see Torres et al. 2010) and can provide masses with an accuracy of better than 3%.

Unfortunately, direct determinations are usually impractical for most stars and indirect methods have to be used. Different empirical and theoretical indirect methods are commonly used to determine the mass of single field stars. The stellar masses can be estimated by using the spectroscopic surface gravity and luminosity of the star, provided the T_{eff} is known. This method can give masses with a precision of 10–20% depending on the precision on $\log g$ and distance (parallax) of stars (Sousa et al. 2011). The masses can be empirically estimated by using mass-luminosity relations with a precision below 10% (e.g., Xia and Fu 2010). Empirical relations between stellar mass and stellar parameters (T_{eff}, $\log g$, and [Fe/H]) by Torres et al. (2010) give a scatter of only ~6% for main-sequence stars with masses above $0.6M_{\odot}$. Finally, stellar masses can be determined by comparing stellar observed properties with stellar evolutionary tracks (e.g., Johnson et al. 2010; Sousa et al. 2015) or by using the power of asteroseismology (e.g., Huber et al. 2012; Chaplin et al. 2014). The latter method can lead to mass uncertainties below 5% (e.g., Chaplin et al. 2014).

We should note that several studies suggested that the masses of planet-hosting evolved stars derived from evolutionary tracks can be largely overestimated (e.g., Lloyd 2011, 2013; Takeda and Tajitsu 2015). However, Ghezzi and Johnson (2015) recently found very good agreement between model-independent masses and the masses estimated using PARSEC evolutionary tracks (Bressan et al. 2012) for a sample of 59 benchmark evolved stars.

Stellar Radii One of the most accurate ways of determining stellar radii is to measure the angular size of stars using interferometry. When precise distances (parallaxes) of these stars are known this method provides a practically direct measurement the radius that reaches 1–3% precision on the angular diameter (e.g., Boyajian et al. 2013, 2014). Until now, distances (parallaxes) of only nearby stars were known with high precision thanks to the *Hipparcos* satellite. However, *Gaia* will improve the situation, providing very precise distances for stars with much larger distances than *Hipparcos* could reach. We note that angular sizes and, consequently, linear radii of stars can also be determined using lunar occultations. This method has clear limitations (e.g., the Moon does not cover all the stars in the sky), but can provide radii with a precision of down to 3% (e.g., Richichi 1997).

Another direct technique to derive accurate stellar radii is to use double-lined eclipsing binary systems. The measured light curve and derived radial velocities of the two components can be used to estimate the radii of the two stars with accuracies of better than ~1% (e.g., Lacy et al. 2005; Southworth et al. 2007). A catalog of about 170 detached eclipsing binary systems with precise mass and radius measurements is presented in Southworth (2015).

Accurate direct measurements of stellar radii with interferometry and/or using eclipsing binaries can be used to develop empirical relations between radius and photometric colors (Boyajian et al. 2014), or else radius and stellar parameters (Torres et al. 2010). These empirical relations can be used to predict radii of stars with errors less than 5%.

Finally, distant stars are inaccessible for direct angular diameter measurements and so indirect methods are necessary to estimate their radii. Stellar radii can be derived from stellar evolution models by using the luminosity and effective temperature of the stars (e.g., Santos et al. 2004b; Torres et al. 2006) or, for transiting systems, by using the stellar density[5] (directly derived from the light curve) and $T_{\rm eff}$ (e.g., Sozzetti et al. 2007). Radii of exoplanet hosts can also be derived by using asteroseismic quantities combined with $T_{\rm eff}$ and stellar metallicity. This technique provides stellar radii with a typical precision of 2–4% (e.g., Chaplin et al. 2014).

Stellar Ages Determination of accurate stellar ages is not an easy task, especially for field stars. Unlike stellar mass and radius, stellar age cannot be directly measured and the use of stellar models is usually necessary to estimate ages. In exceptional cases, stellar ages can be determined without involving stellar models, namely, for young groups of stars through their kinematics (e.g., Makarov 2007) and for old metal-poor stars by using nucleocosmochronometry (e.g., Ludwig et al. 2010).

The most common ways of estimating ages of exoplanet hosts are isochrone placement (e.g., Pont and Eyer 2004; Takeda et al. 2007) and asteroseismology (e.g., Silva Aguirre et al. 2013; Campante et al. 2015). We note that both methods require a knowledge of stellar atmospheric parameters. Whereas the uncertainties on ages derived from stellar isochrones are typically not better than ∼20–30% (e.g., Jørgensen and Lindegren 2005; Maxted et al. 2015), asteroseismology can provide ages with a relative precision of about 10–20% (e.g., Kjeldsen et al. 2009; Silva Aguirre et al. 2016).

Alternatively, stellar ages can be derived by using empirical relations, calibrated between age and rotation period (e.g., Barnes 2007), age and chromospheric activity (e.g., Lyra and Porto de Mello 2005; Mamajek and Hillenbrand 2008), as well as age and chemical abundance ratios (e.g., Nissen 2015). While these empirical relations can provide relative high precision (depending on the calibration), their absolute values are difficult to establish. For an excellent review on stellar age derivation with different techniques we refer the reader to Soderblom (2010).

Summarizing, we can state that, when direct measurements are possible, masses and radii of individual stars can be derived with a precision of better than 1–3%, whereas stellar ages can be estimated with an accuracy of better than a few percent. For large numbers of exoplanet hosts, stellar fundamental properties can be derived with a precision of ∼10–20% for mass and radius, and with a precision of 20–30% for ages, assuming stellar atmospheric parameters are derived with high precision (see sections below).

[5]Note that the derivation of this parameter also depends on the limb-darkening coefficient and orbital eccentricity of the transiting planet (e.g., Mortier et al. 2013).

3 Stellar Atmospheric Parameters

Accurate derivation of stellar atmospheric parameters ($T_{\rm eff}$, $\log g$, and metallicity/-chemical abundances) is very important to fully characterize exoplanet-host stars. We need only to remember that the first interesting hint observed for exoplanet hosts was the correlation between giant-planet occurrence and stellar metallicity (e.g., Gonzalez 1997; Santos et al. 2001), which had crucial importance for the advance of exoplanet formation theories. For individual stars, direct measurements of stellar sizes and masses can be used to determine effective temperature and surface gravity without using stellar models. Stellar metallicity and chemical abundances of stars cannot be directly measured and stellar atmospheric models need to be used.

As discussed in the previous section, the direct determination of radii and masses, and hence $T_{\rm eff}$ and $\log g$, is not possible for most stars. Hence, indirect methods need to be used. Stellar atmospheric parameters ($T_{\rm eff}$, $\log g$, and [Fe/H] as a proxy for overall metallicity) can be derived with different methods and techniques. Photometric calibrations (e.g., Önehag et al. 2009; Casagrande et al. 2010; Brown et al. 2011), depending on the photometric systems, can provide stellar parameters with reasonably high precision (Smalley 2014). Profiles of individual lines (e.g., Catanzaro et al. 2004, 2013; Cayrel et al. 2011) and spectral line depth/equivalent width (EW) ratios (e.g., Gray and Johanson 1991; Sousa et al. 2012) can also be used to determine different stellar parameters. Some of these methods can provide parameters with very high precision (e.g., $\sim$2K in $T_{\rm eff}$; Gray and Livingston 1997), but with significantly less accuracy. Nevertheless, the most used and accurate techniques of deriving stellar parameters are provided by stellar spectroscopy. For a comprehensive description of different methods for atmospheric parameter derivation we refer the reader to Gray (2005) and Niemczura et al. (2014).

The main spectral analysis techniques for the determination of stellar parameters can be divided into two main groups: the EW method and the spectral synthesis method. In classical EW methods, measurements of EWs of isolated individual metallic lines are used to derive stellar parameters assuming excitation equilibrium and ionization balance (e.g., Santos et al. 2004a; Sousa 2014). Spectral synthesis methods yield stellar parameters by fitting the observed spectrum—all, selected parts of the spectrum, or even a selection of lines—with a synthetic one (Valenti and Piskunov 1996; Malavolta et al. 2014), with a library of pre-computed synthetic spectra (Recio-Blanco et al. 2006), or a library of EWs (Boeche and Grebel 2016). Today there are many automatic tools designed to derive stellar parameters with the EW method (e.g., Magrini et al. 2013; Tabernero et al. 2013; Sousa 2014), spectral synthesis techniques (e.g., Allende Prieto et al. 2006; Sbordone et al. 2014), as well as tools that integrate different techniques, models of atmospheres and radiative transfer codes (Blanco-Cuaresma et al. 2014). For further details about these techniques we refer the reader to Niemczura et al. (2014) and Allende Prieto (2016).

Both EW and spectral synthesis techniques have their advantages, disadvantages and limitations. The EW method is usually fast and relies on well selected lines. However, this method cannot be applied to fast-rotating stars or to stars with severe line-blended spectra. For these stars, spectral synthesis methods should be used. Synthesis techniques typically require more complicated computations for the generation of synthetic spectra and heavily depend on the line list and atomic/molecular data. A common limitation of spectroscopic methods is that they cannot constrain stellar surface gravity well (e.g., Sozzetti et al. 2007; Mortier et al. 2013; Tsantaki et al. 2014). The impact of an unconstrained $\log g$ on the derivation of other stellar parameters ($T_{\rm eff}$ and [Fe/H]) is minimal for the EW-based curve-of-growth approach, while it has a significant impact for spectral-synthesis-based methods (Torres et al. 2012; Mortier et al. 2013). Luckily, surface gravity can be derived with high precision using asteroseismology (e.g., Huber et al. 2013), as well as for transiting systems from their light curves combined with spectroscopic $T_{\rm eff}$ and metallicity (e.g., Seager and Mallén-Ornelas 2003). From these two estimates, asteroseismic $\log g$'s are preferable, since transit-based $\log g$'s might be less accurate when the eccentricity or the impact parameter of the transiting planet is not well constrained (Huber et al. 2013). Mortier et al. (2014) proposed an empirical correction—based on the comparison of spectroscopic and asteroseismic $\log g$'s—for the spectroscopic surface gravity that depends only on the effective temperature. A word of caution should be voiced here. It is not advisable to fix the surface gravity—derived from other, non-spectroscopic method—when doing spectral analyses (Mortier et al. 2014; Smalley 2014). Fixing the $\log g$ can bias the results and derivation of other atmospheric parameters.

The spectroscopic determination of stellar parameters is affected by different factors, many of which are briefly discussed in Smalley (2014). The influence of many of these factors (e.g., model atmosphere physics and input data) can be minimized when the spectral analysis is done in a homogeneous way. Consequently, when homogeneous and high-quality data are used, an extremely high precision in stellar parameters can be achieved. For example, the latest works on solar twins that are based on differential line-by-line analysis report a precision (internal error) in atmospheric parameters of $\sim$10 K for $T_{\rm eff}$, $\sim$0.02 dex for $\log g$, and $\sim$0.01 dex for [Fe/H] (e.g., Ramírez et al. 2014; Adibekyan et al. 2016b). However, one should note that when analyzing spectra of the same star obtained with different instruments and at different epochs, dispersion of stellar parameters larger than the aforementioned precision can be obtained (e.g., Bensby et al. 2014; Adibekyan et al. 2016b). Systematic errors, due to the model atmospheres, analysis method and atomic data are much larger than the random errors. Comparison of the results obtained with different methods for very large numbers of stars (e.g., Bensby et al. 2014; Smiljanic et al. 2014), as well as comparison of results with model-independent values for benchmark stars (Jofré et al. 2014; Heiter et al. 2015) show that realistic typical errors on stellar parameters are not less than 50–100 K for $T_{\rm eff}$, 0.1–0.2 dex for $\log g$, and 0.05–0.1 dex for metallicity. Further discussion on the impact of using different atmosphere models and different analysis strategies on the derivation of stellar parameters is presented in Lebzelter et al. (2012).

Homogeneous derivation of stellar parameters is crucial for characterizing exoplanet-host stars. The internal (relative) precision of atmospheric parameters can be as good as ~10 K for $T_{\rm eff}$, ~0.02 dex for $\log g$, and ~0.01 dex for [Fe/H], but the overall precision of these parameters will be considerably smaller.

4 Chemical Abundances of Exoplanet-Host Stars

Exoplanet-related research always requires high precision and accuracy. If very high-precision measurements are needed to detect planets, likewise, finding possible abundance differences between stars with and without planets also requires accurate and homogeneous abundance determinations. Many studies aimed at clarifying whether the planet-hosting stars are different from stars without planets in their content of individual heavy elements other than iron (e.g., Fischer and Valenti 2005; Robinson et al. 2006; Delgado Mena et al. 2010; Adibekyan et al. 2012b, 2015a; Suárez-Andrés et al. 2016). In particular, it was shown that metal-poor hosts tend to show systematic enhancement in α elements (Haywood 2009; Adibekyan et al. 2012a,c). Accurate knowledge of abundances of individual heavy elements and specific elemental ratios (e.g., Mg/Si and C/O) in stars with planets are also very important because they are expected to control the structure and composition of terrestrial planets (e.g., Grasset et al. 2009; Thiabaud et al. 2014; Dorn et al. 2015).

Once the atmospheric parameters of stars are known, chemical abundances of individual elements can be derived spectroscopically by EW or spectral synthesis techniques. Many research groups are intensively working on the derivation of chemical abundances in stellar atmospheres of stars with and without planets. The derivation of chemical abundances may seem very trivial, however, a simple comparison of the (discrepant) results obtained for the same elements from the same data in the same stars, but with different methods, shows that there are important factors (e.g., line list and atomic data, continuum normalization, hyperfine structure, damping, microturbulence, NLTE effects, atmospheric model) that need to be deeply investigated. Intensive and comprehensive discussion about the possible issues can be found in several recent articles (e.g., Smiljanic et al. 2014; Jofré et al. 2015; Hinkel et al. 2016) that had as a common goal to *open the black box of stellar element abundance determination* (Jofré et al. 2016).

As for the stellar parameters, when studying solar twins and solar analogs (i.e., stars that are very similar to our Sun in terms of stellar parameters) extremely precise—accuracy still can be an issue—chemical abundances at the level of ~0.01 dex can be obtained (e.g., Ramírez et al. 2010; González Hernández et al. 2013; Adibekyan et al. 2016a; Saffe et al. 2016). High-precision abundances (at the level of ~0.05–0.10 dex) can be also obtained for large samples of cool stars if high-quality data are used and, importantly, if the spectral analysis is done in a homogeneous way (e.g., Adibekyan et al. 2012b; Bensby et al. 2014; Adibekyan et al. 2015b; Mikolaitis et al. 2016). However, if the data are compiled from different sources, or different methods were used to derive abundances, then the results should be taken with caution. Method-to-method or study-to-study dispersion of chemical abundances can be larger than 0.10–0.20 dex (e.g., Hinkel et al. 2014; Smiljanic et al. 2014).

As for the atmospheric parameters, homogeneous derivation of chemical abundances is important to achieve high precision. Elemental abundances for large samples of cool stars can be derived with a typical internal (relative) precision of ~0.05 dex, but the accuracy of these derivations will be smaller.

5 Other Properties of Exoplanet-Host Stars

Kinematics Kinematics, or Galactic space-velocity components of stars, can be computed when a star's proper motion, radial velocity and parallax are known (e.g., Johnson and Soderblom 1987). The kinematics of exoplanet-host stars and their relation to different stellar populations and moving groups have been discussed in several works (e.g., Barbieri and Gratton 2002; Reid 2002; Ecuvillon et al. 2007; Adibekyan et al. 2012c; Gaidos et al. 2017). Most papers have not reported any significant kinematic peculiarity of planet-hosting stars (e.g., Gonzalez 1999; Barbieri and Gratton 2002). Conversely, Haywood (2008, 2009), combining the chemical and kinematic properties of exoplanet hosts, concluded that most metal-rich stars that host giant planets originate from the inner Galactic disk. The same scenario for the origin of metal-rich planet hosts is explored in a few other works (e.g., Ecuvillon et al. 2007; Santos et al. 2008; Adibekyan et al. 2014).

Activity Understanding stellar magnetic activity phenomena (such as spots, faculae, plages) is very important for different fields of stellar physics and exoplanetary science, as well as for planetary climate studies. Studying magnetic activity in stars of different stellar parameters and activity levels provides an opportunity for detailed tests of stellar/solar dynamo models. From the exoplanetary side, it is well known that stellar active regions, combined with the stellar rotation, can induce signals in high-precision photometric and radial-velocity observations. These activity-induced signals may lead to masking or mimicking of exoplanet signals (e.g., Queloz et al. 2001; Dumusque et al. 2012; Oshagh et al. 2013; Santos et al. 2014). Moreover, these signals constitute one of the main obstacles to the detection and precise characterization of low-mass/small-radius planets, the major goal of future instruments. Several indices (re-emission in the Ca II H and K lines, Mg II h and k lines, Ca infrared triplet, Na I D doublet, Hα) exist to characterize the activity of stars (e.g., Baliunas et al. 1995; Kürster et al. 2003; Mamajek and

Hillenbrand 2008; Gomes da Silva et al. 2011; Haswell et al. 2012; Mathur et al. 2014). The applicability of these indices is restricted, as it depends on the spectral type of the stars and spectral coverage of the used spectrograph. The dependence of stellar activity on the planet-star interaction was discussed in several observational and theoretical studies (e.g., Figueira et al. 2016; and references therein) yielding contradictory results.

Rotation The most common ways of measuring stellar rotation are through spectroscopy (e.g., Benz and Mayor 1981; Donati et al. 1997) and photometry (e.g., Irwin et al. 2009; McQuillan et al. 2013). These techniques—depending on the quality of the data and properties of the stars—can provide rotation velocities[6] with a precision of better than ∼10% (for the limitations and advantages of either technique, see Bouvier 2013). Recent studies show that the stars with planets (or with planet candidates in the case of *Kepler*) rotate more slowly than stars without known planets (e.g., Takeda et al. 2010; Gonzalez 2015). Moreover, it appears that only slow-rotating stars host close-in planets. The slow rotation of exoplanet host stars—if not a selection and/or detection bias—can be caused by early star-disk interactions (Bouvier 2008).

6 Conclusion

Precise and accurate characterization of exoplanet-host stars is crucial to the detailed investigation of exoplanets themselves. Moreover, precise determination of stellar parameters is important to study the star-planet connection. There are different ways of characterizing stars with and without planets. Some of these methods are independent of stellar models, hence fundamental, although most are not. The combination of different methods can provide precise, and even accurate, stellar parameters and chemical abundances of exoplanet hosts.

When studying statistical properties of exoplanets or of their hosts it is very important to use information (parameters) as homogeneous as possible. A catalog of exoplanet hosts with stellar parameters derived and compiled in a homogeneous way is presented in Santos et al. (2013) and, for transiting systems, in Southworth (2012).

Acknowledgements VA would like to thank the SOC for inviting him to *IV*th *Azores International Advanced School in Space Sciences* held in the Azores Islands, Portugal. VA, NCS and SGS acknowledge the support from Fundação para a Ciência e a Tecnologia (FCT) through national funds and from FEDER through COMPETE2020 by the following grants UID/FIS/04434/2013 and POCI-01-0145-FEDER-007672, PTDC/FIS-AST/7073/2014 and POCI-01-0145-FEDER-016880, and PTDC/FIS-AST/1526/2014 and POCI-01-0145-FEDER-016886. VA, NCS and SGS acknowledge the support from FCT through Investigador FCT contracts IF/00650/2015/CP1273/CT0001, IF/00169/2012/CP0150/CT0002, and IF/00028/2014/CP1215/CT0002.

[6]Note that spectroscopy usually provides the $v \sin i$ and photometry gives the rotational period of the stars.

References

Adibekyan, V.Z., Delgado Mena, E., Sousa, S.G., et al.: Astron. Astrophys. **547**, A36 (2012a)
Adibekyan, V.Z., Sousa, S.G., Santos, N.C., et al.: Astron. Astrophys. **545**, A32 (2012b)
Adibekyan, V.Z., Santos, N.C., Sousa, S.G., et al.: Astron. Astrophys. **543**, A89 (2012c)
Adibekyan, V.Z., González Hernández, J.I., Delgado Mena, E., et al.: Astron. Astrophys. **564**, L15 (2014)
Adibekyan, V., Santos, N.C., Figueira, P., et al.: Astron. Astrophys. **581**, L2 (2015a)
Adibekyan, V.Z., Benamati, L., Santos, N.C., et al.: Mon. Not. R. Astron. Soc. **450**, 1900 (2015b)
Adibekyan, V., Delgado-Mena, E., Figueira, P., et al.: Astron. Astrophys. **592**, A87 (2016a)
Adibekyan, V., Delgado-Mena, E., Figueira, P., et al.: Astron. Astrophys. **591**, A34 (2016b)
Allende Prieto, C.: Living Rev. Sol. Phys. **13**, 1 (2016)
Allende Prieto, C., Beers, T.C., Wilhelm, R., et al.: Astrophys. J. **636**, 804 (2006)
Almenara, J.M., Díaz, R.F., Mardling, R., et al.: Mon. Not. R. Astron. Soc. **453**, 2644 (2015)
Baliunas, S.L., Donahue, R.A., Soon, W.H., et al.: Astrophys. J. **438**, 269 (1995)
Barbieri, M., Gratton, R.G.: Astron. Astrophys. **384**, 879 (2002)
Barnes, S.A.: Astrophys. J. **669**, 1167 (2007)
Basu, S., Verner, G.A., Chaplin, W.J., Elsworth, Y.: Astrophys. J. **746**, 76 (2012)
Bensby, T., Feltzing, S., Oey, M.S.: Astron. Astrophys. **562**, A71 (2014)
Benz, W., Mayor, M.: Astron. Astrophys. **93**, 235 (1981)
Blanco-Cuaresma, S., Soubiran, C., Heiter, U., Jofré, P.: Astron. Astrophys. **569**, A111 (2014)
Boeche, C., Grebel, E.K.: Astron. Astrophys. **587**, A2 (2016)
Bouvier, J.: Astron. Astrophys. **489**, L53 (2008)
Bouvier, J.: In: Hennebelle, P., Charbonnel, C. (eds.) EAS Publications Series, vol. 62, pp. 143–168 (2013). doi:10.1051/eas/1362005
Boyajian, T.S., von Braun, K., van Belle, G., et al.: Astrophys. J. **771**, 40 (2013)
Boyajian, T.S., van Belle, G., von Braun, K.: Astrophys. J. **147**, 47 (2014)
Bressan, A., Marigo, P., Girardi, L., et al.: Mon. Not. R. Astron. Soc. **427**, 127 (2012)
Brown, T.M., Latham, D.W., Everett, M.E., Esquerdo, G.A.: Astrophys. J. **142**, 112 (2011)
Campante, T.L., Barclay, T., Swift, J.J., et al.: Astrophys. J. **799**, 170 (2015)
Campante, T.L., Schofield, M., Kuszlewicz, J.S., et al.: Astrophys. J. **830**, 138 (2016)
Casagrande, L., Ramírez, I., Meléndez, J., Bessell, M., Asplund, M.: Astron. Astrophys. **512**, A54 (2010)
Catanzaro, G., Leone, F., Dall, T.H.: Astron. Astrophys. **425**, 641 (2004)
Catanzaro, G., Ripepi, V., Bruntt, H.: Mon. Not. R. Astron. Soc. **431**, 3258 (2013)
Cayrel, R., van't Veer-Menneret, C., Allard, N.F., Stehlé, C.: Astron. Astrophys. **531**, A83 (2011)
Cegla, H.M., Lovis, C., Bourrier, V., et al.: Astron. Astrophys. **588**, A127 (2016)
Chaplin, W.J., Basu, S., Huber, D., et al.: Astrophys. J. Suppl. Ser. **210**, 1 (2014)
Collier Cameron, A., Bruce, V.A., Miller, G.R.M., Triaud, A.H.M.J., Queloz, D.: Mon. Not. R. Astron. Soc. **403**, 151 (2010)
Coughlin, J.L., Mullally, F., Thompson, S.E., et al.: Astrophys. J. Suppl. Ser. **224**, 12 (2016)
de Wit, J., Seager, S.: Science **342**, 1473 (2013)
Delgado Mena, E., Israelian, G., González Hernández, J.I., et al.: Astrophys. J. **725**, 2349 (2010)
Donati, J.-F., Semel, M., Carter, B.D., Rees, D.E., Collier Cameron, A.: Mon. Not. R. Astron. Soc. **291**, 658 (1997)
Dorn, C., Khan, A., Heng, K., et al.: Astron. Astrophys. **577**, A83 (2015)
Dumusque, X., Pepe, F., Lovis, C., et al.: Nature **491**, 207 (2012)
Ecuvillon, A., Israelian, G., Pont, F., Santos, N. C., Mayor, M.: Astron. Astrophys. **461**, 171 (2007)
Figueira, P., Santerne, A., Suárez Mascareño, A., et al.: Astron. Astrophys. **592**, A143 (2016)
Fischer, D.A., Valenti, J.: Astrophys. J. **622**, 1102 (2005)
Gaidos, E., Mann, A.W., Rizzuto, A., et al.: Mon. Not. R. Astron. Soc. **464**, 850 (2017)

Ghezzi, L., Johnson, J.A.: Astrophys. J. **812**, 96 (2015)
Gomes da Silva, J., Santos, N.C., Bonfils, X., et al.: Astron. Astrophys. **534**, A30 (2011)
Gonzalez, G.: Mon. Not. R. Astron. Soc. **285**, 403 (1997)
Gonzalez, G.: Mon. Not. R. Astron. Soc. **308**, 447 (1999)
Gonzalez, G.: Mon. Not. R. Astron. Soc. **450**, 3227 (2015)
González Hernández, J.I., Delgado-Mena, E., Sousa, S.G., et al.: Astron. Astrophys. **552**, A6 (2013)
Grasset, O., Schneider, J., Sotin, C.: Astrophys. J. **693**, 722 (2009)
Gray, D.F.: The Observation and Analysis of Stellar Photospheres (Cambridge University Press, Cambridge, 2005)
Gray, D.F., Johanson, H.L.: Publ. Astron. Soc. Pac. **103**, 439 (1991)
Gray, D.F., Livingston, W.C.: Astrophys. J. **474**, 802 (1997)
Haswell, C.A., Fossati, L., Ayres, T., et al.: Astrophys. J. **760**, 79 (2012)
Haywood, M.: Astron. Astrophys. **482**, 673 (2008)
Haywood, M.: Astrophys. J. **698**, L1 (2009)
Heiter, U., Jofré, P., Gustafsson, B., et al.: Astron. Astrophys. **582**, A49 (2015)
Hinkel, N.R., Timmes, F.X., Young, P.A., Pagano, M.D., Turnbull, M.C.: Astrophys. J. **148**, 54 2014
Hinkel, N.R., Young, P.A., Pagano, M.D., et al.: Astrophys. J. Suppl. Ser. **226**, 4 (2016)
Howell, S.B., Sobeck, C., Haas, M., et al.: Publ. Astron. Soc. Pac. **126**, 398 (2014)
Huber, D., Ireland, M.J., Bedding, T.R., et al.: Astrophys. J. **760**, 32 (2012)
Huber, D., Chaplin, W.J., Christensen-Dalsgaard, J., et al.: Astrophys. J. **767**, 127 (2013)
Irwin, J., Aigrain, S., Bouvier, J., et al.: Mon. Not. R. Astron. Soc. **392**, 1456 (2009)
Jofré, P., Heiter, U., Soubiran, C., et al.: Astron. Astrophys. **564**, A133 (2014)
Jofré, P., Heiter, U., Soubiran, C., et al.: Astron. Astrophys. **582**, A81 (2015)
Jofré, P., Heiter, U., Worley, C.C., et al.: ArXiv e-prints, arXiv:1612.05013 (2016)
Johnson, D.R.H., Soderblom, D.R.: Astrophys. J. **93**, 864 (1987)
Johnson, J.A., Aller, K.M., Howard, A.W., Crepp, J.R.: Publ. Astron. Soc. Pac. **122**, 905 (2010)
Jørgensen, B.R., Lindegren, L.: Astron. Astrophys. **436**, 127 (2005)
Kipping, D.M.: Mon. Not. R. Astron. Soc. **440**, 2164 (2014)
Kjeldsen, H., Bedding, T.R., Christensen-Dalsgaard, J.: In: Pont, F., Sasselov, D., Holman, M.J. (eds.) IAU Symposium, Transiting Planets, vol. 253, pp. 309–317 (2009)
Kürster, M., Endl, M., Rouesnel, F., et al.: Astron. Astrophys. **403**, 1077 (2003)
Lacy, C.H.S., Torres, G., Claret, A., Vaz, L.P.R.: Astrophys. J. **130**, 2838 (2005)
Lebzelter, T., Heiter, U., Abia, C., et al.: Astron. Astrophys. **547**, A108 (2012)
Lloyd, J.P.: Astrophys. J. **739**, L49 (2011)
Lloyd, J.P.: Astrophys. J. **774**, L2 (2013)
Ludwig, H.-G., Caffau, E., Steffen, M., Bonifacio, P., Sbordone, L.: Astron. Astrophys. **509**, A84 (2010)
Lyra, W., Porto de Mello, G.F.: Astron. Astrophys. **431**, 329 (2005)
Magrini, L., Randich, S., Friel, E., et al.: Astron. Astrophys. **558**, A38 (2013)
Makarov, V.V.: Astrophys. J. Suppl. Ser. **169**, 105 (2007)
Malavolta, L., Sneden, C., Piotto, G., et al.: Astrophys. J. **147**, 25 (2014)
Mamajek, E.E., Hillenbrand, L.A.: Astrophys. J. **687**, 1264 (2008)
Mathur, S., García, R.A., Ballot, J., et al.: Astron. Astrophys. **562**, A124 (2014)
Maxted, P.F.L., Serenelli, A.M., Southworth, J.: Astron. Astrophys. **575**, A36 (2015)
Mayor, M., Queloz, D.: Nature **378**, 355 (1995)
McQuillan, A., Mazeh, T., Aigrain, S.: Astrophys. J. **775**, L11 (2013)
Mikolaitis, Š., de Laverny, P., Recio-Blanco, A., et al.: ArXiv e-prints, arXiv:1612.07622 (2016)
Mortier, A., Santos, N.C., Sousa, S.G., et al.: Astron. Astrophys. **558**, A106 (2013)
Mortier, A., Sousa, S.G., Adibekyan, V.Z., Brandão, I.M., Santos, N.C.: Astron. Astrophys. **572**, A95 (2014)

Niemczura, E., Smalley, B., Pych, W.: Determination of Atmospheric Parameters of B-, A-, F- and G-Type Stars. Springer International Publishing, Cham (2014). doi:10.1007/978-3-319-06956-2
Nissen, P.E.: Astron. Astrophys. **579**, A52 (2015)
Önehag, A., Gustafsson, B., Eriksson, K., Edvardsson, B.: Astron. Astrophys. **498**, 527 (2009)
Oshagh, M., Boisse, I., Boué, G., et al.: Astron. Astrophys. **549**, A35 (2013)
Perryman, M.A.C., de Boer, K.S., Gilmore, G., et al.: Astron. Astrophys. **369**, 339 (2001)
Pont, F., Eyer, L.: Mon. Not. R. Astron. Soc. **351**, 487 (2004)
Queloz, D., Henry, G.W., Sivan, J.P., et al.: Astron. Astrophys. **379**, 279 (2001)
Ramírez, I., Asplund, M., Baumann, P., Meléndez, J., Bensby, T.: Astron. Astrophys. **521**, A33 (2010)
Ramírez, I., Meléndez, J., Bean, J., et al.: Astron. Astrophys. **572**, A48 (2014)
Rauer, H., Catala, C., Aerts, C., et al.: Exp. Astron. **38**, 249 (2014)
Recio-Blanco, A., Bijaoui, A., de Laverny, P.: Mon. Not. R. Astron. Soc. **370**, 141 (2006)
Reid, I.N.: Publ. Astron. Soc. Pac. **114**, 306 (2002)
Richichi, A.: In: Bedding, T.R., Booth, A.J., Davis, J. (eds.) IAU Symposium, vol. 189, pp. 45–50 (1997)
Ricker, G.R.: J. Am. Assoc. Var. Star Obs. **42**, 234 (2014)
Robinson, S.E., Laughlin, G., Bodenheimer, P., Fischer, D.: Astrophys. J. **643**, 484 (2006)
Saffe, C., Flores, M., Jaque Arancibia, M., Buccino, A., Jofré, E.: Astron. Astrophys. **588**, A81 (2016)
Santos, N.C., Israelian, G., Mayor, M.: Astron. Astrophys. **373**, 1019 (2001)
Santos, N.C., Israelian, G., Mayor, M.: Astron. Astrophys. **415**, 1153 (2004a)
Santos, N.C., Bouchy, F., Mayor, M., et al.: Astron. Astrophys. **426**, L19 (2004b)
Santos, N.C., Melo, C., James, D.J., et al.: Astron. Astrophys. **480**, 889 (2008)
Santos, N.C., Sousa, S.G., Mortier, A., et al.: Astron. Astrophys. **556**, A150 (2013)
Santos, N.C., Mortier, A., Faria, J.P., et al.: Astron. Astrophys. **566**, A35 (2014)
Sbordone, L., Caffau, E., Bonifacio, P., Duffau, S.: Astron. Astrophys. **564**, A109 (2014)
Seager, S., Mallén-Ornelas, G.: Astrophys. J. **585**, 1038 (2003)
Silva Aguirre, V., Basu, S., Brandão, I.M., et al.: Astrophys. J. **769**, 141 (2013)
Silva Aguirre, V., Lund, M.N., Antia, H.M., et al.: ArXiv e-prints, arXiv:1611.08776 (2016)
Smalley, B.: Stellar Parameters from Photometry, pp. 111–120. Springer International Publishing, New York (2014)
Smiljanic, R., Korn, A.J., Bergemann, M., et al.: Astron. Astrophys. **570**, A122 (2014)
Soderblom, D.R.: Annu. Rev. Astron. Astrophys. **48**, 581 (2010)
Sousa, S.G., Santos, N.C., Israelian, G., et al.: Astron. Astrophys. **526**, A99 (2011)
Sousa, S.G., Santos, N.C., Israelian, G.: Astron. Astrophys. **544**, A122 (2012)
Sousa, S.G.: ARES + MOOG: A Practical Overview of an Equivalent Width (EW) Method to Derive Stellar Parameters, pp. 297–310. Springer International Publishing, New York (2014)
Sousa, S.G., Santos, N.C., Mortier, A., et al.: Astron. Astrophys. **576**, A94 (2015)
Southworth, J.: Mon. Not. R. Astron. Soc. **426**, 1291 (2012)
Southworth, J.: In: Rucinski, S.M., Torres, G., Zejda, M. (eds.) Living Together: Planets, Host Stars and Binaries. Astronomical Society of the Pacific Conference Series, vol. 496, p. 164. Astronomical Society of the Pacific, San Francisco (2015)
Southworth, J., Wheatley, P.J., Sams, G.: Mon. Not. R. Astron. Soc. **379**, L11 (2007)
Sozzetti, A., Torres, G., Charbonneau, D., et al.: Astrophys. J. **664**, 1190 (2007)
Suárez-Andrés, L., Israelian, G., González Hernández, J.I., et al.: Astron. Astrophys. **591**, A69 (2016)
Tabernero, H.M., González Hernández, J.I., Montes, D.: In: Guirado, J.C., Lara, L.M., Quilis, V., Gorgas, J. (eds.) Highlights of Spanish Astrophysics VII, pp. 673–673 (2013). http://adsabs.harvard.edu/abs/2013hsa7.conf..673T
Takeda, Y., Tajitsu, A.: Mon. Not. R. Astron. Soc. **450**, 397 (2015)
Takeda, G., Ford, E.B., Sills, A., et al.: Astrophys. J. Suppl. Ser. **168**, 297 (2007)
Takeda, Y., Honda, S., Kawanomoto, S., Ando, H., Sakurai, T.: Astron. Astrophys. **515**, A93 (2010)

Thiabaud, A., Marboeuf, U., Alibert, Y., et al.: Astron. Astrophys. **562**, A27 (2014)
Torres, G., Lacy, C.H., Marschall, L.A., Sheets, H.A., Mader, J.A.: Astrophys. J. **640**, 1018 (2006)
Torres, G., Andersen, J., Giménez, A.: Astron. Astrophys. Rev. **18**, 67 (2010)
Torres, G., Fischer, D.A., Sozzetti, A., et al.: Astrophys. J. **757**, 161 (2012)
Tsantaki, M., Sousa, S.G., Santos, N.C., et al.: Astron. Astrophys. **570**, A80 (2014)
Valenti, J.A., Piskunov, N.: Astron. Astrophys. Suppl. **118**, 595 (1996)
Wolszczan, A., Frail, D.A.: Nature **355**, 145 (1992)
Xia, F., Fu, Y.-N.: Chin. Astron. Astrophys. **34**, 277 (2010)

Noise Sources in Photometry and Radial Velocities

Mahmoudreza Oshagh

Abstract The quest for Earth-like, extrasolar planets (exoplanets), especially those located inside the habitable zone of their host stars, requires techniques sensitive enough to detect the faint signals produced by those planets. The radial velocity (RV) and photometric transit methods are the most widely used and also the most efficient methods for detecting and characterizing exoplanets. However, presence of astrophysical "noise" makes it difficult to detect and accurately characterize exoplanets. It is important to note that the amplitude of such astrophysical noise is larger than both the signal of Earth-like exoplanets and state-of-the-art instrumentation limit precision, making this a pressing topic that needs to be addressed. In this chapter, I present a general review of the main sources of noise in photometric and RV observations, namely, stellar oscillations, granulation, and magnetic activity. Moreover, for each noise source I discuss the techniques and observational strategies which allow us to mitigate their impact.

1 Introduction

Exoplanetology is a vigorous and exciting new area of astrophysics. Since the revolutionary discovery of a planet orbiting the solar-like star 51 Peg (Mayor and Queloz 1995), over 3500 exoplanets have been discovered in about 2600 planetary systems,[1] which places our unique Solar System into context through the new field of comparative planetology. The radial velocity (RV) and photometric

[1] http://exoplanet.eu.

M. Oshagh (✉)
Institut für Astrophysik, Georg-August Universität Göttingen, Friedrich-Hund-Platz 1, 37077 Göttingen, Germany

Instituto de Astrofísica e Ciências do Espaço, Universidade do Porto, CAUP, Rua das Estrelas, 4150-762 Porto, Portugal
e-mail: moshagh@astro.physik.uni-goettingen.de

T.L. Campante et al. (eds.), *Asteroseismology and Exoplanets: Listening to the Stars and Searching for New Worlds*, Astrophysics and Space Science Proceedings 49, DOI 10.1007/978-3-319-59315-9_13

transit methods are the most widely used—and most successful techniques—in the detection and characterization of exoplanets.

Exoplanet-host stars are also the source of astrophysical "noise" with different amplitudes and timescales that can hamper the detection of accurate numbers of planets in a system and the accurate characterization of the detected planets. In this chapter, I provide a general review of the different sources of noise which are present in high-precision photometric and RV observations. Since the timescales of these noise signals are diverse, I have thus organized this chapter in such a way that the timescales of the noise signals increase as we move along. Moreover, for each noise source I discuss several proposed observational strategies and data analysis techniques which could help eliminate their impact.

2 Stellar Oscillations

Due to the presence of pressure waves in the interiors of stars, stellar surfaces often exhibit oscillations. The typical amplitude and timescale of oscillation modes increase with stellar mass along the main sequence. For that reason, the measurement of stellar oscillations has been used to extract crucial information about the interior structure of stars as part of a field of stellar astrophysics known as asteroseismology. As a consequence, asteroseismology has enabled us to characterize stellar properties with extremely high precision (Christensen-Dalsgaard 2016). Nevertheless, in the field of exoplanets the stellar oscillation signal is regarded as a source of noise that can hamper the detection of weak exoplanet-induced signals.

2.1 Radial Velocities

Several studies have attempted to estimate the exact timescale and amplitude of RV signals induced by stellar oscillations. For example, Bazot et al. (2007) used the HARPS spectrograph to perform extensive high-precision RV measurements of α Cen A during five consecutive nights with very short exposure times. Their observations revealed the timescale of the oscillations to be of the order of 5–15 min and the RV amplitude to be in the range 0.2–3 $\mathrm{m\,s^{-1}}$.

2.1.1 Eliminating Stellar Oscillation Noise in RV

Santos et al. (2007) explored various observing strategies to reduce the induced RV signal due to stellar oscillations. They reached to the conclusion that the appropriate strategy is to use long exposure times (minimum of 15 min per exposure) so as to significantly average out the stellar oscillation noise. Subsequently, this strategy has been used in performing RV measurements using stable spectrographs such as HARPS.

2.2 *Photometry*

Short-cadence and high-precision photometric observations obtained with the *Kepler* space telescope (Borucki et al. 2010) allowed us to estimate the timescale and amplitude of the stellar oscillation noise in photometry. The range of timescales closely matches the values obtained from RV observations (5–15 min). The amplitude of photometric variations due to stellar oscillations was seen to lie in the range 100–300 parts-per-million (ppm) (e.g., Carter et al. 2012).

2.2.1 Eliminating Stellar Oscillation Noise in Photometry

Similarly to the RV case, long-exposure photometric observations would mitigate the impact of stellar oscillations. However, since transits can have durations of only a few hours, a long-exposure strategy would negatively impact on the detection of transiting exoplanets. Consequently, large surveys such as *Kepler* and *CoRoT* have provided high-precision photometric measurements at short cadences (60 s cadence in the case of *Kepler*) in order to enable the detection of transits by Earth-size planets. Moreover, short-cadence observations provide more data points during the transit, and hence facilitate accurate estimation of a planet's parameters through the analysis of the transit light curve. Note that it is not uncommon to bin the transit light curve (with bin sizes of 15–30 min) in order to cancel out the effect of stellar oscillations (e.g., Barclay et al. 2013).

3 Granulation

Stars with convective envelopes exhibit a granulation pattern at their surfaces. The granulation pattern[2] manifests itself as the upward flow of bright and hot material from deeper layers followed by a downward flow of darker material after being cooled off at the surface. Stellar granulation adds substantial correlated noise to RV and photometric time-series observations, with larger amplitudes and longer timescales than the noise due to stellar oscillations.

[2]Granulation patterns at the surfaces of stars can only be observed through the analysis of spatially-resolved images, which are currently only possible for the Sun.

3.1 Radial Velocities

Dumusque et al. (2011) used HARPS to obtain long-term, continuous and high-precision RV measurements for five stars of different spectral types. They modeled the resulting RV power density spectra using a functional form that had been previously introduced to describe the granulation signal in the Sun (the so-called Harvey-like profile; Harvey 1985). As a result, Dumusque et al. (2011) estimated the timescale of the granulation noise to lie in the range from 15 min to 24 h, and the amplitude to be in the range 1–30 m s^{-1}, depending on the spectral type of the star.

3.1.1 Eliminating Granulation Noise in RV

Dumusque et al. (2011) also evaluated several observational strategies to reduce the RV noise due to granulation. They established that the best observational strategy is to obtain three RV measurements per night for each star separated by 1–2 h. They demonstrated that this approach can significantly reduce the granulation noise. This strategy has ever since been used when performing RV observations with spectrographs such as HARPS, HARPS-N, and SOPHIE.

3.2 Photometry

Based on solar observations obtained with the *SOHO* spacecraft, Jenkins (2002) demonstrated that the granulation of the quiet Sun can produce photometric variability of up to 50 ppm. More recently, several studies have used short-cadence photometric observations obtained with space telescopes such as *Kepler* and *CoRoT* and proceeded with the analysis of the corresponding power density spectra. The amplitude and timescale of the granulation noise in photometric observations across different spectral types and evolutionary states have been constrained as a result. Gilliland et al. (2011) provide scaling relations for the estimation of the amplitude and timescale of the granulation noise in photometric observations:

$$\sigma_{\mathrm{gran}} = (75\,\mathrm{ppm})\left(\frac{M}{\mathrm{M}_{\odot}}\right)^{-0.5}\left(\frac{R}{\mathrm{R}_{\odot}}\right)\left(\frac{T_{\mathrm{eff}}}{T_{\mathrm{eff},\odot}}\right)^{0.25} \quad (1)$$

and

$$\tau_{\mathrm{gran}} = (220\,\mathrm{s})\left(\frac{M}{\mathrm{M}_{\odot}}\right)^{-1}\left(\frac{R}{\mathrm{R}_{\odot}}\right)^{2}\left(\frac{T_{\mathrm{eff}}}{T_{\mathrm{eff},\odot}}\right)^{0.5}. \quad (2)$$

The granulation noise amplitude is close to the expected amplitude of a transit signal of an Earth-size planet, hence it could become a serious obstacle for the detection and characterization of small planets via the transit method.

3.2.1 Eliminating Granulation Noise in Photometry

Just as in Sect. 2.2.1, averaging photometric observations in order to reduce the impact of granulation will directly hamper the detection of a transiting planet's signal. Therefore, light curves are ideally obtained with short cadence to ensure that no transit signals are missed. Once the transit signal has been detected, the light curve is then binned to average out the granulation noise.

4 Stellar Magnetic Activity

Stellar magnetic activity manifests itself in the form of various contrasting structures at the stellar surface (e.g., dark spots and bright faculae), commonly known as stellar active regions. The combination of active regions present at the stellar surface and stellar rotation generates RV and photometric signals with amplitudes and periods commensurate with those of exoplanet-induced signals. Basri et al. (2013) found that more than 30% of the 150,000 stars observed by *Kepler* possess significantly higher levels of magnetic activity than the Sun. Therefore, one realizes how crucial it is to estimate the impact of stellar activity on exoplanet-induced signals as well as to mitigate its effect.

4.1 Radial Velocities

Stellar active regions, due to their temperature contrast, affect the shape of spectral lines and as a consequence deform the cross-correlation function (CCF) profile. Since radial velocities are measured by fitting a Gaussian function to the CCF, a deformation of the CCF profile may be compensated by an offset in the mean of the fitted Gaussian. Therefore, presence of active regions may lead to inaccurate and incorrect RV measurements. Due to stellar rotation, this incorrect RV estimate will exhibit a variation with a period close[3] to the stellar rotation period.

Convective motion at the stellar surface generates a net blueshifted RV signal. In active regions, however, convective motion is significantly reduced due to the presence of strong magnetic fields. The inhibition of convective blueshifts in these

[3]Depending on the latitude of active regions and the stellar differential rotation, different active regions would induce different periodicities.

Table 1 Mass estimates of planets in the CoRoT-7 system

CoRoT-7b	CoRoT-7c	References
$4.8 \pm 0.8\, M_\oplus$	$8.4 \pm 0.9\, M_\oplus$	Queloz et al. (2009)
$6.9 \pm 1.4\, M_\oplus$	$12.4 \pm 0.42\, M_\oplus$	Hatzes et al. (2010)
$7.42 \pm 1.21\, M_\oplus$	...	Hatzes et al. (2011)
$2.3 \pm 1.8\, M_\oplus$	...	Pont et al. (2011)
$5.7 \pm 2.5\, M_\oplus$	$13.2 \pm 4.1\, M_\oplus$	Boisse et al. (2011)
$8.0 \pm 1.2\, M_\oplus$	$13.6 \pm 1.4\, M_\oplus$	Ferraz-Mello et al. (2011)
$4.8 \pm 2.4\, M_\oplus$	$11.8 \pm 3.4\, M_\oplus$	Tuomi et al. (2014)
$4.73 \pm 0.95\, M_\oplus$	$13.56 \pm 1.08\, M_\oplus$	Haywood et al. (2014)
$5.52 \pm 0.78\, M_\oplus$	...	Barros et al. (2014)
$5.53 \pm 0.86\, M_\oplus$	$12.62 \pm 0.77\, M_\oplus$	Faria et al. (2016)

regions thus leads to extra RV variations (Meunier et al. 2010; Dumusque et al. 2014).

Consequently, it is a challenging task to assure that the observed RV variations are purely due to the Doppler reflex motion caused by the presence of exoplanets. The presence of activity-induced RV noise has been known since the very beginning of Doppler exoplanet searches (Saar and Donahue 1997; Hatzes and Cochran 1999; Santos et al. 2000; Queloz et al. 2001). Later on, with the emergence of high-precision RV measurements,[4] it became clearer how crucial it is to correct for the activity-induced noise in order to be able to detect the signal due to low-mass planets in the habitable zones of solar-like stars (Boisse et al. 2009, 2011; Dumusque et al. 2012). Moreover, determining the exact number of planets in a system and estimating their masses has been a challenging task whenever in the presence of stellar magnetic activity. The CoRoT-7 system best demonstrates this. Depending on the methods used to model the stellar activity noise as well as on the techniques employed to disentangle the activity and exoplanet signals, several teams have obtained conflicting results on the number of planets in the system and on their masses. In Table 1, I summarize the number of planets and their mass estimates as obtained in different studies.

4.1.1 Eliminating the Activity-Induced Signal in RV

There are two steps in the elimination process of activity-induced signals in RV (also known as RV jitter). The first, and main, step is to assess the presence of RV jitter. The second step is to predict its signal profile and to attempt its removal from the RV measurements.

[4]A precision of $0.5\ \mathrm{m\,s^{-1}}$ was achieved by the HARPS spectrograph, which enabled the detection of signals due to low-mass/Earth-size planets.

Activity Indicators: The first type of stellar activity indicators aim at quantifying the spectral line (or the CCF) asymmetry, e.g., the full width at half maximum (*FWHM*) of the CCF (Queloz et al. 2009), the bisector span[5] (*BIS*; Queloz et al. 2001; Santos et al. 2002), V_{span}[6] (Boisse et al. 2011), V_{asy}[7] (Figueira et al. 2013), and the bi-Gaussian method[8] (Figueira et al. 2013).

The second type of stellar activity indicators carry information directly about the magnetic activity of the star, e.g., the average magnetic field (B) estimated by measuring the Zeeman splitting of spectral lines (Reiners 2012; e.g.,), the Mount Wilson S-index[9] (Wilson 1978), and the $\log(R'_{HK})$ index[10] (Noyes et al. 1984).

To assess whether the observed RV signal is contaminated by RV jitter, researchers usually look for a correlation between any of the above activity indicators and the RV measurements. Presence of strong correlation means that the RV measurements need to be corrected for the RV jitter, which I describe next.

Modeling Activity: Two main approaches have been used to model RV jitter. One approach is based on using the information provided by the activity indicators and to employ empirical proxies to predict the RV jitter. This approach has been used in detecting low-mass planets around active stars, e.g., CoRoT-7 (Queloz et al. 2009; Hatzes et al. 2010; Boisse et al. 2011; Pont et al. 2011; Haywood et al. 2014; Faria et al. 2016), GJ 674 (Bonfils et al. 2007), and HD 189733 (Boisse et al. 2009; Aigrain et al. 2012).

Another approach is based on the numerical simulation of active regions at the stellar surface, including computation of all observables (e.g., activity indicators). The synthetic RVs are then simultaneously fitted to the observed RVs and to any activity indicator measurements. There are several numerical tools available to the community capable of performing this analysis, e.g., SOAP (Boisse et al. 2012), SOAP2.0 (Dumusque et al. 2014), and STARSIM (Herrero et al. 2016). This approach has been used in correcting RV observations of, e.g., HD 189733 (Boisse et al. 2012) and α Cen B (Dumusque et al. 2012, 2014). Although the use of numerical simulations has been shown to be the more robust and accurate of the two approaches described, it is also the more time-consuming from a computational perspective.

[5] $BIS = V_{high} - V_{low}$, where V_{high} and V_{low} are the velocity average of the points at the top and bottom of the CCF profile, respectively.

[6] $V_{span} = RV_{high} - RV_{low}$, where RV_{high} and RV_{low} are Gaussian fits to the upper and lower parts of the CCF, respectively.

[7] V_{asy} estimates the unbalance between the red and blue wings of the CCF.

[8] This approach consists in fitting a Gaussian with wings characterized by two different values of the *HWHM* (half width at half maximum) to the CCF.

[9] The S-index is based on the measurement of the emission in the cores of the Ca II H and K lines, and reflects the non-thermal chromospheric heating associated with the magnetic field.

[10] $\log(R'_{HK})$ is closely related to the S-index, giving the emission in the narrow bands normalized by the bolometric brightness of the star.

I would like to note that stellar active regions vary spatially and temporally, further evolving over several stellar rotation periods, which makes RV jitter not a strictly periodic and stable signal. Therefore, most of the correction techniques fail to explain the real observed RV jitter, which points to the necessity of developing models that take into account physical processes related to the active regions' formation and evolution. For instance, a recent effort by Dumusque et al. (2015) aimed at observing the Sun as a star with the HARPS-N spectrograph and trying to model the solar RV variation using observables that could be obtained through analysis of the resolved images of the Sun from solar satellites.

4.2 Photometry

The temperature contrast of active regions also produces photometric variations, which can be periodic due to the stellar rotation. This noise signal influences the detection and characterization of planets via the transit method. One can split activity-induced photometric noise into two main types depending on their source, i.e., active regions unocculted by the transiting planet during transit and occulted regions. These two types of active regions affect the transit light curve in different ways.

4.2.1 Unocculted Stellar Active Regions

Unocculted stellar active regions lead to periodic photometric modulation due to stellar rotation. The influence of such light-curve modulation on the planetary parameter estimates has been explored in several observational and simulation studies. For instance, Czesla et al. (2009) demonstrated that the planet radius can be overestimated by up to 4%.

4.2.2 Occulted Stellar Active Regions

In case the transiting planet occults the stellar active regions, this produces anomalies in the transit light curve that may lead to an inaccurate estimation of the planetary parameters, e.g., the planet radius and orbital inclination. Through simulations, Oshagh et al. (2013b, 2015b) showed that the planet radius can be underestimated by 5% due to stellar active region occultation. Moreover, Oshagh et al. (2014) demonstrated that the planet radius underestimation can be as large as 10% if the light curve is obtained at short wavelengths.

Analysis of high-precision, transit light curves allows us to accurately measure the transit times. The variation of transit times—known as transit-timing variation or TTV—may indicate the presence of other non-transiting planets in the system, which perturb the orbit of the transiting planet. As shown by Oshagh et al. (2013a), however, the anomalies caused by occulted active regions can mimic a TTV signal with an amplitude of 200 s, similar to the TTV signal induced by an Earth-mass planet in a mean-motion resonance with a Jovian body transiting a solar-mass star in a 3-day orbit (Boué et al. 2012).

A study by Oshagh et al. (2015a) also showed that the occultation by a transiting planet of a large, polar stellar spot can smear out the transit light curve. It should be noted that large, cool (dark) and long-lived stellar spots located near the stellar rotational axis are common features in stars regardless of the stellar rotational velocity and spectral type (e.g., Strassmeier et al. 1991). Furthermore, the occultation of active regions can affect the estimation of the spin-orbit angle based on measurements of the Rossiter–McLaughlin effect (e.g., Oshagh et al. 2016; and references therein).

4.2.3 Eliminating the Activity-Induced Signal in Photometry

The most promising strategy for estimating and eliminating the impact of stellar active regions is to use state-of-the-art models—e.g., SOAP-T (Oshagh et al. 2013a), MACULA (Kipping 2012), and SPOTROD (Béky et al. 2014)—to reproduce the noise signal generated by these regions and to subsequently remove it from the observational data. However, this approach faces several issues. First, the models require that assumptions be made concerning the values taken by their parameters, and there exists strong degeneracy[11] in determining the properties of the stellar active regions. Second, running numerical models is a time-consuming process from a computational perspective. Similarly to the RV jitter correction (Sect. 4.1.1), the evolution of stellar active regions makes accurate modeling of the photometric variation a challenging and difficult task. In this regard, MACULA is the only tool which takes the evolution of stellar active regions into account by implementing a linear stellar-active region evolution model.

5 Conclusion

In this chapter, I reviewed the sources and characteristics of astrophysical "noise" signals that contaminate RV and photometric observations in exoplanet searches. These noise signals have distinct timescales and amplitudes and, therefore, the strategies and techniques used to eliminate them will differ. In Table 2, I present a summary of the timescales and amplitudes of the several noise signals described above, as well as the most efficient way of eliminating them from our observations.

[11]For instance, a stellar spot's temperature contrast and filling factor are strongly degenerate, and cannot thus be estimated independently.

Table 2 Characteristics of astrophysical "noise" signals in RV and photometric exoplanet searches

Noise source	Timescale	RV ($m\,s^{-1}$)	Photometry (ppm)	Treatment
Oscillations	5–15 min	0.2–3	100–300	**RV:** at least 15 min exposure **Photometry:** binning the light curve into 15-min bins after detection of transit signal
Granulation	15 min to 24 hr	1–30	50–500	**RV:** three measurements per night with 1–2 h separation and averaging them **Photometry:** binning the light curve into 1-h bins after detection of transit signal
Magnetic activity	Several days	1–200	50–10,000	**RV:** finding correlation between measured RVs and activity indicators; if any correlation found, remove RV jitter by modeling **Photometry:** model out-of- and in-transit portions of light curve

More details can be obtained from the slides presented at the School (available at http://www.iastro.pt/research/conferences/faial2016/files/presentations/CE6.pdf).

Acknowledgements I would like to thank the members of the SOC for inviting me to the *IV*th *Azores International Advanced School in Space Sciences* held in the Azores Islands, Portugal. I acknowledge funding from the Deutsche Forschungsgemeinschaft (DFG, German Research Foundation): OS 508/1-1.

References

Aigrain, S., Pont, F., Zucker, S.: Mon. Not. R. Astron. Soc. **419**, 3147 (2012)
Barclay, T., Rowe, J.F., Lissauer, J.J., et al.: Nature **494**, 452 (2013)
Barros, S.C.C., Almenara, J.M., Deleuil, M., et al.: Astron. Astrophys. **569**, A74 (2014)
Basri, G., Walkowicz, L.M., Reiners, A.: Astrophys. J. **769**, 37 (2013)
Bazot, M., Bouchy, F., Kjeldsen, H., et al.: Astron. Astrophys. **470**, 295 (2007)
Béky, B., Kipping, D.M., Holman, M.J.: Mon. Not. R. Astron. Soc. **442**, 3686 (2014)
Boisse, I., Moutou, C., Vidal-Madjar, A., et al.: Astron. Astrophys. **495**, 959 (2009)
Boisse, I., Bouchy, F., Hébrard, G., et al.: Astron. Astrophys. **528**, A4 (2011)
Boisse, I., Bonfils, X., Santos, N.C.: Astron. Astrophys. **545**, A109 (2012)
Bonfils, X., Mayor, M., Delfosse, X., et al.: Astron. Astrophys. **474**, 293 (2007)
Borucki, W.J., Koch, D., Basri, G., et al.: Science **327**, 977 (2010)
Boué, G., Oshagh, M., Montalto, M., Santos, N.C.: Mon. Not. R. Astron. Soc. **422**, L57 (2012)
Carter, J.A., Agol, E., Chaplin, W.J., et al.: Science **337**, 556 (2012)
Christensen-Dalsgaard, J.: ArXiv e-prints, arXiv:1602.06838 (2016)
Czesla, S., Huber, K.F., Wolter, U., Schröter, S., Schmitt, J.H.M.M.: Astron. Astrophys. **505**, 1277 (2009)

Dumusque, X., Udry, S., Lovis, C., Santos, N.C., Monteiro, M.J.P.F.G.: Astron. Astrophys. **525**, A140 (2011)
Dumusque, X., Pepe, F., Lovis, C., et al.: Nature **491**, 207 (2012)
Dumusque, X., Boisse, I., Santos, N.C.: Astrophys. J., **796**, 132 (2014)
Dumusque, X., Glenday, A., Phillips, D.F., et al.: Astrophys. J. **814**, L21 (2015)
Faria, J.P., Haywood, R.D., Brewer, B.J., et al.: Astron. Astrophys. **588**, A31 (2016)
Ferraz-Mello, S., Tadeu Dos Santos, M., Beaugé, C., Michtchenko, T.A., Rodríguez, A.: Astron. Astrophys. **531**, A161 (2011)
Figueira, P., Santos, N.C., Pepe, F., Lovis, C., Nardetto, N.: Astron. Astrophys. **557**, A93 (2013)
Gilliland, R.L., Chaplin, W.J., Dunham, E.W., et al.: Astrophys. J. Suppl. Ser. **197**, 6 (2011)
Harvey, J.: In: Rolfe, E., Battrick, B. (eds.) Future Missions in Solar, Heliospheric and Space Plasma Physics. ESA Special Publication, vol. 235. European Space Agency, Paris (1985)
Hatzes, A.P., Cochran, W.D.: Mon. Not. R. Astron. Soc. **304**, 109 (1999)
Hatzes, A.P., Dvorak, R., Wuchterl, G., et al.: Astron. Astrophys. **520**, A93 (2010)
Hatzes, A.P., Fridlund, M., Nachmani, G., et al.: Astrophys. J. **743**, 75 (2011)
Haywood, R.D., Collier Cameron, A., Queloz, D., et al.: Mon. Not. R. Astron. Soc. **443**, 2517 (2014)
Herrero, E., Ribas, I., Jordi, C., et al.: Astron. Astrophys. **586**, A131 (2016)
Jenkins, J.M.: Astrophys. J. **575**, 493 (2002)
Kipping, D.M.: Mon. Not. R. Astron. Soc. **427**, 2487 (2012)
Mayor, M., Queloz, D.: Nature **378**, 355 (1995)
Meunier, N., Desort, M., Lagrange, A.-M.: Astron. Astrophys. **512**, A39 (2010)
Noyes, R.W., Hartmann, L.W., Baliunas, S.L., Duncan, D.K., Vaughan, A.H.: Astrophys. J. **279**, 763 (1984)
Oshagh, M., Boisse, I., Boué, G., et al.: Astron. Astrophys. **549**, A35 (2013a)
Oshagh, M., Santos, N.C., Boisse, I., et al.: Astron. Astrophys. 556, A19 (2013b)
Oshagh, M., Santos, N.C., Ehrenreich, D., et al.: Astron. Astrophys. **568**, A99 (2014)
Oshagh, M., Santos, N.C., Figueira, P., et al.: Astron. Astrophys. **583**, L1 (2015a)
Oshagh, M., C., S. N., Boisse, I., et al.: in European Physical Journal Web of Conferences, Vol. 101, European Physical Journal Web of Conferences, 05003 (2015b)
Oshagh, M., Dreizler, S., Santos, N.C., Figueira, P., Reiners, A.: Astron. Astrophys. **593**, A25 (2016)
Pont, F., Aigrain, S., Zucker, S.: Mon. Not. R. Astron. Soc. **411**, 1953 (2011)
Queloz, D., Henry, G.W., Sivan, J.P., et al.: Astron. Astrophys. **379**, 279 (2001)
Queloz, D., Bouchy, F., Moutou, C., et al.: Astron. Astrophys. **506**, 303 (2009)
Reiners, A.: Living Rev. Sol. Phys. **9**, 1 (2012)
Saar, S.H., Donahue, R.A.: Astrophys. J. **485**, 319 (1997)
Santos, N.C., Mayor, M., Bouchy, F., et al.: Astron. Astrophys. **474**, 647 (2007)
Santos, N.C., Mayor, M., Naef, D., et al.: Astron. Astrophys. **361**, 265 (2000)
Santos, N.C., Mayor, M., Naef, D., et al.: Astron. Astrophys. **392**, 215 (2002)
Strassmeier, K.G., Rice, J.B., Wehlau, W.H., et al.: Astron. Astrophys. **247**, 130 (1991)
Tuomi, M., Anglada-Escude, G., Jenkins, J.S., Jones, H.R.A.: ArXiv e-prints, arXiv:1405.2016 (2014)
Wilson, O.C.: Astrophys. J. **226**, 379 (1978)

Atmospheres of Exoplanets

David Ehrenreich

Abstract Atmospheres of exoplanets are our only window into the physical and chemical processes occurring in these distant worlds. These processes are important tracers of the origins and evolution of planetary systems, including our Solar System. In this broader context, we could better understand how common (or unique) are the conditions leading to the emergence of life, which could leave its spectroscopic imprints precisely into planetary atmospheres. This lecture is focused on one of the available techniques to study exoplanetary atmospheres: transit transmission spectroscopy. After describing some basics about this technique, I will illustrate through some case studies how it can practically bring observational constraints on these remote and exotic atmospheres.

1 Our Observation Window into (Exo)Planets

Except from being slightly closer to the Sun than our Earth, how much different from Earth could Venus be? With similar mass and size, hence density and surface gravity, surely Venus would make for a great and exotic, if not a tad too warm, destination for tourists. Its atmosphere, revealed as it refracted sunlight during observations of the 1761 transit of Venus by Mikhaïl Lomonosov, is known to be covered by thick clouds, casting doubts about how sunny the cytherean weather really is (and adding to the surface mystery). Apart form this, what could possibly go wrong (except wildlife; Fig. 1)? The previous text summons the popular view, before the 1950s, that Venus could be habitable; but there was no way to tell. Only from the moment radio observations determined an apparent black-body temperature >500 K (Mayer et al. 1958), did it become clear that Venus was utterly different and harsher than the Earth.

D. Ehrenreich (✉)
Observatoire astronomique de l'Université de Genève, 51 chemin des Maillettes, 1290, Versoix, Switzerland
e-mail: david.ehrenreich@unige.ch

T.L. Campante et al. (eds.), *Asteroseismology and Exoplanets: Listening to the Stars and Searching for New Worlds*, Astrophysics and Space Science Proceedings 49, DOI 10.1007/978-3-319-59315-9_14

Fig. 1 An artist's impression of Venus in the late 1940s. Before the surface temperature of Venus was measured, science-fiction writers and artists could base their stories on the then-reasonable belief that the planet could host advanced life forms. Extract from *Planet Comics* #51 (1947)

Exoplanets with masses and sizes, hence bulk densities, similar to Earth's or Venus's are being discovered at an accelerating pace, including near the habitable zones of their systems. How can we find out whether these worlds are hospitable or hellish? The answer lies in their atmospheres. The reader interested in the theoretical foundations of (exo)planetary atmospheres is referred to Heng (2017).

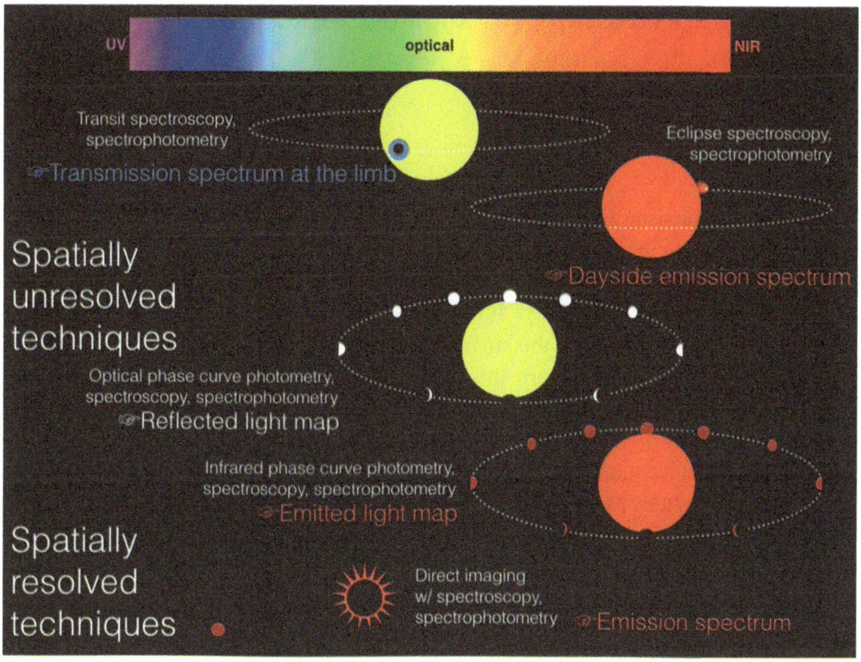

Fig. 2 Sketch of the different techniques for characterising exoplanetary atmospheres

Different techniques exist for characterising exoplanet atmospheres. These techniques, sketched in Fig. 2, can be separated into two broad categories, depending on whether the planet is spatially resolved from the star or not. The earlier category involves high-contrast imaging coupled to spectroscopy or spectrophotometry. It can provide the emission spectra of young giant planets at large separations around early main-sequence stars. Spatially unresolved techniques use temporal or spectroscopic differentiation to separate the planetary contribution from the (overwhelming) stellar contribution. Transit spectroscopy reveals the transmission spectrum of the atmospheric limb, at the day-night terminator of the transiting planet. Eclipse spectroscopy yields the emission spectrum of the planet's illuminated dayside. Photometric observations of the planet phase curve in the optical or in the infrared provide us with longitudinal maps of reflected or emitted light, respectively. Although challenging because of the small signals that are sought after, all these techniques have produced spectacular results (for a review, see, e.g., Crossfield 2015); importantly, they are also complementary to each other. Addressing all of them in depth is beyond the scope of this single lecture. Here, I will focus on transit spectroscopy.

2 Transit Spectroscopy of Exoplanet Atmospheres

2.1 Basics

2.1.1 Method

The method consists in retrieving the spectroscopic imprint of the planetary atmosphere on the star light that filters through it during a transit. As illustrated in Fig. 3, the transit depth δ is the normalised difference of the stellar fluxes out of transit ($F_{\rm out}$) and in-transit ($F_{\rm in}$) . It does also simply correspond (when neglecting the limb-darkening effect) to the surface ratio between the planet and the star, i.e., $(R_p/R_\star)^2$, where R_p and $R_\star$ are the planet and star radii, respectively. Different chemical species within the planetary atmosphere extinct (absorb or scatter) star light as a function of wavelength λ and also depending at what altitude z they are concentrated. Consequently, it is possible to build the transmission spectrum of the planetary atmosphere by measuring the planet radius at different wavelengths. The opacity τ along the light of sight crossing the planetary limb can be expressed as:

$$\tau(z,\lambda) = \int n(z)\sigma(\lambda)\mathrm{d}l\,. \tag{1}$$

The number density $n(z)$ (units of cm^{-3}) denotes the dependence on the atmospheric structure, whereas the extinction cross-section $\sigma(\lambda)$ (units of cm^2) represents the

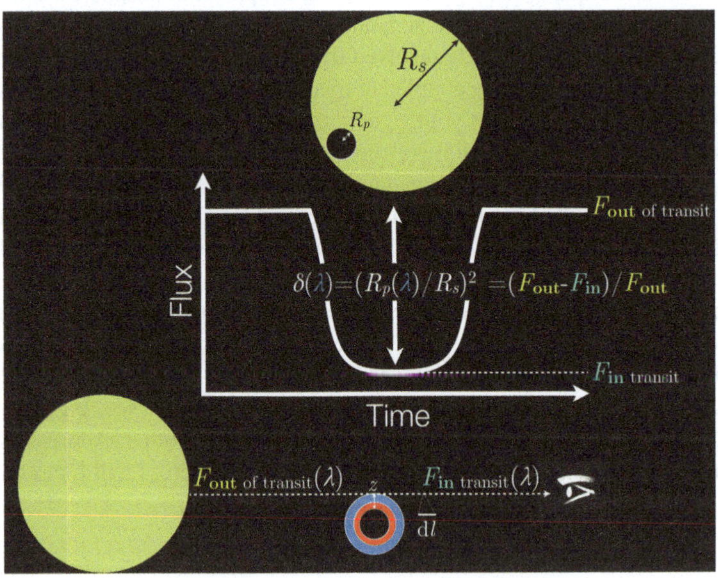

Fig. 3 Transit light curve and transit spectroscopy geometric parameters

dependence on the atmospheric composition. Assuming that (i) the planet is small with respect to the star and (ii) the atmosphere of the planet is isothermal, allows one to approximate this integral as the length of the chord throughout the limb (Fortney 2005):

$$\tau(z, \lambda) \approx n(z)\sigma(\lambda)\sqrt{2\pi R_p H}\,, \tag{2}$$

where H is an important quantity called the atmospheric scale height, which I will formally introduce below.

2.1.2 Radiative Transfer

Following the notation from Fig. 3, the in-transit flux can be expressed as the solution of the simplified (no source term) radiative transfer equation:

$$F_{\mathrm{in}}(\lambda) = F_{\mathrm{out}}(\lambda)\exp(-\tau)\,. \tag{3}$$

This allows relating the transit depth to the opacity as:

$$\delta(\lambda) = \exp\left(-n\sigma\sqrt{2\pi R_p H}\right) - 1\,. \tag{4}$$

The two physically interesting ingredients in this equation, regarding the planetary atmosphere, are the extinction cross-section, σ, and atmospheric density profile, n. The extinction term includes:

- the photo-absorption by atoms and molecules,
- the scattering of light by atoms, molecules and small particles (Rayleigh scattering),
- the scattering of light by large particles such as dust grains or droplets in hazes or clouds (Mie scattering).

Some examples of these different extinction components are plotted as cross-sections in Fig. 4. It is worth noticing that despite the extinction properties of atoms and molecules (in particular their photo-absorption cross-sections) can be measured in the lab, it is only recently that substantial efforts have been started in order to complete line lists for important molecules such as water or methane, at temperatures relevant to irradiated exoplanets. Also relevant to these atmospheres or those of brown dwarfs, the pressure-broadened wings of prominent resonant lines from alkaline metals (sodium and potassium) are still not well known (Burrows et al. 2000).

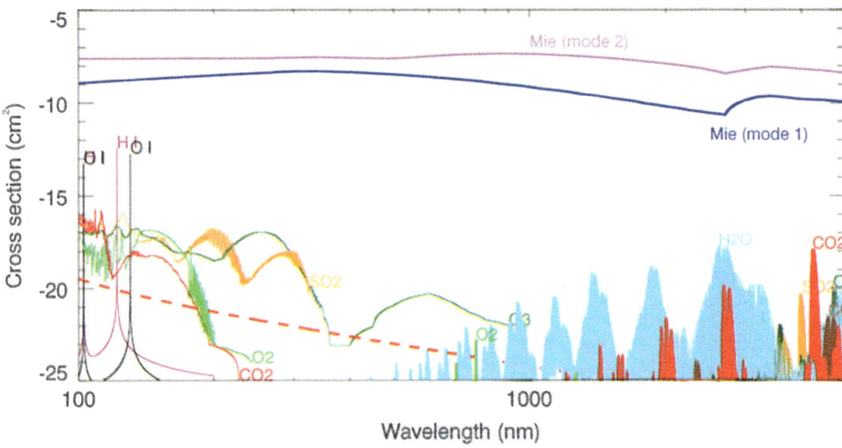

Fig. 4 Absorption cross-sections of different atmospheric gases present on Venus, including CO_2 (*red*), for which the Rayleigh scattering cross-section is also plotted (*thick, dashed red line*). The Mie scattering cross-sections of haze particles with two different size distributions are shown by the *thick violet* and *blue lines*. After Ehrenreich et al. (2012)

2.1.3 Atmospheric Structure

The density (ρ) structure of a planetary atmosphere pertains to its pressure-temperature (p-T) profile. A good description for well mixed, lower atmospheres of planets is that of a perfect gas in hydrostatic equilibrium. Hence, the continuity equation and the microscopic perfect gas law below are extremely useful:

$$\frac{\mathrm{d}p}{\mathrm{d}z} = -\rho g\,, \tag{5}$$

$$p = n\,k_{\mathrm{B}}T\,, \tag{6}$$

where g is the acceleration of gravity and k_{B} is Boltzmann's constant. Introducing μ, the mean molecular mass of the atmosphere (units of g molecule^{-1}—this is not the molar mass!) to relate the mass and number densities as $\rho = n\mu$, playing with the two equations above leads to express the pressure as:

$$p = -\frac{\mathrm{d}p}{\mathrm{d}z}H\,, \tag{7}$$

$$H = \frac{k_{\mathrm{B}}T}{\mu g}\,. \tag{8}$$

The atmospheric scale height, H, measures the 'compacity' of an atmosphere. Values of H for different planets are provided in Table 1. More precisely, it represents the height above which the pressure decreases by a factor of e. The exponential

Table 1 Atmospheric scale height, transit depth and atmospheric signal for different typical objects

	Radius ($R_\oplus$)	Temperature (K)	Gravity (m s^{-2})	Molar mass (g mol^{-1})	H (km)	Transit depth	Atmospheric signal[a] (H^{-1})
Earth	1	300	10	29	8	80 ppm	0.2 ppm
Venus	1	740	9	44	16	80 ppm	0.4 ppm
Jupiter	11	110	25	2.2	16	1%	5 ppm
Titan	0.4	90	1.4	29	20	14 ppm	0.2 ppm
Hot Jupiter	17	1100	25	2.2	160	2.5%	70 ppm

[a]The atmospheric signal is for absorption by one atmospheric scale height during transit

decrease of the pressure with altitude can be easily seen when integrating Eq. (7):

$$p(z) = p_0 \exp\left(-\int_{z_0}^{z} \frac{dz'}{H(z')}\right), \tag{9}$$

where $p_0 = p(z_0)$. Because T depends on z, H should stay within the integral. Only when the temperature is constant with altitude (isothermal profile), could we take H out and use the well-known expression $p = p_0 \exp\left[-(z - z_0)/H\right]$.

2.2 *Amplitude of the Expected Atmospheric Signal*

With the scale height, it is now possible to approximate δ_H, the magnitude of the absorption signal due to the planet atmosphere during a transit: for this, let us simply assume that the atmospheric absorption is equivalent to that caused by an optically-thick annulus, of thickness H, around the planet. This is represented in Fig. 5, and from this one can write:

$$\delta_H = \left(\frac{R_p + H}{R_\star}\right)^2 - \left(\frac{R_p}{R_\star}\right)^2 . \tag{10}$$

After some manipulations and assuming that the scale height is small with respect to the radius of the planet ($H \ll R_p$, hence $(H/R_p)^2 \approx 0$), this simply expresses as:

$$\delta_H \approx 2\left(\frac{R_p}{R_\star}\right)^2 \frac{H}{R_p}, \tag{11}$$

where $(R_p/R_\star)^2$ is the transit depth δ (see some examples in Table 1). While this expression would leave one thinking that the planetary radius is an important factor, developing H, then g, surprisingly leads to R_p being eliminated from the equation!

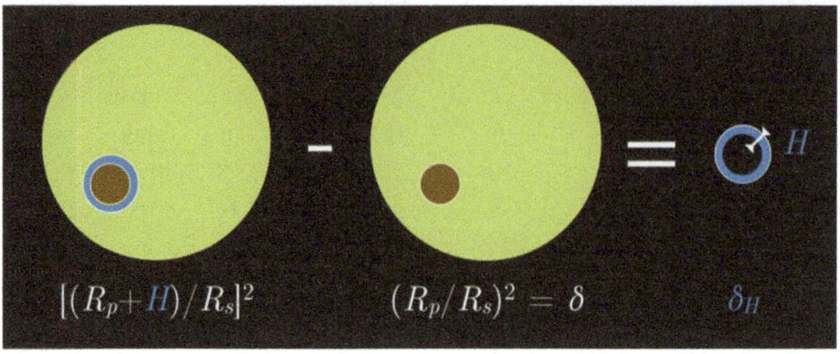

Fig. 5 How to approximate the transit transmission signal of a planetary atmosphere

In fact,

$$\delta_H \approx 2\frac{\cancel{R_p^2}}{R_\star^2}\frac{k_{\rm B}T}{\frac{4}{3}\pi G\mu\rho \cancel{R_p^2}}\,, \tag{12}$$

$$\delta_H \propto (R_\star)^{-2}(\rho)^{-1}(\mu)^{-1}(T)^{+1}\,. \tag{13}$$

The only relevant bulk parameter of the planet is its mean density, ρ. Thus, hot planets mainly made of gas or ices (low densities), with primordial atmospheres rich in hydrogen and helium (low μ) in transit across red dwarfs (small $R_\star$) are the best targets for transmission spectroscopy. This quantity can guide the selection of targets for surveys dedicated to compare the atmospheric properties of exoplanets. The other critical quantity is the stellar magnitude; in fact, transmission spectroscopy is a photon-starved technique relying on the stellar flux, hence the brighter, the better. The stellar magnitude of planet-host stars is plotted as a function of the planet δ_H in Fig. 6. Around bright stars, the signal-to-noise ratio of atmospheric signatures increases. Transitioning from the exoplanet detection census (e.g., from the *Kepler* space telescope) to exoplanet characterisation surveys, it is thus essential to collect more exoplanets amenable to atmospheric characterisation, hence in transit across bright (or small) stars. This is the objective of the refurbishment of the *Kepler* mission into K2 and the goal of the three next space missions dedicated to exoplanets:

- The *Transiting Exoplanet Survey Satellite* (*TESS*) is a medium-class explorer (MIDEX) NASA mission that will be launched in 2018 (Ricker et al. 2015).
- The *CHaracterising ExOPlanet Satellite* (*CHEOPS*) is a small (S) ESA mission developed jointly with a consortium of European countries led by Switzerland, that will also be launched in 2018 (Broeg et al. 2013).
- *PLATO* (*PLAnetary Transits and Oscillations of stars*) is a medium (M) ESA mission that will be launched in 2025 (Rauer et al. 2014).

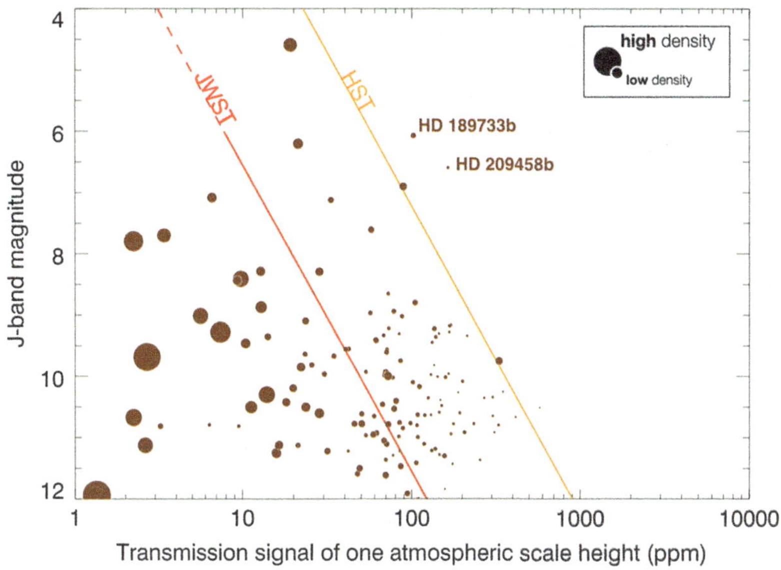

Fig. 6 Detectability of exoplanet atmospheres. The signal of one atmospheric scale height seen in transmission during transit is plotted against the stellar J magnitude. The signal is calculated in ppm following Eq. (11). The size of the *circles* scales with the planet bulk density. Approximate *HST* and *JWST* 3σ detection limits (*orange* and *red lines*, respectively) are shown. After Pepe et al. (2014)

3 Case Studies

In order to illustrate how the technique of transit spectroscopy works, three case studies of atmospheric signatures are presented and discussed below. These are:

- The detection of sodium at optical wavelengths and high-spectral resolution in the upper atmosphere of the hot gas giant HD 189733b, observed from the ground with the HARPS instrument at the ESO 3.6-m telescope (Wyttenbach et al. 2015).
- The detection of hydrogen atoms escaping from the warm Neptune GJ 436b, observed in the UV with the Space Telescope Imaging Spectrograph on the *Hubble Space Telescope* (Ehrenreich et al. 2015).
- The multiple detections and non-detections of water from several exoplanets, observed in the near-IR with the Wide-Field Camera 3 (WFC3) on the *Hubble Space Telescope* (Deming et al. 2013; Ehrenreich et al. 2014; Fraine et al. 2014; Knutson et al. 2014).

For each case, I present below (Figs. 7, 8 and 9) what the raw data and the 'end product' (i.e., the transmission spectrum) look like.

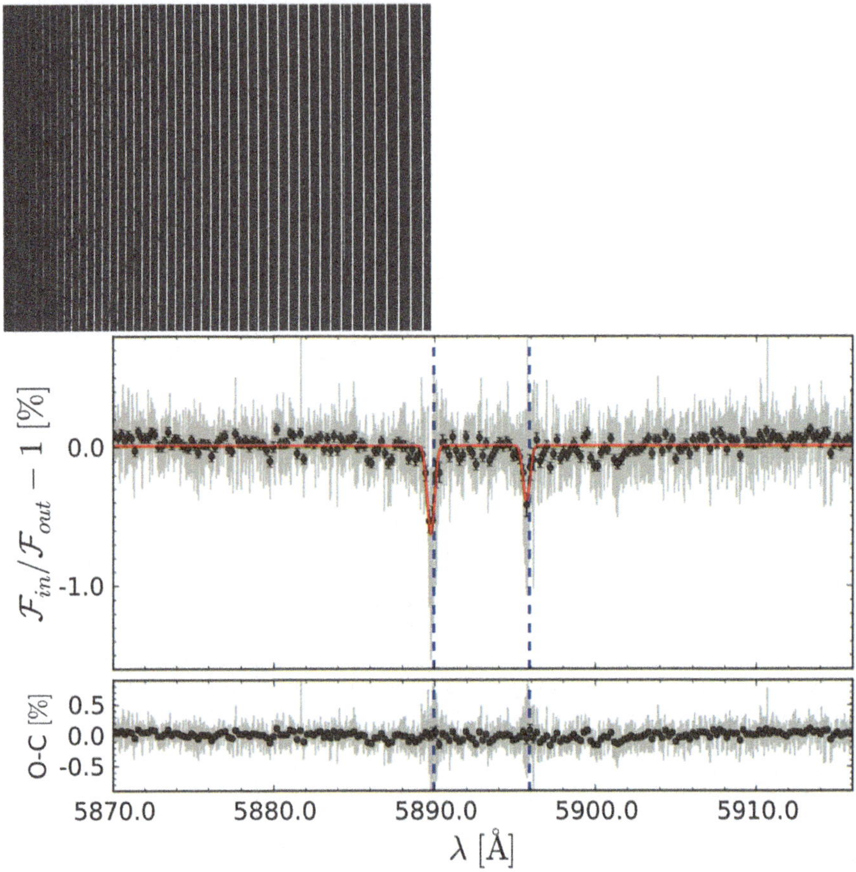

Fig. 7 *Top panel:* A raw 2D echelle spectrum of HD 189733 (the star) from HARPS. Spectral orders appear as *vertical stripes* and the dispersion for each order is along the y-axis. Obtaining a time series of such spectra over the planetary transit allows calculating the transit transmission spectrum of the planet, which is shown in the *bottom panel*. *Bottom panel:* The transmission spectrum of HD 189733b (the planet) is shown unbinned (*grey*) and binned (*black dots*). A Gaussian fit to both sodium lines (indicated by the *vertical blue dashed lines*) is shown in *red*. The *lower panel* shows the residuals after subtracting this Gaussian fit. After Wyttenbach et al. (2015)

3.1 Sodium Sky of a Hot Jupiter

The signature of atomic sodium was the first atmospheric signal revealed in an exoplanet (Charbonneau et al. 2002) with the Space Telescope Imaging Spectrograph (STIS) on board *HST*. It has since then been found in other hot gas giants, thanks not only to *HST*, but also with high-resolution spectrographs installed at ground-based observatories. As two, easy-to-identify resonant transitions, the line cores of the sodium doublet at 589 nm represent a powerful probe: they are so

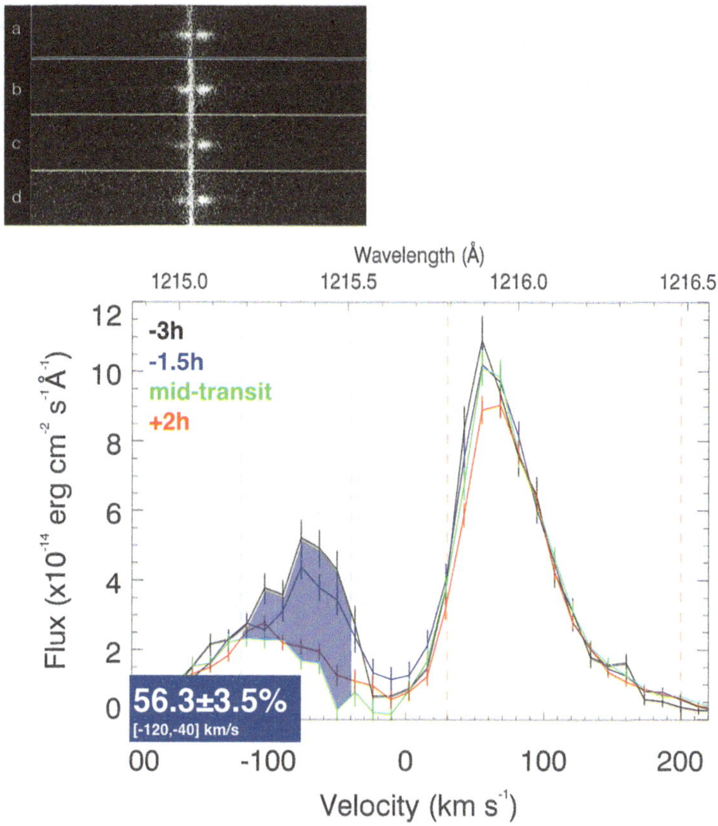

Fig. 8 *Top panel*: Four raw 2D spectra obtained after 50-min exposures of M dwarf GJ 436 in the far-UV. Each spectrum is obtained within 40 min of the previous one. The dispersion direction is along the x-axis. The *vertical stripe* crossing all exposures is the geocoronal emission line (airglow). The wings of the stellar Lyα emission line are visible on both sides of the airglow. Exposure (*c*) is obtained during the optical transit of the Neptune-size planet GJ 436b: the blue wing (*left*) appears fainter than in out-of-transit exposures (*a*) and (*b*). *Bottom panel*: Reduced and averaged 1D spectrum of GJ 436. The different colours refer to exposures obtained −3h (*black*; *a*), −1.5h (*blue*; *b*), 0h (*green*; *c*), and +2h (*red*; *d*) with respect to the mid-transit time. The *hatched region* indicates the extent of the airglow-contaminated region

intense that, observed at high-spectral resolution, they can be detected very high up in the planetary atmospheres, typically above the cloud layers that damp spectral signatures from the lower parts of the atmosphere.

Planet-hunting spectrographs like HARPS at the ESO 3.6-m telescope in La Silla, Chile, acquire high-resolution spectra ($\lambda/\Delta\lambda = 115{,}000$) from which radial-velocity measurements are produced. It turns out that these spectra could also be used to search for atmospheric signatures during transits. In Wyttenbach et al. (2015), we searched existing HARPS spectra of the prototypical hot gas giant HD 189733b for signs of sodium. Figure 7 shows a raw stellar spectrum obtained

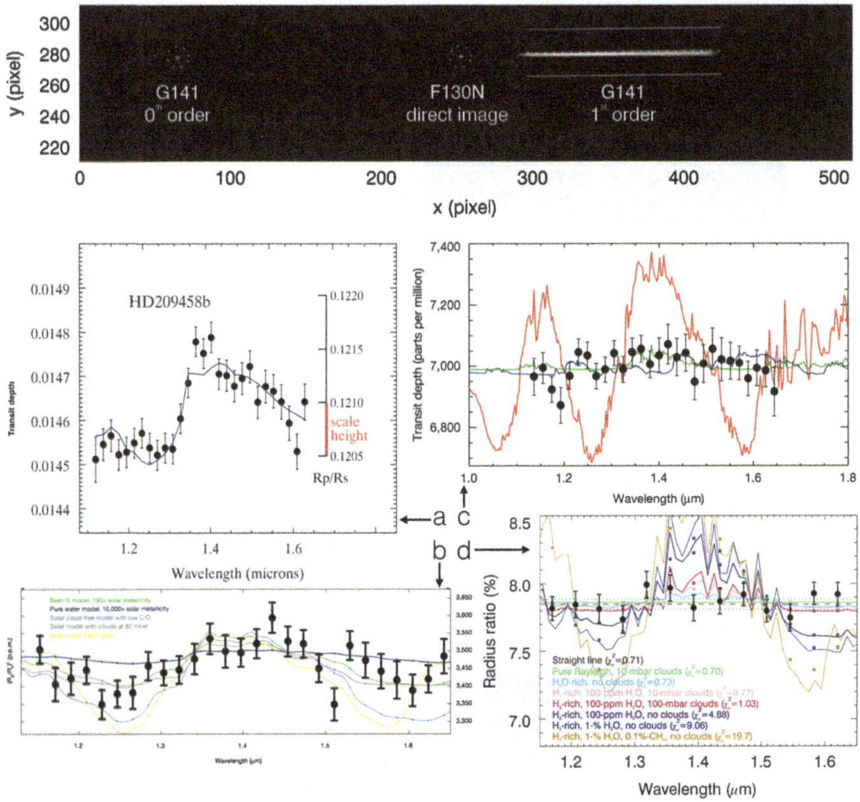

Fig. 9 *Top panel*: A raw WFC3/G141 stare-mode spectral image of GJ 3470 (including orders 0—the undispersed image—and +1). A direct image of the star, obtained with a narrow-band filter (F130N) is superimposed. *Bottom panels*: Reduced WFC3/G141 transmission spectra of (**a**) HD 209458b (Deming et al. 2013), (**b**) HAT-P-11b (Fraine et al. 2014), (**c**) GJ 436b (Knutson et al. 2014), and (**d**) GJ 3470b (Ehrenreich et al. 2014). Water is detected in the leftmost panels (**a**,**b**), not in the rightmost ones (**c**,**d**)

with HARPS on this target. As there are no reference stars, the idea is to perform differential spectroscopy to retrieve the transit spectrum and light curve of the planet, implementing corrections for telluric contamination and planetary orbital motion. Only then, the normalised spectra obtained out of transit and in-transit are compared. In the final transit spectrum obtained, each line of the sodium doublet arising from absorption in the planetary atmosphere is resolved, and large line contrasts are measured: $0.64 \pm 0.07\%$ (D_2) and $0.40 \pm 0.07\%$ (D_1) for full widths at half maximum (FWHMs) of 0.52 ± 0.08 Å. Comparing the data to synthetic atmospheric models, Wyttenbach et al. (2015) measure temperatures of ~3000 K in the upper atmosphere and derive a temperature gradient ~$0.2\,K\,km^{-1}$. Hot and high-altitude winds blowing from the hot dayside to the cooler nightside of the planet could also be detected as a blueshift in the line position. This result demonstrates the relevance of studying exoplanet atmospheres with high-resolution spectrographs mounted on 4-m-class telescopes and paves the way for an in-depth characterisation of physical conditions in the atmospheres of many exoplanetary systems with future spectrographs such as ESPRESSO on the VLT or HiReS and METIS on the E-ELT.

3.2 Hydrogen Corona Evaporating from a Warm Neptune

Exoplanets orbiting close to their parent stars may lose some fraction of their atmospheres because of the extreme irradiation. Atmospheric mass loss primarily affects low-mass exoplanets, leading to the suggestion that hot rocky planets might have begun as Neptune-like, but subsequently lost all of their atmospheres; however, no confident measurements have hitherto been available. The signature of this loss could be observed as absorption signatures in the Lyman-α line (121 nm) of neutral hydrogen in the ultraviolet, when the planet and its escaping atmosphere transit the star, giving rise to deeper and longer transit signatures than the ones detected in the optical spectrum.

The hydrogen Lyα line is the only source of stellar flux at these far-UV wavelengths, where there is no black-body continuum but only chromospheric and coronal stellar emission lines. The hydrogen Lyα line can only be observed from space with *HST* because Lyα photons are readily absorbed by the Earth atmosphere. Further difficulties involve the contamination of the line centre by the emission from atomic hydrogen of the Earth geocorona, within which *HST* orbits. This contamination can be subtracted; however, the core of the stellar Lyα line remains inaccessible: neutral hydrogen in the interstellar medium on the line of sight absorbs all the flux, even for the closest stars. Transit signatures are thus sought after in the remaining wings of the Lyα line.

Figure 8 shows what the raw *HST*/STIS UV data look like. There are four raw spectra of the M dwarf GJ 436, which is host to a transiting Neptune-mass planet. The planet is fully transiting in one of these spectra, causing the blue wing of the stellar emission line to become perceptibly fainter by eye, even in these raw data. Ehrenreich et al. (2015) indeed report that the ultraviolet transit of this Neptune-mass exoplanet GJ 436b has a depth of $56.3 \pm 3.5\%$, far beyond the 0.69% optical transit depth. It is inferred from this that the planet is surrounded and trailed by a large exospheric cloud composed mainly of hydrogen atoms. With this information, it is possible to estimate the mass-loss rate of the planet atmosphere in the range of about 10^8–10^9 g s^{-1}, which is far too small to deplete the atmosphere of a Neptune-like planet in the lifetime of the parent star, but would have been much greater in the past.

3.3 Water Vapour in (More-or-Less) Cloudy Planets

Water vapour has been firmly detected in the atmosphere of an exoplanet (HD 209458b) by Deming et al. (2013), who used the Wide-Field Camera 3 on board *HST* to observe planetary transits and spot the signatures of the water band centred at 1.38 μm (see Fig. 9a). This led to a flurry of observations and observation surveys using similar settings, sometimes devoting large amounts of telescope time (∼50 *HST* orbits) for a promising planet (Kreidberg et al. 2014). While spectroscopic signatures have been identified in several hot giant planets (Sing et al. 2016), the atmospheres of lower-mass planets turned out to be more tricky to explore. The atmospheric composition of super-Earths and Neptunes is, indeed, the object of intense observational and theoretical investigations. Meanwhile, the transmission spectra obtained for many such exoplanets are featureless as seen with *HST*/WFC3 (Ehrenreich et al. 2014; Knutson et al. 2014; Kreidberg et al. 2014). This flat signature is attributed to the presence of optically-thick clouds or translucent hazes. A noticeable exception is the warm-Neptune HAT-P-11b, for which the 1.38 μm water band has been detected (Fraine et al. 2014).

The planet GJ 3470b is a warm Neptune detected in transit across a bright, late-type star (Bonfils et al. 2012). The transit of this planet has already been observed in several band passes from the ground and space, allowing observers to draw an intriguing yet incomplete transmission spectrum of the planet atmospheric limb (see Chen et al. 2017; and references therein for an update). In particular, published data in the visible suggest the existence of a Rayleigh scattering slope making GJ 3470b a unique case among the known Neptunes, while data obtained beyond 2 μm are consistent with a flat infrared spectrum. The unexplored near-infrared spectral region between 1 and 2 μm is thus the key to understanding the atmospheric nature of GJ 3470b. Ehrenreich et al. (2014) report on the first space-borne spectrum of GJ 3470, obtained during one transit of the planet with the Wide-Field Camera 3 (WFC3) on board *HST*, operated in stare mode. An example of a raw spectral image, superimposed over a direct image, is presented in Fig. 9 (upper panel). Because WFC3 was not initially designed to observe bright objects such as exoplanet-host stars, this kind of observations have poor duty cycles. However, a more efficient observation mode called 'scanning mode' is now the standard for exoplanet observations. It consists in scanning the detector with the stellar spectrum to allow for increased exposure times, better duty cycles, and eventually better measurement precision. This precision increase can be critical for detecting the water feature, as in another warm Neptune, HAT-P-11b (Fraine et al. 2014) (see Fig. 9b); yet it may not be sufficient in other cases, such as GJ 436b (Knutson et al. 2014) (see Fig. 9c), which we met in the previous section.

As for GJ 436b, the reduced transmission spectrum of GJ 3470b, shown in Fig. 9d, appears flat within the uncertainties: the WFC3 spectrum obtained with the G141 grism covers the 1.1–1.7 μm region with a resolution of ∼300. The transmission spectrum of GJ 3470b is retrieved with a chromatic planet-to-star radius ratio precision of 0.09% (about half a scale height) per 40 nm bins. At this precision, the spectrum indeed appears featureless. This is actually in good agreement with ground-based and *Spitzer* infrared data at longer wavelengths, pointing to a flat transmission spectrum from 1 to 5 μm. Simulations of possible theoretical transmission spectra for GJ 3470b allow showing that the *HST*/WFC3 observations rule out cloudless hydrogen-rich atmospheres ($>10\sigma$) as well as hydrogen-rich atmospheres with tholin haze ($>5\sigma$). Considering the full set of available measurements as of 2014 in 0.3–5 μm supports a cloudy, hydrogen-rich atmosphere (Ehrenreich et al. 2014). The tentative Rayleigh slope in the visible has since been confirmed from the ground, hinting at a low-metallicity atmosphere (Chen et al. 2017).

4 Future Landscape

Atmospheric characterisation of exoplanets will benefit from a combination of powerful instruments located in space and on the ground. Figure 10 synthesises all relevant and accepted experiments. Capturing the most amenable targets for atmospheric characterisation will be performed as a continuous observational effort,

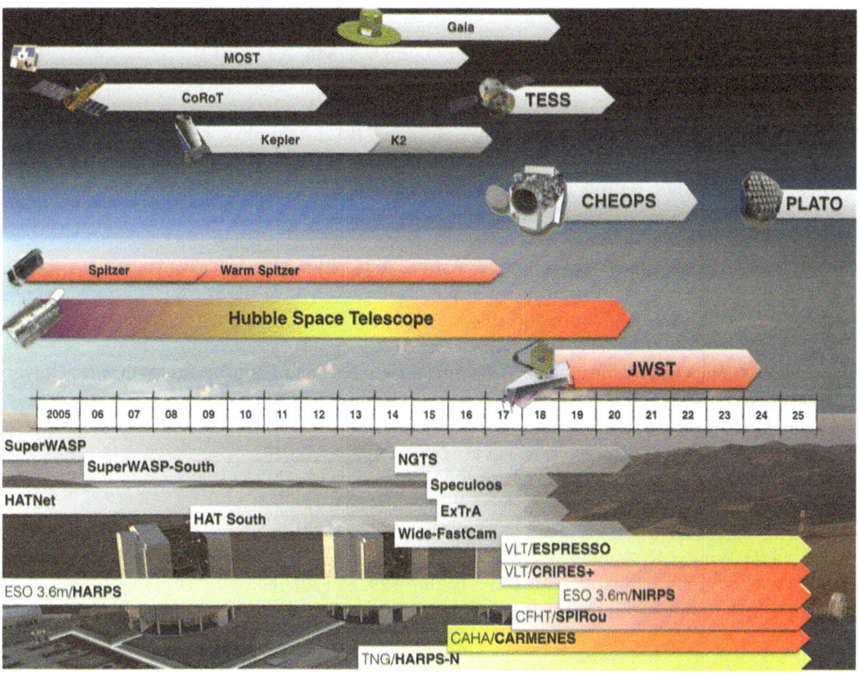

Fig. 10 Timeline of current and future instruments and missions relevant to exoplanets and characterising exoplanetary atmospheres

already started (K2) and which will be pursued throughout the 2020 decade with *TESS*, *CHEOPS*, and *PLATO*. It is mesmerising to imagine that after *PLATO*, the exoplanet on which traces of life could eventually be found will be among our known sample.

A strong emphasis on infrared characterisation will result from the exploitation of the *James Webb Space Telescope*. Because the UV window will close with the end of *HST*, an overlap with *JWST* as long as possible will be strategic.

High-resolution spectroscopy from the ground in the optical and near-infrared will not only complement, but also possibly exceed space-borne observations, in particular when the instruments are coupled to a large collecting area. This is for instance the case of ESPRESSO and CRIRES+ at the 8-m VLT, or by the end of the 2020s, HIRES or METIS at the 39-m European Extremely Large Telescope.

There will not be enough telescope time on the most powerful facilities to characterise all interesting objects. Therefore, a reconnaissance or triage approach will be essential to down-select the top candidates. This could be accomplished with high-resolution spectrographs on intermediate-size ground-based telescopes (4-m class now and 8-m class in the ELT era) and by dedicated space-borne surveys.

References

Bonfils, X., Gillon, M., Udry, S., et al.: Astron. Astrophys. **546**, A27 (2012)
Broeg, C., Fortier, A., Ehrenreich, D., et al.: EPJ Web Conf. **47**, 03005 (2013)
Burrows, A., Marley, M.S., Sharp, C.M.: Astrophys. J. **531**, 438 (2000)
Charbonneau, D., Brown, T.M., Noyes, R.W., Gilliland, R.L.: Astrophys. J. **568**, 377 (2002)
Chen, G., Guenther, E.W., Palle, E., et al.: ArXiv e-prints, arXiv:1703.01817 (2017)
Crossfield, I.J.M.: Publ. Astron. Soc. Pac. **127**, 941 (2015)
Deming, D., Wilkins, A., McCullough, P., et al.: Astrophys. J. **774**, 95 (2013)
Ehrenreich, D., Vidal-Madjar, A., Widemann, T., et al.: Astron. Astrophys. **537**, L2 (2012)
Ehrenreich, D., Bonfils, X., Lovis, C., et al.: Astron. Astrophys. **570**, A89 (2014)
Ehrenreich, D., Bourrier, V., Wheatley, P.J., et al.: Nature **522**, 459 (2015)
Fortney, J.J.: Mon. Not. R. Astron. Soc. **364**, 649 (2005)
Fraine, J., Deming, D., Benneke, B., et al.: Nature **513**, 526 (2014)
Heng, K.: Exoplanetary Atmospheres: Theoretical Concepts and Foundations. Princeton University Press, Princeton (2017)
Knutson, H.A., Benneke, B., Deming, D., Homeier, D.: Nature **505**, 66 (2014)
Kreidberg, L., Bean, J.L., Désert, J.-M., et al.: Nature **505**, 69 (2014)
Mayer, C.H., McCullough, T.P., Sloanaker, R.M.: Astrophys. J. **127**, 1 (1958)
Pepe, F., Ehrenreich, D., Meyer, M.R.: Nature **513**, 358 (2014)
Rauer, H., Catala, C., Aerts, C., et al.: Exp. Astron. **38**, 249 (2014)
Ricker, G.R., Winn, J.N., Vanderspek, R., et al.: J. Astron. Telesc. Instrum. Syst. **1**, 014003 (2015)
Sing, D.K., Fortney, J.J., Nikolov, N., et al.: Nature **529**, 59 (2016)
Wyttenbach, A., Ehrenreich, D., Lovis, C., Udry, S., Pepe, F.: Astron. Astrophys. **577**, A62 (2015)

Tutorial: Detecting Planetary Transits and Radial-Velocity Signals

Susana Barros and João P. Faria

Abstract Since the discovery of the first exoplanet orbiting a main-sequence star in 1995, more than 3500 planets have been discovered. Most of these were detected and characterized by means of radial-velocity and/or photometric observations. In this tutorial we present examples of exoplanet detection and characterization using these two methods. We start by showing an application of the Box-fitting Least Squares (BLS) algorithm to search for a transiting planet signal in the photometric light curve of EPIC 211089792 recorded by K2. Once the transits are detected we then show how to fit a model to the transit light curve and derive the transit parameters using a least-squares minimization algorithm. In a second exercise we show how to detect the same planet based on radial-velocity data and using the Lomb–Scargle periodogram. Finally, we exemplify how to combine the radial velocities and transits of the exoplanet EPIC 211089792b using a Bayesian approach to derive the planetary parameters. This tutorial hence covers some of the state-of-the-art methods of exoplanet search and characterization.

1 Introduction

The radial-velocity (RV) and transit methods are the workhorses of exoplanet discovery, with a combined yield of more than 90% of all the planets discovered up to now. The presence of an orbiting planet imprints a periodic signal in the two observables: the RV and brightness of the host star. Analyzing these data to search for planets therefore requires searching for periodicities in unevenly-spaced time series and the further characterization of the detected signals.

In this hands-on tutorial we describe and implement a collection of methods to perform this search and to fit the relevant models to observed data. To that end, we use K2 data of EPIC 211089792 (K2–29, WASP-152), which has also been observed with the SOPHIE spectrograph and is known to host a hot Jupiter with

S. Barros (✉) • J.P. Faria
Instituto de Astrofísica e Ciências do Espaço, Universidade do Porto, CAUP, Rua das Estrelas, 4150-762 Porto, Portugal
e-mail: susana.barros@astro.up.pt; joao.faria@astro.up.pt

T.L. Campante et al. (eds.), *Asteroseismology and Exoplanets: Listening to the Stars and Searching for New Worlds*, Astrophysics and Space Science Proceedings 49, DOI 10.1007/978-3-319-59315-9_15

an orbital period of 3.25 days (Santerne et al. 2016). The tutorial is presented in a very straightforward manner here, where we have decided to list the same exercises that were presented to the students during the School. The first task is to search for periodic, transit-like signals with the BLS algorithm, after which we fit a full transit model to the K2 data. We then use the Lomb–Scargle periodogram to search for sinusoidal signals in the RV measurements. Finally, we perform a joint Bayesian analysis of both data sets. We cite all necessary software packages and make the data and code available online at https://github.com/iastro-pt/AzoresTE1.

2 Transit Search

The methods used to estimate periodicities in time-series data can be broadly divided into two types: Fourier analysis (e.g., Deeming 1975; Lomb 1976; Scargle 1982; van der Klis 1988; Press and Rybicki 1989) and epoch-folding techniques (e.g., Stellingwerf 1978; Leahy et al. 1983; Schwarzenberg-Czerny 1989; Davies 1990). Fourier methods are based on the decomposition of the signal into sinusoidal functions of a given frequency, i.e., the calculation of Fourier transforms. They are useful when we want to detect a sinusoidal signal, even in data of low signal-to-noise ratio. Folding techniques consist in folding the data over trial periods and then analyzing the scatter of the resulting profile with an appropriate statistic. These methods can perform better than Fourier transforms for non-sinusoidal periodic signals, but may fail in the presence of multiple periods.

When searching for transit signals in stellar light curves, one deals with non-sinusoidal signals that are further localized in time. A large number of harmonics is needed to describe the signal, which can lead to leakage of the Fourier power to higher harmonics. To avoid this problem, Kovács et al. (2002) proposed the Box-fitting Least Squares (BLS) algorithm that uses box-shaped functions to fit the signal instead of sinusoids. The box shape is the superposition of two step functions with opposite signs representing a low and a high state, with the low state lasting much less time. The algorithm works as a folding technique, calculating the power for a defined set of frequencies with frequency separation `df`, a minimum frequency `fmin` and a number `nf` of frequency bins to test. The light curve is folded at each trial frequency and binned in phase into `nb` bins. Then the algorithm searches for the low state within a fractional duration range (`qmi`-`qma`) and fits the depth, duration and start/end of the transit, with the power corresponding to the goodness of fit. The result is a periodogram (power as a function of each trial frequency or period) and estimation of the parameters for the best period.

Problem 1 Search for transits in the K2 Campaign 4 light curve of EPIC 211089792 using a `Python` implementation of the BLS algorithm.[1] Consider periods in the

[1]The package is available at https://github.com/dfm/python-bls.

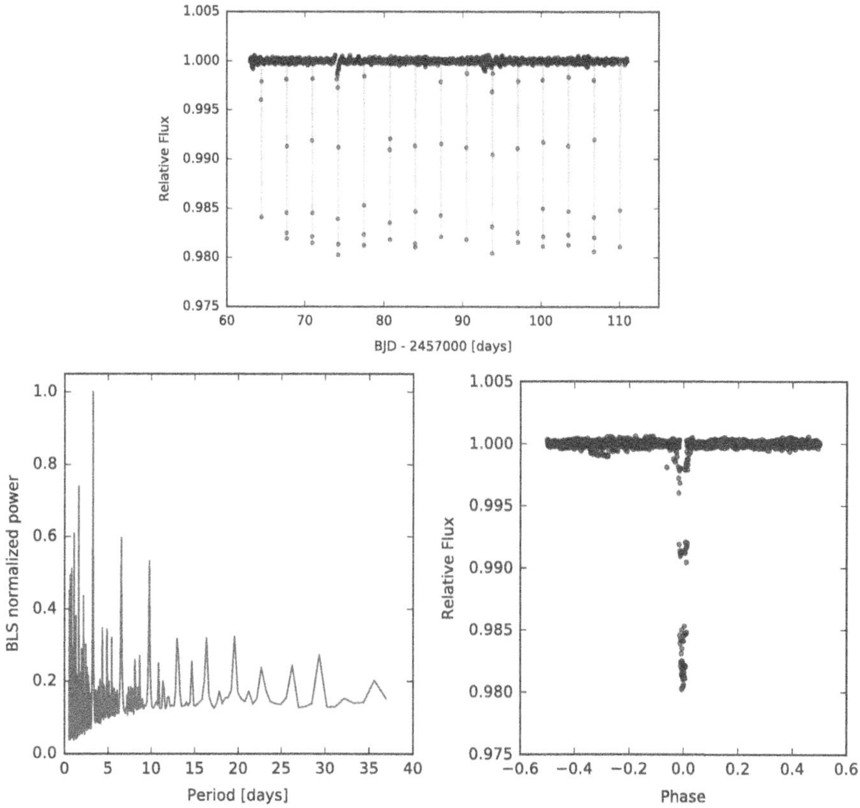

Fig. 1 Partial solution to Problem 1: K2 light curve (*top*), the BLS periodogram (*bottom left*) and the folded light curve at the period of maximum power (*bottom right*)

range from 0.5 days to 70% of the full duration of the observations with a frequency resolution of 0.001, and choose the values for the remaining parameters `nb`, `qmi` and `qma`. Plot the resulting BLS periodogram. Identify the frequency of the highest peak and its corresponding parameters: power, depth, duration and epoch. Phase fold the light curve using the period found. Try out different input parameters and compare your results with the panels in Fig. 1. Take note of all the parameters you obtained.

3 Transit Fitting

A transit event occurs when a planet passes in front of its parent star as seen by the observer. In its simplest form, a transit is a geometrical effect and has five observables: the orbital period P, mid-transit time T_0, depth d, transit duration, and duration of ingress/egress as the planet crosses the stellar limb.

For circular orbits it was shown by Seager and Mallén-Ornelas (2003) that these observables lead to a unique solution of the transit parameters: r_p/r_*, a/r_* and inclination, where r_p and r_* are the radius of the planet and star, respectively, and a is the semi-major axis of the orbit. For eccentric orbits, two more parameters affect the transit shape: the eccentricity, e, and the longitude of periastron, ω. These affect both the velocity of the planet and the distance between the star and planet at the time of transit. Here we will be assuming circular orbits. This corresponds to setting $e = 0$ and $\omega = 90°$ (any ω would be valid, but this choice guarantees that the reference time will coincide with the transit moment).

Real stars show a limb-darkening effect, which also changes the transit shape. Here we will be using a quadratic limb-darkening law, setting $u_1 = 0.4983$ and $u_2 = 0.2042$ (cf. Santerne et al. 2016). You will also need to use the following relations (e.g., Seager and Mallén-Ornelas 2003):

$$d = \left(\frac{r_p}{r_*}\right)^2 \tag{1}$$

and

$$\rho_* = \frac{4\pi^2}{GP^2}\left(\frac{a}{r_*}\right)^3, \tag{2}$$

where ρ_* is the stellar density.

Problem 2 Plot the light curve of EPIC 211089792. Make an initial guess for the transit parameters using the equations above and the parameters obtained from the BLS output. You can also assume that the star has the same density as the Sun. Use these initial guesses of the transit parameters to obtain a transit model with the `batman` package.[2] Overplot the transit model.

Problem 3 Fit the transits using the `batman` transit models and the `lmfit` package,[3] which implements non-linear least-squares fitting routines. Assume a circular orbit and the limb-darkening parameters given above. Hint: start by creating a function that calls the transit model given the model parameters and returns the difference between the model and the data points. The parameters need to be defined according to the `lmfit` requirements. Call the `minimise` function from `lmfit` that will minimize the above function and estimate the best-fit transit parameters. Plot the phase-folded data using the new derived parameters and overplot the best transit model. Cut the out-of-transit parts in order to have a total of three transit durations for each transit, centered at the mid-transit time. Save the resulting time series.

[2]The package is available from https://www.cfa.harvard.edu/~lkreidberg/batman and is described in Kreidberg (2015).

[3]Available from https://lmfit.github.io/lmfit-py.

4 The (Generalized) Lomb–Scargle Periodogram

The Lomb–Scargle periodogram is commonly used for period search and frequency analysis of time series. It is equivalent to fitting sine waves of the form

$$y = a\cos\omega t + b\sin\omega t \tag{3}$$

for a range of frequencies. This method has an analytical solution and is thus both convenient to use and efficient. A formula for the periodogram was given by Barning (1963). Lomb (1976) and Scargle (1982) further investigated its statistical behavior, especially the statistical significance of the detection of a signal.

Because the detection of periodicities in time-series data is such a common problem, many generalizations of the Lomb–Scargle periodogram have been proposed. These take into account individual measurement errors, the inclusion of a constant offset, or other types of periodic signals (Ferraz-Mello 1981; Cumming et al. 1999; Zechmeister and Kürster 2009; Mortier et al. 2015). Implementations of these algorithms are also common and mature in many programming languages.

Problem 4 Use the `astropy` implementation of the Lomb–Scargle periodogram to search for periodic variations in the radial velocities of EPIC 211089792 (see Fig. 2). Take into account each individual RV uncertainty. Plot the periodogram and calculate the period of maximum power. Fold the RV observations at the best period.

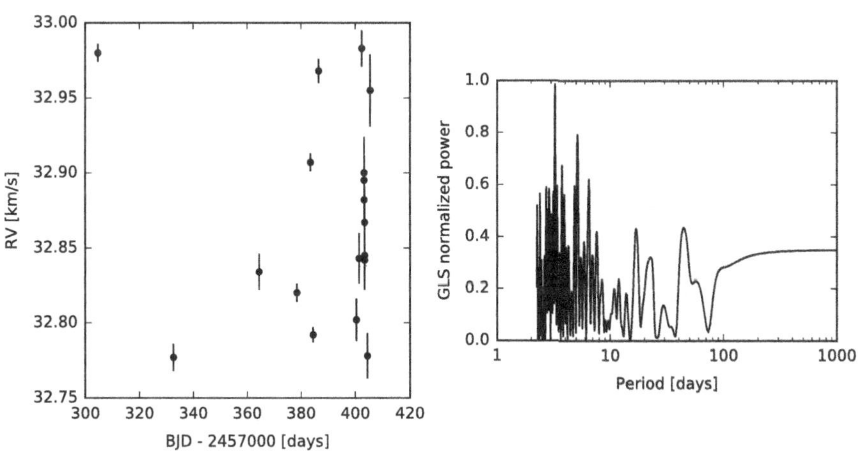

Fig. 2 Partial solution to Problem 4: RV observations (*left*) and their generalized Lomb–Scargle periodogram (*right*)

5 Transit and RV Fit: Bayesian Approach

Sometimes the fit of non-linear models by least-squares minimization can lead to wrong results because the minimization algorithm gets trapped in local minima. In the Bayesian context, the problem becomes that of sampling from the posterior distribution for the model parameters. Markov chain Monte Carlo (MCMC) is an efficient technique to solve this sampling problem.

Almost every Bayesian analysis has at least two ingredients: the likelihood, $p(D|\boldsymbol{\theta})$, and the prior distribution, $p(\boldsymbol{\theta})$. They are related through Bayes' theorem:

$$p(\boldsymbol{\theta}|D) \propto p(\boldsymbol{\theta})\, p(D|\boldsymbol{\theta}), \tag{4}$$

where $\boldsymbol{\theta}$ represents the set of model parameters and D the data. The term $p(\boldsymbol{\theta}|D)$ is the posterior distribution for the parameters, i.e., the distribution we want to sample from.

Problem 5 What is the least informative likelihood we can come up with?

In this tutorial we actually make use of two datasets, D_1 and D_2, respectively coming from transit and RV observations. The two should be combined in order to obtain an orbital solution which agrees with both sets of observations.

Problem 6 What is the combined likelihood of the two datasets? How does it depend on the individual likelihoods of D_1 and D_2?

We can use a Gaussian likelihood and include in our model an extra source of noise, s, which accounts for effects that are not included in the measurement uncertainties:

$$\begin{aligned} p(D|\boldsymbol{\theta}) &= \prod_{i=1}^{N} p(D_i|\boldsymbol{\theta}) \\ &= \prod_{i=1}^{N} \mathcal{N}(d_i|f(x_i), \sigma_i + s) \\ &= \prod_{i=1}^{N} \frac{1}{\sqrt{2\pi\,(\sigma_i^2 + s^2)}} \exp\left\{ -\frac{[d_i - f(x_i)]^2}{2(\sigma_i^2 + s^2)} \right\} \\ &= (2\pi)^{-N/2} \left(\prod_{i=1}^{N} \frac{1}{\sqrt{\sigma_i^2 + s^2}} \right) \exp\left\{ -\frac{1}{2} \sum_{i=1}^{N} \frac{[d_i - f(x_i)]^2}{2(\sigma_i^2 + s^2)} \right\}, \end{aligned} \tag{5}$$

so that the log-likelihood is

$$\ln p(D|\boldsymbol{\theta}) = -\frac{N}{2} \ln 2\pi - \sum_{i=1}^{N} \left\{ \frac{\ln(\sigma_i^2 + s^2)}{2} + \frac{[d_i - f(x_i)]^2}{2(\sigma_i^2 + s^2)} \right\}. \tag{6}$$

Problem 7 Write a function that computes the combined log-likelihood given in Eq. (6).

Problem 8 Using the distributions included in `scipy`, define priors for each parameter and write a function to produce random samples from those prior distributions and/or from the joint prior distribution. Create also a function that calculates the value of the log-prior.

Problem 9 Write a function that computes the posterior and use the `emcee` package[4] to sample from this posterior.

Problem 10 Make a corner plot to check the correlations between parameters and the posterior distributions of the model parameters. Plot your MCMC chains to check if they have converged.

Problem 11 Important: You should oversample the model light curve. What happens when you do not? Run your code using different priors and check if the results change. Fix the orbital period of the planet and see what happens (you can use the values from the BLS or the generalized Lomb–Scargle periodogram). Assume you know the stellar density and consider using the Gaussian prior

$$\frac{a}{r_*} = \mathcal{N}(10.51, 0.15)\,. \tag{7}$$

Calculate the mass and radius of the planet.

References

Barning, F.J.M.: Bull. Astron. Inst. Neth. **17**, 22 (1963)
Cumming, A., Marcy, G.W., Butler, R.P.: Astrophys. J. **526**, 890 (1999)
Davies, S.R.: Mon. Not. R. Astron. Soc. **244**, 93 (1990)
Deeming, T.J.: Astrophys. Space Sci. **36**, 137 (1975)
Ferraz-Mello, S.: Astron. J. **86**, 619 (1981)
Foreman-Mackey, D., Hogg, D.W., Lang, D., Goodman, J.: Publ. Astron. Soc. Pac. **125**, 306 (2013)
Kovács, G., Zucker, S., Mazeh, T.: Astron. Astrophys. **391**, 369 (2002)
Kreidberg, L.: Publ. Astron. Soc. Pac. **127**, 1161 (2015)
Leahy, D.A., Elsner, R.F., Weisskopf, M.C.: Astrophys. J. **272**, 256 (1983)
Lomb, N.R.: Astrophys. Space Sci. **39**, 447 (1976)
Mortier, A., Faria, J.P., Correia, C.M., Santerne, A., Santos, N.C.: Astron. Astrophys. **573**, A101 (2015)
Press, W.H., Rybicki, G.B.: Astrophys. J. **338**, 277 (1989)
Santerne, A., Hébrard, G., Lillo-Box, J., et al.: Astrophys. J. **824**, 55 (2016)
Scargle, J.D.: Astrophys. J. **263**, 835 (1982)

[4]Available from http://dan.iel.fm/emcee and described in Foreman-Mackey et al. (2013).

Schwarzenberg-Czerny, A.: Mon. Not. R. Astron. Soc. **241**, 153 (1989)
Seager, S., Mallén-Ornelas, G.: Astrophys. J. **585**, 1038 (2003)
Stellingwerf, R.F.: Astrophys. J. **224**, 953 (1978)
van der Klis, M.B.M.: NATO Advanced Study Institutes Series. Series C, Mathematical and Physical Sciences (1988)
Zechmeister, M., Kürster, M.: Astron. Astrophys. **496**, 577 (2009)

Tutorial: Measuring Stellar Atmospheric Parameters with ARES+MOOG

Sérgio G. Sousa and Daniel T. Andreasen

Abstract The technical aspects of using an Equivalent Width (EW) method for the derivation of spectroscopic stellar parameters with ARES+MOOG are described herein. While the science background to this method can be found in numerous references, the goal here is to provide a user-friendly guide to the several codes and scripts used in the tutorial presented at the School. All the required data have been made available online at the following repository: https://github.com/sousasag/school_codes.

1 Introduction

Several methods are used for the derivation of stellar spectroscopic parameters. These can be divided into two main groups. One set of methods is based on spectral synthesis, for which synthetic spectra are created and compared with the observed spectrum to find the best fit. The second group adopts a line-by-line analysis strategy, measuring the strength of observed spectral lines, and then estimating abundances and applying relevant criteria—such as the ionization and excitation balance—to find the best spectroscopic stellar parameters that match the observations. The description of the physics behind both groups of methods can be found in several textbooks (e.g., Gray 2008).

The workflow of the EW method can be briefly described as follows: For a high-quality, observed stellar spectrum, we start by measuring the strength of several spectral lines (e.g., with ARES). These measurements are then converted into individual line abundances, which are computed (e.g., with MOOG) using stellar atmospheric models, normally assuming a Local Thermodynamical Equilibrium (LTE) approximation. The model parameters are then adjusted until the individual line abundances show evidence of excitation and ionization

S.G. Sousa (✉) • D.T. Andreasen
Instituto de Astrofísica e Ciências do Espaço, Universidade do Porto, CAUP, Rua das Estrelas, 4150-762 Porto, Portugal
e-mail: Sergio.Sousa@astro.up.pt; Daniel.Andreasen@astro.up.pt

T.L. Campante et al. (eds.), *Asteroseismology and Exoplanets: Listening to the Stars and Searching for New Worlds*, Astrophysics and Space Science Proceedings 49, DOI 10.1007/978-3-319-59315-9_16

balance. This step can be made automatic if a proper minimization method is adopted.

The rest of the document will focus on the technical aspects of the codes provided for the tutorial and on how a student can use them to actually derive spectroscopic stellar parameters. A more in-depth description of ARES+MOOG can be found in Sousa (2014). The next sections are based on the codes and test data made available online.

2 Requirements

The tutorial was implemented so as to be executed on a Linux machine. Note that even without access to a Linux machine, today it is very easy to emulate one by using virtual machines (e.g., VirtualBox). All the instructions provided both in this document and in the repository are compatible for Debian/Ubuntu systems, but can be easily adapted for other Linux flavors.

2.1 Main Codes

ARES ARES is a `C` code that allows a fast and automatic measurement of EWs of spectral absorption lines. The ARES code is a submodule in the repository and therefore easily updated via the root ARES repository,[1] which contains more information. For a more detailed description of the code please see the ARES papers (Sousa et al. 2007, 2015). Note that the latest version of ARES is already able to deal with in-situ radial-velocity correction and automatic parameterization for the continuum level, which makes the spectral analysis easier and more consistent. In order to compile ARES, some external libraries (e.g., `CFITSIO`, `GSL` etc.) are required, which can be easily installed in Linux machines. The compilation of the code is then handled by the `Makefile` provided in the repository.

MOOG MOOG is a code that performs a variety of LTE spectral analyses. The original code can be found in its own website.[2] The code available in the repository is an adapted version of MOOG2014, modified to neglect its (nonfree) plot library dependency. In the tutorial only the silent version of the code is used when making abundance computations for our EW method. The compilation of the code is also handled by the `Makefile`.

---

[1]https://github.com/sousasag/ARES.

[2]http://www.as.utexas.edu/~chris/moog.html.

2.2 Spectral Test Data

Observed Spectra The spectra available for analysis should be of good quality in terms of resolution ($R > 30{,}000$) and signal-to-noise ratio ($S/N > 100$). Three HARPS spectra are provided for the tutorial, having $R \sim 110{,}000$ and $S/N > 300$. These are standard `FITS` 1D spectra. In order to be compatible with ARES, they need to have the standard keywords `CDELTA1` and `CRVAL1`, and wavelengths should be given in angstroms. Alternatively, ARES can also read `ASCII` spectra containing two columns, the first with the wavelengths and the second with the flux. In this tutorial only the standard `FITS` 1D files are used.

Line List The strength of specific absorption spectral lines should be measured during the analysis. As in most EW methods, iron lines are mostly used because they strongly populate the solar-type spectrum and because iron can be used as a proxy for the stellar metallicity. The line list provided in the repository (`ironlines_parameters.dat`) compiles 263 Fe I and 36 Fe II lines. These lines were carefully selected (cf. Sousa et al. 2008) and their atomic data were revised using a solar spectrum as reference, thus making it possible to perform a differential analysis (Sousa 2014). This line list is used both for ARES and MOOG. For ARES the important information is only the wavelength of the lines. The format should be such that wavelengths (in angstroms) are provided in the first column. For the individual abundance computations, MOOG needs the atomic data, Excitation Potential (E.P.), and oscillator strength, log(gf). The format of this file should be kept fixed so that the scripts work without a problem.

2.3 Scripts

Input Line List for MOOG The EW measurements and the atomic data in the line list require a special format to be used by MOOG. To facilitate this task, a `Python` script is provided (`make_moog_lines.py`) that reads the atomic data from the line list as well as the output file from ARES, and compiles the input file to MOOG in the required format.

Creating a Stellar Atmospheric Model For the computation of the individual line abundances, the use of an atmospheric stellar model is required. MARCS models[3] are used in the tutorial. For this task a script is provided that inter-

[3]http://marcs.astro.uu.se.

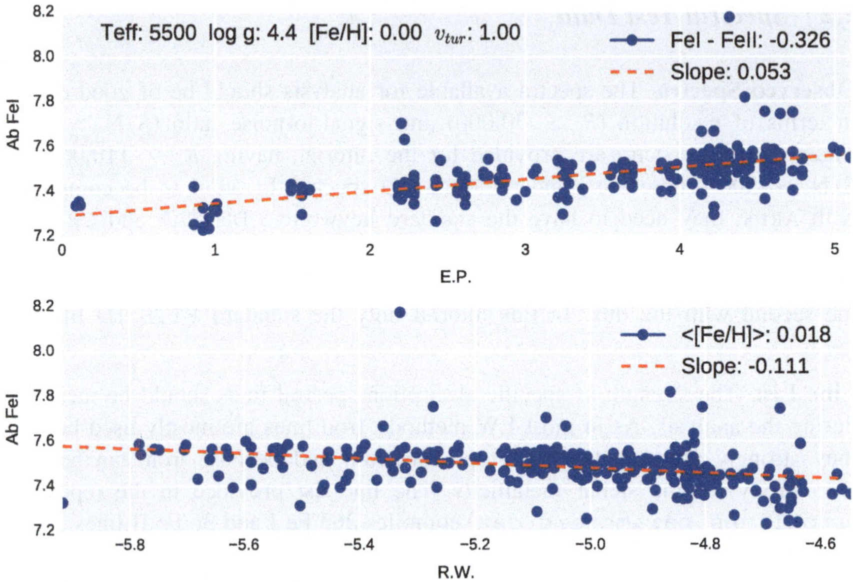

Fig. 1 Result of the first iteration of the analysis of the `TestA.fits` spectrum

polates a grid of MARCS models for specific stellar parameters. The folder in the repository named `interpol_models_marcs` includes a script for this task. In order to make it work, it is first required to follow the instructions to download and extract the grid of MARCS models. The use of the script to get a specific atmospheric stellar model is straightforward and is described below.

Excitation and Ionization Correlation To understand how the individual line abundances depend on the stellar atmospheric parameters and how to find the best stellar parameters, a plotting script is provided that allows the visualization of the correlations/indicators needed to constrain the parameters. These include the abundance vs. E.P., abundance vs. reduced EW, the information on $|\langle \mathrm{Fe\,I} \rangle - \langle \mathrm{Fe\,II} \rangle|$ and the average [Fe/H] compared with the model [M/H]. For more details about these indicators see Sect. 4 and Sousa (2014). The `Python` script `running_dir/read_moog_plot.py` allows to plot these correlations as well as to display the information of these indicators (see Figs. 1 and 2).

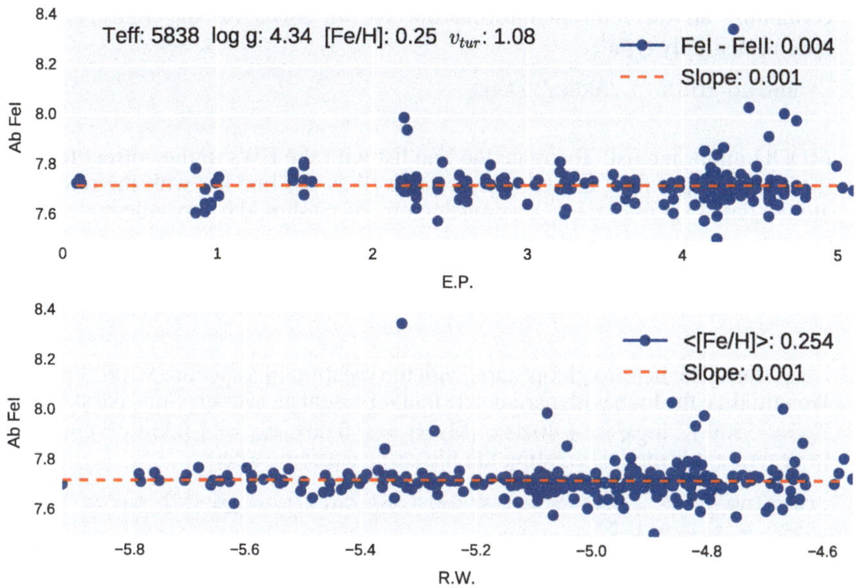

Fig. 2 An optimized model derived for the `TestA.fits` spectrum

3 Step-by-Step Tutorial

The criteria to derive good spectroscopic parameters with this method rely on the model that enforces the excitation and ionization balance for the line abundances. Therefore, the final step of this iterative occurs when the same abundance for all the lines is found, which will translate into null slopes for abundance vs. E.P., abundance vs. reduced EW, and $\langle \text{Fe I} \rangle = \langle \text{Fe II} \rangle$. How to reach this? The steps to derive stellar parameters—where the spectrum `TestA.fits` is used for this example—are as follows:

1. Location: To start the process let us first define a reference folder. Let the folder `running_dir` be our reference folder. To confirm this, when displaying the spectra available in the repository the result should be:

```
running_dir$ ls ../spectra/
TestA.fits  TestB.fits  TestC.fits
```

2. Measuring EWs: The required ARES input file (`mine.opt`) is already present in the reference folder and contains the recommended parameters for the analysis of high-quality spectra. Details about these parameters can be found in Sousa (2007, 2014). Running ARES will display plenty of information in the terminal. An output file (`TestA.ares`) will be created as well as a log file (`logARES.txt`)

containing all the relevant information. To run ARES for the `TestA.fits` spectrum simply type[4]:

```
running_dir$ ../ARES/./ARES
```

3. MOOG input line list: To create the line list with the EWs in the correct format, the ARES output file (`TestA.ares`) as well as the line list with the reference atomic data (`ironlines_parameters.dat`) should be provided:

```
running_dir$ ../. /make_moog_lines.py TestA.ares ../ironlines_p
arameters.dat
Saved in: lines.TestA.ares
```

4. Loop start: The iteration loop starts with the creation of a specific MARCS model. We initialize the loop with parameters that represent an average Sun-like star (i.e., $T_{\rm eff} = 5500$ K, $\log g = 4.40$ dex, [M/H] $= 0.0$ dex, $v_{\rm tur} = 1.0$ km s^{-1}) in order to generate this model (creating the file `out_marcs.atm`):

```
running_dir$ ../interpol_models_marcs/./make_model_marcs.bash
5500 4.4 0.0 1.0
```

5. MOOG: The next iterative step is to compute the individual line abundances assuming the stellar atmospheric model. In this step, MOOG will use a predefined input file (`batch.par`) and will then create an output file with the computed line abundances (`output.moog`). To run MOOG in its silent mode:

```
running_dir$ ../MOOG2014/./MOOGSILENT
```

6. Control check: This is the crucial step where the stellar parameters in the model are validated. Again, optimized parameters are found when the same abundance is derived for all the lines. The slopes should therefore be negligible (all indicators having values < 0.005 is a sensible criterion). If significant correlations are found in these indicators, these can be used to adjust the parameters back in step 4. To check the status of these indicators the `Python` plotting script can be used:

```
running_dir$ python read_moog_plot.py output.moog
Model Parameters: Teff logg vtur [M/H]
                  5500 4.4 1.00 0.00
-----------------------------
|   Slope  E.P. :0.053
|   Slope  R.W. :-0.111
|  Fe I   - Fe II:-0.326
|      [Fe/H]    :0.018
| [Fe/H]  - [M/H]:0.018
```

Figure 1 shows the plots generated during this last step. The indicators' values for this first iteration are far from zero. Therefore, it is required to go back to step

[4]Note that `mine.opt` should be adapted for the analysis of the other test spectra.

4 and try different parameters. The strategy to minimize these indicators and change the parameters in the correct direction is discussed in the next section.

4 Finding the Best Parameters

In this section the strategy used to find the best parameters is briefly described. A complete description of this strategy can be found in Sousa (2014). From Fig. 1 the four indicators' values provide the necessary hints to obtaining a better model during the next iteration:

- Slope of E.P.: This indicator strongly depends on the temperature of the model. Negative values of the slope mean that the correct temperature should be lower. This indicator also controls the excitation balance. Given the very high number of iron lines, the temperature is one of the best constrained parameters using this method.
- Slope of reduced EW: This indicator is connected to the microturbulence parameter, which basically controls the abundance determination for the stronger lines where saturation becomes especially significant in the wings of the absorption lines (Gray 2008). Positive values of the slope mean the microturbulence is being overestimated.
- ⟨Fe I⟩–⟨Fe II⟩: Since Fe II lines are more sensitive to surface gravity than Fe I lines (Gray 2008), this indicator can be used to control the $\log g$ of the model. For negative values, $\log g$ should decrease during the next iteration, while for positive values $\log g$ should increase.
- [Fe/H]–[M/H]: More than an indicator, this is actually a logical constraint for the model, meaning that the input metallicity should lead to a compatible global iron abundance. Here it is assumed that the iron abundance is a proxy for the stellar metallicity. Therefore, if the iron abundance is larger than the model metallicity (positive indicator), then it should be increased during the next iteration. If the indicator is negative, then this means that the metallicity of the model is overestimated.

These indicators can be used to find an optimized model for the `TestA.fits` spectrum. Figure 2 shows such a model with $T_{\rm eff} = 5838$ K, $\log g = 4.34$ dex, [M/H] $= 0.25$ dex, and $v_{\rm tur} = 1.08$ km s^{-1}. With this model all the indicators are close to zero (i.e., all obey the < 0.005 criterion). These results can still be improved given the clear presence of outliers in Fig. 2. In the present case most of these outliers are overestimating the iron abundance (compared to the average). The cause for this may be related to less reliable, and overestimated, EW measurements for those specific spectral lines and hence removing such outliers will improve the result.

The stars used in the tutorial are identified in Table 1. The table also summarizes the stellar parameters derived after the removal of outliers. For the `TestA.fits` spectrum, corresponding to HD 128620 (α Cen A), the results presented in Fig. 2

Table 1 Parameters derived with ARES+MOOG for test spectra

File	Star	$T_{\rm eff}$ (K)	$\log g$ (dex)	[Fe/H] (dex)	$v_{\rm tur}$ (km s^{-1})
`TestA.fits`	HD 128620	5832	4.33	0.23	1.11
`TestB.fits`	HD 128621	5234	4.40	0.16	0.90
`TestC.fits`	HD 179949	6287	4.54	0.21	1.36

are quite close to the ones in Table 1. Note that the temperature is within 10 K, whereas the metallicity should be slightly lower, as expected, given that the outliers overestimate the iron abundance. Although no error analysis is discussed here for this method, it is clear from the results that the precision of the method is quite high, by simply considering the very low spread of the derived parameters when disregarding lines with problematic EW measurements.

5 Summary

The technical aspects concerning the ARES+MOOG EW method for deriving spectroscopic stellar parameters were presented in the form of a tutorial. The strategy for finding an optimized model based on a set of indicators provided during each iteration is briefly described. This document acts as a complement to the material provided online. For a more complete understanding of the ARES+MOOG method we direct the reader to Sousa (2007, 2014, 2015).

Acknowledgements SGS and DTA acknowledge the support from Fundação para a Ciência e a Tecnologia (FCT) through national funds and a research grant (project ref. UID/FIS/04434/2013, PTDC/FIS-AST/7073/2014, and grant ref. CAUP-09/2014-BD). SGS also acknowledges the support from FCT through Investigador FCT contract with reference IF/00028/2014 and POPH/FSE (EC) by FEDER funding through Programa Operacional de Factores de Competitividade COMPETE.

References

Gray, D.F.: The Observation and Analysis of Stellar Photospheres. Cambridge University Press, Cambridge (2008)

Sousa, S.G.: ARES + MOOG: A Practical Overview of an Equivalent Width (EW) Method to Derive Stellar Parameters, pp. 297–310. Springer International Publishing, Cham (2014)

Sousa, S.G., Santos, N.C., Israelian, G., Mayor, M., Monteiro, M.J.P.F.G.: Astron. Astrophys. **469**, 783 (2007)

Sousa, S.G., Santos, N.C., Mayor, M., et al.: Astron. Astrophys. **487**, 373 (2008)

Sousa, S.G., Santos, N.C., Adibekyan, V., Delgado-Mena, E., Israelian, G.: Astron. Astrophys. **577**, A67 (2015)

The manufacturer's authorised representative in the EU is Springer Nature Customer Service Centre GmbH, Europaplatz 3, 69115 Heidelberg, Germany. If you have any concerns regarding our products, please contact ProductSafety@springernature.com

Printed and bound by CPI Group (UK) Ltd, Croydon, CR0 4YY
30/09/2026
02225228-0011